T0270958

Engineering Mathematics for Marine Applications

Gaining expertise in marine floating systems typically requires access to multiple resources to obtain the knowledge required, but this book fills the long-felt need for a single cohesive source that brings together the mathematical methods and dynamic analysis techniques required for a meaningful analysis, primarily, of large and small bodies in oceans. You will be introduced to fundamentals such as vector calculus, Fourier analysis, and ordinary and partial differential equations. Then you will be taken through dimensional analysis of marine systems, viscous and inviscid flow around structures, surface waves, and floating bodies in waves. Real-life applications are discussed and end-of-chapter problems help ensure full understanding. Students and practicing engineers will find this book an invaluable resource for developing problem-solving and design skills in a challenging ocean environment through the use of engineering mathematics.

Umesh A. Korde works on wave energy, waves, oscillating systems, and control. His recent interests include dynamics of natural and engineered multi-scale systems with a focus on energy and climate. He is a Fellow of the American Society of Mechanical Engineers.

R. Cengiz Ertekin's research includes nonlinear wave theories, hydroelasticity, wave energy, and storm impact. He is a Fellow of the American Society of Mechanical Engineers and the Society of Naval Architects and Marine Engineers. He is the Founder and editor-in-chief of the *Journal of Ocean Engineering and Marine Energy*.

Engineering Mathematics for Marine Applications

UMESH A. KORDE
Johns Hopkins University

R. CENGIZ ERTEKIN
University of Hawai'i, Manoa

CAMBRIDGE
UNIVERSITY PRESS

CAMBRIDGE
UNIVERSITY PRESS

Shaftesbury Road, Cambridge CB2 8EA, United Kingdom

One Liberty Plaza, 20th Floor, New York, NY 10006, USA

477 Williamstown Road, Port Melbourne, VIC 3207, Australia

314–321, 3rd Floor, Plot 3, Splendor Forum, Jasola District Centre,
New Delhi – 110025, India

103 Penang Road, #05–06/07, Visioncrest Commercial, Singapore 238467

Cambridge University Press is part of Cambridge University Press & Assessment,
a department of the University of Cambridge.

We share the University's mission to contribute to society through the pursuit of
education, learning and research at the highest international levels of excellence.

www.cambridge.org
Information on this title: www.cambridge.org/9781108421041

DOI: 10.1017/9781108363235

© Umesh A. Korde and R. Cengiz Ertekin 2023

First published 2023

A catalogue record for this publication is available from the British Library.

Library of Congress Cataloging-in-Publication Data
Names: Korde, Umesh A., author. | Ertekin, R. Cengiz, author.
Title: Engineering mathematics for marine applications / Umesh A. Korde, R. Cengiz Ertekin.
Description: Cambridge ; New York, NY : Cambridge University Press, 2023.
| Includes bibliographical references and index.
Identifiers: LCCN 2022040981 | ISBN 9781108421041 (hardback) |
ISBN 9781108363235 (ebook)
Subjects: LCSH: Ocean engineering – Mathematical models. | Engineering mathematics.
Classification: LCC TC1650 .K669 2023 | DDC 620/.4162015118–dc23/eng/20221223
LC record available at https://lccn.loc.gov/2022040981

ISBN 978-1-108-42104-1 Hardback

Contents

Preface

Mathematics allows us to describe and interpret the myriad intricacies of nature. Engineers tend to use mathematics to help them design structures and systems that must work with, survive, and sometimes utilize the forces of nature. As we look ahead to an era in which oceans and their effect on life on land command ever greater attention, we might expect that more and more engineers of the future will require some understanding of how engineered structures and systems respond to the forces of the ocean. This may, therefore, be a good time to pause and reflect on the works of brilliant mathematicians and ocean engineers who, over the last three centuries, have put together a rich array of methods and techniques, without which the development of offshore oil, ocean energy, and ocean exploration technologies would have been slower, more painful, and more expensive.

This book is a record of our attempt to assemble in one place the mathematical methods that would serve an engineer well when tackling problems that involve bodies that have to withstand and perhaps exploit the harsh environment on and close to the ocean surface. Forces in the ocean are time and location dependent, so we begin with a study of vectors and their rates of change with time and space, and we learn about surfaces, volumes, and fields so that we could conveniently and compactly describe them and relate them to each other through known laws of physics (such as conservation mass, conservation of momentum, conservation of energy, viscous dissipation and energy loss, etc.). We review principles of system dynamics, so that we could relate the rates of change of our field variables to other variables and any external forces, particularly as and when they impact an engineered structure in the ocean. The review of vector calculus, complex algebra, and Fourier analysis provides the foundation for material on ordinary and partial differential equations. While our goal has been to frame these discussions in the context of marine, and/or in-water applications, we have also been mindful of the more traditional mechanical engineering treatments of diffusion, conduction, and wave propagation on/over solids. A brief treatment of analytical dynamics is also included, given the importance of variational calculus in engineering optimization, control, finite-element analysis, etc.

It is our hope that the material in Parts I and II will whet the reader's appetite for the broader and deeper treatments of Parts III and IV.

In Part III, we discuss the mathematics of scaling, a topic which is so important in experiments that are generally used to understand how closely we predict what happens in nature. After all, mother nature is the ultimate judge of the theories we develop. In Part IV, the mathematical basis of the book discussed in earlier chapters is used to further develop hydrodynamical theories that can be employed to solve some rather practical problems encountered in marine applications. Both viscous and inviscid theories are covered to the extent possible. We mostly deal with irrotational flows in this book when the fluid is inviscid.

We hope that this book will in some ways provide a sense of continuity to students and practicing engineers who sometimes like to review their basics as they work on advanced problems, and to those who sometimes like to get a preview of the interesting applications and new techniques that lie ahead while working through the basics. It could be used as a text for an advanced undergraduate or early graduate two-semester course sequence. We hope that the self-assessment exercises at the end of each chapter will be helpful to those readers with an interest in practicing their understanding before directly plunging into an application in an engineering situation.

Acknowledgments

UAK:

I would like to express my immense gratitude to all my mathematics teachers to date: starting with my sister Meena and brother Shri, my parents, my teachers from school through university, and the teachers and colleagues I work with. I owe particular thanks to Professor Hisaaki Maeda (then at the University of Tokyo) who was the first person to make the mathematics of floating bodies come alive for me. Indeed, it may have been his love of the subject and clarity of understanding that kept me close to mathematics at a time in life when I was gravitating toward everything nonmathematical. Others who have taught me mathematics through their works include the many authors whose books we have used heavily and referenced in this text.

I am particularly grateful to Professor R. Cengiz Ertekin for agreeing to collaborate on this project. This volume would never have materialized without his involvement and his unwavering motivation.

I want to thank the Johns Hopkins University and the Environmental Health and Engineering (EHE) department for giving me the time and opportunity to work on this book. I greatly appreciate the support and encouragement of Dr. Marsha Wills-Karp, Dr. Ben Hobbs, and Dr. Alan Goldberg of the EHE department. On a related note, I cannot thank enough Professor Mike McCormick and Mr. Bob Murtha for their wonderful friendship and for every way in which they have helped me through the last several years. Finally, I want to thank my wife Michelle for her understanding and support of my pursuits and for the personal sacrifices she has made over the years.

RCE:

I am grateful for the teachings of many of my teachers and their kind support and advice throughout my education and career. In particular, my deep appreciation goes to Prof. M. Cengiz Dökmeci of the Technical University of Istanbul, Professors William C. Webster, Ronald W. Yeung, Joe L. Hammack, Marshall Tulin, Paul M. Naghdi of the University of California at Berkeley, Prof. Theodore Y. Wu of California Institute of Technology, and Prof. Ying Z. Liu of Shanghai Jia Tong University. It was very kind of Prof. Michael Isaacson of the University of British Columbia to allow us to adapt and use a number of figures from his book with Prof. Turgut Sarpkaya. Many of my former students have helped to improve my lecture notes that I developed for my hydrodynamics courses at the University of Hawaii for almost 30 years, and some of the chapters are based on those notes. In particular, I thank my former Ph.D. students, Drs. Murthy Chitrapu, Zhengmin Qian, Suqin Wang, Douglas R. Neill, Yingfan Xu, Charles Liu, Bala Padmanbhan, Hari Sundararaghavan, Dingwu Xia, Richard W. Carter, Masoud Hayatdavoodi, and Betsy R. Seiffert. In particular I am indebted to Prof. Masoud Hayatdavoodi of the University of Dundee who kindly created a number of figures used in this book among his busy schedule.

I owe who I am to Prof. John V. Wehausen of the University of California at Berkeley who was my Ph.D. advisor at Berkeley. His teachings have been followed by so many hydrodynamicists since the early 1960s, and I am no exception. A number of chapters in this book would show the influence of his class notes at U.C. Berkeley. Not only was he a brilliant mathematician but also a patient, kind, and gentle human being who is dearly missed; it is impossible to fill his shoes. Finally, I am very thankful to my wife Juliet who has supported me without any reservations since my junior year in college. Without her sacrifices, I could not have done much, šhat šhnorhakalut'yun Juliet.

UAK and RCE:

We owe particular gratitude to Mr. Steven Eliot of Cambridge University Press for his constant encouragement and support. Our thanks also to other staff at Cambridge, notably, Ms. Julia Ford, who kept the book on track. We are also grateful to Ms. Naomi Chopra for her thoughtful handling of this book, as well as for all of her kindness, help, and patience through the final stages of this book. Mr. Santhamurthy Ramamoorthy and his staff did an excellent job with the copy-editing, typesetting, and proof-checking steps.

Finally, we thank you, the reader, for your attention to the topics in this book. As you read this book, if you discover any errors or if you have any questions, please feel free to contact us, so that we may correct our errors and/or answer any of your questions.

Part I

The Foundations

Here we review foundational topics such as vector mechanics, vector algebra and calculus, and indicial notation. We discuss differential operators such as gradient, divergence, and curl and their physical significance. We also review line integrals and surface integrals. We discuss some of the celebrated results often used in mechanics, such as the Gauss divergence theorem, Stokes' theorem, and Green's identities, and consider how they could be used to restructure problems in ways that simplify the path to their solutions.

We next discuss complex variables, which we use frequently in marine applications, looking closely at analytical functions and complex maps, branch cuts, and contour integration. Fourier analysis follows next, starting with Fourier series on finite domains and moving on to Fourier transforms for functions defined over infinite domains. Much of our later work in irregular waves will utilize this background. Some solution techniques for partial differential equations describing linear systems will also utilize Fourier transformation.

1   Vector Calculus-I

Marine applications encompass a wide spectrum of engineering problems involving solid rigid bodies, solid flexible bodies, and fluids; sometimes solids within fluids and sometimes fluids within solids. Most situations require analysis of the underlying statics and dynamics that reveal important design drivers and methods that could be used in evaluating our designs for successful operation. Insightful and efficient analysis depends on physically and mathematically consistent representations of variables and phenomena. Vectors and matrices are the fundamental building blocks of such representations. We begin with a brief treatment of vectors in two and three dimensions. Vectors are intuitive and quantitative. Vectors have magnitudes, represented by a number (having the same units as the quantity represented, e.g., Newton or N for force), and a direction [represented by another number between 0 and 2π rad (360°)]. Thus, a force pulling an object in a certain direction may be specified using a magnitude F and a direction θ. An opposite force pushing (rather than pulling) the object has an opposite direction $\theta + \pi$. Vectors with a magnitude 1 are unit vectors and are convenient for defining directions. Vectors can be interrelated to each other and to scalars within the consistent frameworks of algebra and calculus, ultimately enabling design-critical decisions involving forces, masses, stiffnesses, velocities, accelerations, volumes, areas, and time. Our goal in this chapter is to review some of the algebra and differential calculus of vectors. We attempt informally to understand more than we formally prove. More detailed treatments of topics covered in this chapter can be found in Kreyszig, Kreyszig, and Norminton (2011) and Greenberg (1978, 1998), and so forth.

1.1   Basic Properties of Vectors

Consider a ship in calm sea conditions that is tied to bollards at 2 or 3 points around the vessel (see, for instance, Figure 1.1). Clearly, the net force on the vessel will depend on the locations of these points relative to the vessel as well as the magnitudes of the forces applied by the cables. In other words, the force directions as determined by the locations of the bollards relative to the vessel are important. For N cables pulling at the vessel from N points, the net force on the vessel is given by

$$\mathbf{F} = \sum_{n=1}^{N} \mathbf{F}_n. \tag{1.1}$$

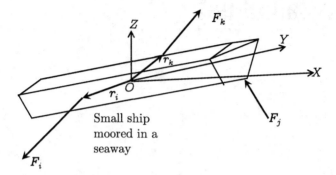

Figure 1.1 A small ship moored at sea, experiencing forces from different directions. Forces may include tension in the mooring lines between the buoys or platforms to which the ship may be moored, as well as any forces from the current flow past the ship at that site. Also included, though not shown here, among the forces and the effective moments would be the wave forces and moments due to wave effects.

Equation (1.1) represents vector summation in three-dimensional Cartesian space. Denoting the directions of the three axes (x, y, z) as chosen in our particular example using the unit vectors $\mathbf{i}$, $\mathbf{j}$, and $\mathbf{k}$, we have

$$F_x\mathbf{i} + F_y\mathbf{j} + F_z\mathbf{k} = \left(\sum_{n=1}^{N} F_{xi}\right)\mathbf{i} + \left(\sum_{n=1}^{N} F_{yi}\right)\mathbf{j} + \left(\sum_{n=1}^{N} F_{zi}\right)\mathbf{k}, \qquad (1.2)$$

where F_x, F_y, and F_z represent the components of the resultant vector along the three coordinate directions. The resultant force vector can thus be obtained simply by adding the respective components of the N force vectors, and

$$F_x = \sum_{n=1}^{N} F_{xi},$$

$$F_y = \sum_{n=1}^{N} F_{xi},$$

$$F_z = \sum_{n=1}^{N} F_{zi}. \qquad (1.3)$$

The components of each vector are related to its direction cosines. Thus, the magnitude of a vector $\mathbf{F}$ in terms of its components F_x, F_y, and F_z is given by

$$F = \sqrt{F_x^2 + F_y^2 + F_z^2} = \left[F_x^2 + F_y^2 + F_z^2\right]^{1/2}. \qquad (1.4)$$

The direction cosines $[\alpha \ \beta \ \gamma]$ can then be expressed as

$$\cos \alpha = \frac{F_x}{F},$$
$$\cos \beta = \frac{F_y}{F},$$
$$\cos \gamma = \frac{F_z}{F}. \tag{1.5}$$

It is not difficult to see that an equal and opposite vector will have the same magnitude F but opposite direction, as given by direction cosines $\pi + \alpha$, $\pi + \beta$, and $\pi + \gamma$.

One property of vector addition that is important to recognize (but easy to pass over lightly) is that it is commutative. Indeed, sometimes it is this property that allows us to decide whether a given quantity is a vector or not. Angular displacement through finite angles (i.e., $\gg 0$) is not a vector since the sequence of rotations matters. This can be illustrated using a book (or another object shaped like a parallelepiped). A rotation through 90° about the long axis followed by a rotation through 90° about the short axis does not place the book in the same orientation as when an opposite sequence is employed. On the other hand, infinitesimal rotations do commute, for which reason, angular velocities also commute and therefore can be considered vectors.

The property of commutativity also plays an important role in vector multiplication. Returning to the aforementioned example of the ship pulled from different directions, so far we have tacitly assumed that the lines of action of all forces pass through the ship's center of mass. In such a situation, there is no net moment generated by the forces that would cause the ship to rotate. However, if the line of action of a force does not pass through the center of mass, then the force will also produce a moment that acts to rotate the body in the direction of the moment. Thus, if different forces $\mathbf{F}_n$ pass through points P_n on the body where each P_n is at a position vector $\mathbf{r}_n$ from the center of mass, then the moment produced by $\mathbf{F}_i$ about the mass center is

$$\mathbf{M}_n = \mathbf{r}_n \times \mathbf{F}_n. \tag{1.6}$$

Note that, as expected, moment is a vector. If both force and position vector lie in a plane, positive moment is counterclockwise and is thought to be out of the plane ("right-hand screw rule"), while negative moment is clockwise and into the plane. The net moment on the ship above is simply the vector sum of the individual moments $\mathbf{M}_i$, such that

$$\mathbf{M} = \sum_{n=1}^{N} \mathbf{r}_n \times \mathbf{F}_n. \tag{1.7}$$

The product above is referred to as "cross product." It is also known as "vector product," since the result of this multiplication of one vector with another is also a vector (see, for instance, Figure 1.2). The magnitude of this product can be evaluated as, continuing the example in equation (1.6),

$$M_n = r_n F_n \sin \theta_n, \tag{1.8}$$

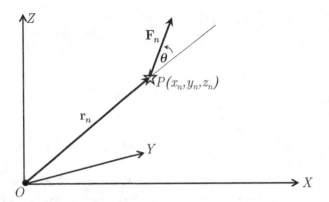

Figure 1.2 A schematic showing the relative disposition of the two vectors $\mathbf{r}_n$ and $\mathbf{F}_n$ involved in the cross product. The result will be a vector, perpendicular to the plane containing the two vectors being multiplied together.

where θ_n is the angle between $\mathbf{r}_n$ and $\mathbf{F}_n$. The vector $\mathbf{M}_n$ resulting from this multiplication is perpendicular to the plane that contains both $\mathbf{r}_n$ and $\mathbf{F}_n$. Further, if one could take the first vector in the product (e.g., here $\mathbf{r}_n$) and literally turn it toward the second vector as if turning a screw, a right-hand screw would advance out of the plane containing $\mathbf{r}_n$ and $\mathbf{F}_n$. This is intuitively consistent with what one expects of moments. It is also easy to see from equation (1.8) that when the two vectors involved in a cross product are collinear (i.e., the angle between them is zero or π radians), their cross product is zero. On the other hand, when the two vectors are perpendicular (i.e., the angle between them is $\pi/2$ or $3\pi/2$ radians), their cross product is the greatest it can be for the given vector magnitudes. Note that the product $\mathbf{F}_n \times \mathbf{r}_n$ also has the same magnitude as shown in equation (1.8). However, if one could now turn $\mathbf{F}_n$ toward $\mathbf{r}_n$ as if turning a screw, a right-hand screw recedes into the plane containing $\mathbf{r}_n$ and $\mathbf{F}_n$. Thus,

$$\mathbf{r}_n \times \mathbf{F}_n = -\mathbf{F}_n \times \mathbf{r}_n. \tag{1.9}$$

We see thus that order is important in a cross product, and that a cross product is not commutative. A more formal way to evaluate a cross product is to evaluate the following determinant, where $\mathbf{i}$, $\mathbf{j}$, and $\mathbf{k}$ are the unit vectors along the three Cartesian coordinate axes.

$$\mathbf{r}_n \times \mathbf{F}_n = \begin{vmatrix} \mathbf{i} & \mathbf{j} & \mathbf{k} \\ x_n & y_n & z_n \\ F_{nx} & F_{ny} & F_{nz} \end{vmatrix}, \tag{1.10}$$

where $\mathbf{r}_n = x_n\mathbf{i} + y_n\mathbf{j} + z_n\mathbf{k}$ and $\mathbf{F}_n = F_{nx}\mathbf{i} + F_{ny}\mathbf{j} + F_{nz}\mathbf{k}$. The relation in equation (1.10) is more convenient to use in three-dimensional static and dynamic analyses, and also makes it clear why order is important in a cross product.

   In the following section, we examine the inner product or dot product, which leads to a scalar result and is therefore commutative.

1.2   Inner Product

When trying to evaluate the work done by a force in causing a displacement of a body, such as perhaps, the work done by the thrust to be applied on a ship in order to move through a certain trajectory, we may want to know how much energy is spent in performing that operation. Alternatively, when one wants to understand the contribution of a force in a particular direction, or the amount of displacement in a particular direction given a displacement vector, one resorts to the so-called inner product or dot product of vectors (see, for instance, Figure 1.3). Thus, the component of a force $\mathbf{F}$ in a direction $\mathbf{e}$ could be represented as

$$F_e = \mathbf{F} \cdot \mathbf{e}, \tag{1.11}$$

where $\mathbf{e}$ is the unit vector in the direction of interest. Since the direction is already defined, F_e only needs to be, and is, a scalar. The result of an inner product is thus a scalar. It follows then that order can be reversed in an inner product, since either way, the same scalar results. The inner product of two vectors $\mathbf{a}$ and $\mathbf{b}$ can be found using

$$\mathbf{a} \cdot \mathbf{b} = ab \cos \theta, \tag{1.12}$$

where a and b denote the magnitudes of $\mathbf{a}$ and $\mathbf{b}$, and θ is the angle between them. It is easy to see from equation (1.12) that when two vectors are orthogonal, i.e., when $\theta = \pi/2$, the inner product between them is zero. Intuitively, it is not surprising that the component of a vector along a direction normal to itself is zero. While the expression in equation (1.12) is convenient to use when the two vectors are coplanar, a more general expression for evaluating them is

$$\mathbf{a} \cdot \mathbf{b} = a_1 b_1 + a_2 b_2 + a_3 b_3, \tag{1.13}$$

where the subscripts 1, 2, and 3 represent the three coordinate directions x, y, and z, respectively.

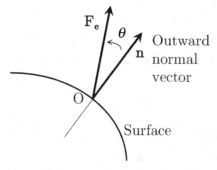

Figure 1.3 A schematic showing the relative disposition of the two vectors involved in the inner product. Here we are interested in knowing the component of the force in a direction normal to the surface.

1.3   General Observations on Cross, Inner Products, and Their Combinations

It is interesting to consider, based on equations such as (1.6) and (1.11), that the cross product of a vector with itself is zero, while the inner product of a vector with itself equals the square of its magnitude. Further, it is also interesting to consider how differently the two products operate on the unit vectors defining the three Cartesian coordinate directions. Thus,

$$\mathbf{i} \cdot \mathbf{j} = \mathbf{i} \cdot \mathbf{k} = \mathbf{j} \cdot \mathbf{k} = 0,$$
$$\mathbf{i} \cdot \mathbf{i} = \mathbf{j} \cdot \mathbf{j} = \mathbf{k} \cdot \mathbf{k} = 1. \tag{1.14}$$

and

$$\mathbf{i} \times \mathbf{j} = \mathbf{k}, \quad \mathbf{j} \times \mathbf{k} = \mathbf{i}, \quad \mathbf{k} \times \mathbf{i} = \mathbf{j},$$
$$\mathbf{i} \times \mathbf{i} = \mathbf{j} \times \mathbf{j} = \mathbf{k} \times \mathbf{k} = 0. \tag{1.15}$$

The vector product relations among the unit vectors can be visualized as a cyclic permutation. Combinations of cross and inner products also lead to interesting results. Thus, an inner product between a vector $\mathbf{A}$ and another vector representing the cross product of two other vectors $\mathbf{B}$ and $\mathbf{C}$ can be represented as $\mathbf{A} \cdot \mathbf{B} \times \mathbf{C}$. It can be seen that this combination results in a scalar. Further, any cyclic permutation of the three vectors leads to the same scalar result. Thus,

$$\mathbf{A} \cdot \mathbf{B} \times \mathbf{C} = \mathbf{C} \cdot \mathbf{A} \times \mathbf{B} = \mathbf{B} \cdot \mathbf{C} \times \mathbf{A}. \tag{1.16}$$

The scalar triple product could arise, for instance, when we need to compute the instantaneous power provided by a force on a body that is constrained to rotate about a fixed point in space. In this case, the cross product of the angular velocity of the body and the radius vector to its center of mass gives the linear velocity vector for the center of mass. The inner product of the force vector and the velocity vector determines the power transmitted by the force. The scalar triple product can be written in determinant form as

$$\mathbf{A} \cdot \mathbf{B} \times \mathbf{C} = \begin{vmatrix} A_x & A_y & A_z \\ B_x & y_n & z_n \\ C_x & C_y & C_z \end{vmatrix}. \tag{1.17}$$

The vector product of a vector $\mathbf{A}$ with a vector product of two other vectors $\mathbf{B}$ and $\mathbf{C}$ is a vector quantity, and therefore sensitive to the order in which the vectors appear. Thus,

$$\mathbf{A} \times \mathbf{B} \times \mathbf{C} = -\mathbf{C} \times \mathbf{A} \times \mathbf{B} = \mathbf{C} \times \mathbf{B} \times \mathbf{A}. \tag{1.18}$$

A useful identity involving the vector triple product is

$$\mathbf{A} \times \mathbf{B} \times \mathbf{C} = \mathbf{B}(\mathbf{A} \cdot \mathbf{C}) - \mathbf{C}(\mathbf{A} \cdot \mathbf{B}). \tag{1.19}$$

Vector triple products often arise in the study of dynamics of bodies in three dimensions, when transforming quantities expressed in a coordinate system at the centroid of a rotating body to quantities written with respect to an inertial reference frame. Recall that an inertial reference frame is either fixed in space or moving at constant velocity (hence by extension, nonrotating).

1.4   Differential Calculus of Vectors

In rigid-body dynamics, one deals with time derivatives of vectors as a routine matter. In flexible-body dynamics, as well as in fluid dynamics, however, quantities can vary in time as well as space. Often the quantities to be observed are scalars. Density, for instance, is a scalar quantity that in an inhomogeneous fluid may be a function of position of a point within a fluid. Similarly, velocity of fluid particles is a vector that may be different at different points. When we need to keep track of the spatial and temporal dependence of quantities such as density, velocity, and pressure, the term "field" is commonly employed. Thus, we have a velocity field described as $\mathbf{v}(x, y, z; t)$ or a pressure field represented as $p(x, y, z; t)$. Note that since pressure at a point within a fluid is independent of direction, it is a scalar. Derivatives of field variables such as pressure and velocity take on a new significance when their rates of change relative to spatial coordinates need to be analyzed. Here, we get our early introduction to the concepts and notations that are frequently used in marine applications, as we will see in Chapters 10 and 11, later in this text. We begin by reviewing an important symbol, ∇, the so-called "del" operator, which is defined as

$$\nabla \equiv \frac{\partial}{\partial x}\mathbf{i} + \frac{\partial}{\partial y}\mathbf{j} + \frac{\partial}{\partial z}\mathbf{k}. \tag{1.20}$$

We have here assumed the Cartesian rectangular coordinate system. Note that ∇ is a vector and combines partial derivatives along the three coordinate axes in a consistent manner, using the unit vectors $\mathbf{i}$, $\mathbf{j}$, and $\mathbf{k}$ to represent the three coordinate directions with which the individual partial derivatives are associated. Note further that ∇ represents an operation, inasmuch as the individual partial derivatives are operations. Since pressure is a scalar, the ∇ operation on pressure p can be represented as

$$\nabla p = \frac{\partial p}{\partial x}\mathbf{i} + \frac{\partial p}{\partial y}\mathbf{j} + \frac{\partial p}{\partial z}\mathbf{k}. \tag{1.21}$$

Here the three partial derivatives represent the rates of change of pressure in the three coordinate directions. When ∇ operates on a scalar, the result is a vector and is referred to as the "gradient" of the scalar pressure field at each point. The gradient is an important operation, because we can use it to find the rates of change of a field variable in particular directions. An arbitrary direction in space can be described using a unit vector $\mathbf{n}$, where $\mathbf{n} = n_x\mathbf{i} + n_y\mathbf{j} + n_z\mathbf{k}$. The rate of change of pressure above, for instance, in the $\mathbf{n}$ direction is simply the component of its gradient vector along the vector $\mathbf{n}$, as given by the inner product

$$\nabla p \cdot \mathbf{n} = \frac{\partial p}{\partial x}n_x + \frac{\partial p}{\partial y}n_y + \frac{\partial p}{\partial z}n_z. \tag{1.22}$$

What if the field variable that ∇ is to operate on is also a vector, such as velocity $\mathbf{v}(x,y,z;t)$? We can perform this operation in two ways: (i) define it as an inner product between ∇ and $\mathbf{v}$, or (ii) define it as a cross product between ∇ and $\mathbf{v}$. In the former case, the result is a scalar, while in the latter case, it is a vector. Both are meaningful physically. The inner product is referred to as the "divergence" of the vector field $\mathbf{v}$, while the cross product is referred to as the "curl" of the vector field $\mathbf{v}$. We begin with a discussion of the inner product:

$$\mathrm{div}\,\mathbf{v} = \nabla \cdot \mathbf{v} = \frac{\partial v}{\partial x} + \frac{\partial v}{\partial y} + \frac{\partial v}{\partial z}. \tag{1.23}$$

We note again that divergence is a scalar quantity, and since it also is a function of the spatial coordinate, it represents a scalar field in its own right. In physical terms, divergence of velocity represents the net outward flow per unit volume at a point represented by the coordinates (x,y,z) (see, for instance, (Greenberg 1978) for a further discussion of divergence). This interpretation of divergence leads naturally into the following discussion.

Another scalar field was mentioned above, namely, the fluid density $\rho(x,y,z;t)$. By combining density ρ with velocity $\mathbf{v}$, one can discuss mass flow rates and examine what conservation of mass means in fluids. If we consider a small hypothetical elemental volume within the fluid, we can argue that the rate of change of mass (increase or decrease) within the small element must equal the net inflow or outflow of mass into or from the element. This can be expressed in the form

$$\frac{\partial \rho}{\partial t} + \mathrm{div}(\rho \mathbf{v}) = 0. \tag{1.24}$$

With the help of equation (1.23), we find that equation (1.24) implies that

$$\frac{\partial \rho}{\partial t} + \nabla \cdot \rho \mathbf{v} = 0. \tag{1.25}$$

Expanding the second term,

$$\frac{\partial \rho}{\partial t} + \nabla \rho \cdot \mathbf{v} + \rho \nabla \cdot \mathbf{v} = 0. \tag{1.26}$$

The ∇ in the second term represents the gradient of the density field, given by

$$\nabla \rho = \frac{\partial \rho}{\partial x}\mathbf{i} + \frac{\partial \rho}{\partial y}\mathbf{j} + \frac{\partial \rho}{\partial z}\mathbf{k}. \tag{1.27}$$

Thus,

$$\nabla \rho \cdot \mathbf{v} = u\frac{\partial \rho}{\partial x} + v\frac{\partial \rho}{\partial y} + w\frac{\partial \rho}{\partial z}. \tag{1.28}$$

Here, we have used $\mathbf{v} = (u,v,w)$, or in other words, utilized the three Cartesian components of velocity $\mathbf{v}$. Equation (1.26) can thus be rewritten as

$$\left[\frac{\partial \rho}{\partial t} + u\frac{\partial \rho}{\partial x} + v\frac{\partial \rho}{\partial y} + w\frac{\partial \rho}{\partial z}\right] + \rho \nabla \cdot \mathbf{v} = 0. \tag{1.29}$$

The terms enclosed by the square bracket represent the so-called "material derivative" or "Lagrangian derivative" in a fluid flow, represented using the notation D/Dt. This is the rate of change of a quantity (here density) that we would see if we followed the fluid, flowing with it at velocity $\mathbf{v}$. Thus, here

$$\frac{D\rho}{Dt} + \rho\nabla \cdot \mathbf{v} = 0. \qquad (1.30)$$

If the fluid is incompressible, then

$$\frac{D\rho}{Dt} = 0, \qquad (1.31)$$

which implies (since ρ is not in general, zero)

$$\nabla \cdot \mathbf{v} = \frac{\partial u}{\partial x} + \frac{\partial v}{\partial y} + \frac{\partial w}{\partial z} = 0. \qquad (1.32)$$

Equation (1.30) represents the continuity equation (or the conservation of mass principle) in compressible fluids, while when the fluid is incompressible, equation (1.31) represents conservation of mass principle. In general, when flow velocities are much smaller than the speed of sound in the fluid (i.e., two or three orders of magnitude smaller), the effect of compressibility becomes relatively negligible, and the fluid can be treated as nearly incompressible.

We mention here that an important application of the material derivative arises when one wants to study the dynamics of flowing fluids, where the first step is the application of Newton's second law on an arbitrary small fluid element of volume dV. With the condition that the mass of the element itself remains constant, we can express its momentum as $\rho \mathbf{v} dV$. The rate of change of momentum, for incompressible fluids, as we track the element through its motion is understood to be

$$\rho\frac{D\mathbf{v}}{Dt} = \rho\left[\frac{\partial \mathbf{v}}{\partial t} + (\mathbf{v} \cdot \nabla)\mathbf{v}\right]. \qquad (1.33)$$

Let us next consider the cross product of the operator ∇ with another vector. Using the fluid velocity $\mathbf{v}$ again as an example, we can express the cross product $\nabla \times \mathbf{v}$ as

$$\nabla \times \mathbf{v} = \begin{vmatrix} \mathbf{i} & \mathbf{j} & \mathbf{k} \\ \frac{\partial}{\partial x} & \frac{\partial}{\partial y} & \frac{\partial}{\partial z} \\ u & v & w \end{vmatrix}. \qquad (1.34)$$

This can be rewritten as

$$\nabla \times \mathbf{v} = \left(\frac{\partial w}{\partial y} - \frac{\partial v}{\partial z}\right)\mathbf{i} + \left(\frac{\partial u}{\partial z} - \frac{\partial w}{\partial x}\right)\mathbf{j}$$
$$\left(\frac{\partial v}{\partial x} - \frac{\partial u}{\partial y}\right)\mathbf{k}. \qquad (1.35)$$

This product thus takes a vector field such as velocity and defines another, closely associated, vector field. The product in equations (1.34) and (1.35) is referred to as the curl of velocity $\mathbf{v}$. It is easier to see the physical significance of curl in a two-dimensional case, where $\mathbf{v} = u\mathbf{i} + v\mathbf{j}$. In this case, $w = 0$, and there is no variation of

u and v in the z direction in purely two-dimensional flow. Thus, the curl of this vector becomes

$$\nabla \times \mathbf{v} = \left(\frac{\partial v}{\partial x} - \frac{\partial u}{\partial y}\right)\mathbf{k}. \tag{1.36}$$

First of all, we notice that the curl in equation (1.36) is perpendicular to the plane occupied by the $\mathbf{v}$ vector. On a little more specific level, we observe that the first term represents the rate of change of the y-component of velocity with respect to x, while the second term denotes the rate of change of the x-component of velocity relative to y. Figure 1.4 explains what this means for a small fluid element of dimensions dx and dy. Counterclockwise rotation is taken to be positive. If the element were to rotate about the fixed point O as a rigid body, the horizontal side would rotate by the same amount as the vertical side, in the same sense/direction. Then, the horizontal side OA would rotate counterclockwise at an angular velocity $\partial v/\partial x = \omega_H$, and the vertical side OC would rotate counterclockwise at an angular velocity $\partial u/\partial y = \omega_V$, where the two are numerically equal, or $\omega_H = \omega_V = \omega_R$, out of the plane of the page. Since u is positive to the left, the angular velocity of the element could then be expressed as

$$\omega = \frac{1}{2}(\omega_H + \omega_V) = \frac{1}{2}\left(\frac{\partial v}{\partial x} + \frac{\partial u}{\partial y}\right)\mathbf{k} = \omega_R\mathbf{k}. \tag{1.37}$$

The first term after the equality denotes an average of the two angular velocities, which in this case happens to be the actual angular velocity of the element. The above, of course, assumes that the element is rotating like a rigid body. At small scales, fluid elements tend to be deformable, however, and the sides OA and OC may not rotate at the same angular velocity, or in the same sense, for that matter. Thus, the side OA could rotate counterclockwise, while the side OC rotates clockwise, and a square-shaped element may not look like a square upon deformation. This is the case of shear

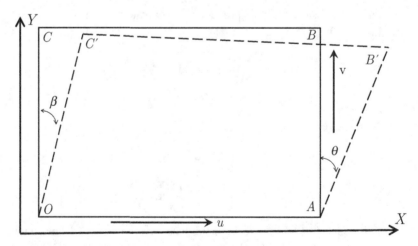

Figure 1.4 A schematic showing the deformation of a small fluid element through non-rigid-body rotation of the element, where its vertical and adjacent sides undergo different rotational deformations.

deformation, as would be caused by viscous effects in a fluid, which we will study in detail in Chapter 10. In that case, we would need to use the original expression in equation (1.36) to define the average angular velocity of the element, such that

$$\omega = \frac{1}{2}\left(\frac{\partial v}{\partial x} - \frac{\partial u}{\partial y}\right)\mathbf{k} = \frac{1}{2}\left(\nabla \times \mathbf{v}\right). \tag{1.38}$$

In either case, we see that

$$\nabla \times \mathbf{v} = \frac{1}{2}\omega. \tag{1.39}$$

In terms of its physical significance, curl of the velocity is thus related to the angular velocity in a fluid and is referred to as "vorticity." An example of flow such as in equation (1.37) can occur in a fluid that is stirred into rotation about a certain axis at a uniform angular velocity. We make a slight digression here to point out that fluid particle motion under a small-amplitude propagating surface gravity wave in deep water is circular. Yet such waves and the particle motion under them can be described using velocity potentials, which presumes irrotational motion. Particle motion here is therefore such that the individual particle itself does not rotate about its own axis as it travels over its circular trajectory. In other words, a hypothetical rectangular particle with a horizontal long dimension travels such that the long horizontal dimension remains horizontal throughout the trajectory. If the particle were to undergo rotational motion at angular ω, all sides of the particle would rotate at ω. As pointed out by Greenberg (1978), particle would undergo irrotational motion if it traveled as an individual seat in a Ferris wheel would. Some interesting observations follow:

$$\nabla \times \nabla f = 0, \tag{1.40}$$

where f denotes a scalar field. In words, this is to say that the curl of a gradient of a scalar field is zero. In addition,

$$\nabla \cdot \nabla \times \mathbf{v} = 0. \tag{1.41}$$

Equation (1.41) states that the divergence of a curl of a vector is zero. Both statements can be verified using the determinant expansion form for cross product and working through the entire operations, realizing that

$$\frac{\partial^2 u}{\partial y \partial z} = \frac{\partial^2 u}{\partial z \partial y}, \text{ etc.} \tag{1.42}$$

Some other statements and interesting interrelationships involving the del operator are included below:

$$\nabla^2 = \nabla \cdot \nabla = \frac{\partial^2}{\partial x^2} + \frac{\partial^2}{\partial y^2} + \frac{\partial^2}{\partial z^2},$$
$$\nabla \cdot (f\mathbf{v}) = \nabla f \cdot \mathbf{v} + f\nabla \cdot \mathbf{v},$$
$$\nabla \times (f\mathbf{v}) = \nabla f \times \mathbf{v} + f\nabla \times \mathbf{v},$$
$$\nabla \cdot (\mathbf{u} \times \mathbf{v}) = \mathbf{v} \cdot (\nabla \times \mathbf{u}) - \mathbf{u} \cdot (\nabla \times \mathbf{v}). \tag{1.43}$$

The ∇^2 operator in the first of equations (1.43) is generally referred to as the "Laplacian" and will be used frequently in the hydrodynamics discussions of the subsequent chapters.

1.5   Indicial Notation

Here we introduce a system for representing three-dimensional (and higher) quantities. The indicial notation is widely used in Physics, Hydrodynamics, and Elasticity and represents a very compact way to represent, manipulate, and understand higher-dimensional quantities (i.e., vectors, tensors, etc.). Indicial notation will be used frequently in this text, for instance, in Chapters 9, 10, 11, and so forth. The representations for inner products (dot products), derivatives, and so forth, require some attention and are also discussed here.

Under the rules of indicial notation, when an index occurs unrepeated in a term, that index is understood to take on the values 1, 2, ..., K, where K is a specified integer. When an index appears twice in a term, that index is understood to take on all the values of its range, and the resulting terms summed. This often is called "Einstein's Summation Convention." In this notation, the repeated indices are called "dummy indices" and the unrepeated ones are called "free indices." For example, the material derivative can be written, in this notation, as

$$\frac{D}{Dt} = \frac{\partial}{\partial t} + u_j \frac{\partial}{\partial x_j}, \quad j = 1,2,3,$$

and material derivative of a vector $\mathbf{u} = [u_1, u_2, u_3]$ can be written as

$$\frac{Du_i}{Dt} = \frac{\partial u_i}{\partial t} + u_j \frac{\partial u_i}{\partial x_j}, \quad i = 1,2,3, \quad j = 1,2,3 \text{ (sum over } j\text{).} \qquad (1.44)$$

Note also that $\partial u_i / \partial x_j$ can be written as $u_{i,j}$ in indicial notation, in other words, comma denotes differentiation.

The most important rules of the indicial notation are:

1. A letter index may occur either once or twice in a given term. If it occurs once, it is a free index. If it occurs twice, it is a dummy index. For example, the term a_{ijj} has one free index, i, and one dummy index, j.
2. If an index is a free index, it is understood to take on the values $1, 2, \ldots, K$, where K is an integer that defines the range of the index. For example, $a_{ijj}, i, j = 1,2,3$, indicates that i and j run from 1 through 3. The tensorial rank of a term is determined by the number of free indices in that term. For example, a_{ijj} indicates that the term is a tensor of rank 1, since there is only one free index, i. This will be discussed in Section 2.6.
3. When an index is a dummy index, i.e., it occurs twice in a given term, it is understood that the term is summed over the dummy index from 1 through K, the entire range of the index. For example, $a_{ijj} = 0, i, j = 1,2,3$, represents $K = 3$

equations, one for each free index i, and each equation has 3 terms on the left-hand side, e.g., $a_{1jj} = a_{111} + a_{122} + a_{133} = 0$, for $i = 1$, $j = 1,2,3$, and so forth.

4. A dummy (or repeated) index can be replaced by another dummy index in any one or more terms in an equation since the meaning of the dummy index does not change, e.g., $a_{ijj} = a_{ikk}$ is acceptable or is equivalent. But a free index can only be replaced by another letter index if the same free index in all other terms in the equation is replaced with the same letter index. For example, $a_{ijj} + b_{ill} = 0$ can also be written as $a_{kjj} + b_{kll} = 0$.

1.5.1   Examples

(**a**) Velocity vector:

$$V = u_1 e_1 + u_2 e_2 + u_3 e_3 \quad \text{or} \quad V = u_j e_j.$$

(**b**) Equation of a plane in Ox_1, x_2, x_3:

$$ax_1 + bx_2 + cx_3 = d \quad \text{or} \quad a_1 x_1 + a_2 x_2 + a_3 x_3 = d \quad \text{or} \quad \sum_{j=1}^{3} a_j x_j = d \quad \text{or} \quad a_j x_j = d.$$

(**c**) Dot product:

$$q \cdot q = (q_1 e_1 + q_2 e_2 + q_3 e_3) \cdot (q_1 e_1 + q_2 e_2 + q_3 e_3),$$

$$= q_1^2 + q_2^2 + q_3^2 = \sum_{j=1}^{3} q_j q_j$$

$$= q_j q_j, \qquad j = 1, 2, 3.$$

(**d**) Derivative of a scalar:

$$\frac{\partial^2 \phi}{\partial x_1{}^2} + \frac{\partial^2 \phi}{\partial x_2{}^2} + \frac{\partial^2 \phi}{\partial x_3{}^2} = \nabla^2 \phi = \Delta\phi = \phi_{,jj} = 0,$$

which is the Laplace equation. Note that $\nabla^2 \equiv \nabla \cdot \nabla \equiv \Delta$ (Delta).

We now introduce the "Kronecker delta" and the "permutation symbol." Kronecker delta is defined by:

$$\delta_{ij} = \begin{cases} 1 & \text{for } i = j, \\ 0 & \text{for } i \neq j. \end{cases} \tag{1.45}$$

We also introduce the "permutation symbol," defined by

$$\epsilon_{ijk} = \begin{cases} 1 & \text{if } i, j, k \text{ are an even permutation of } 1, 2, 3, \\ -1 & \text{if } i, j, k \text{ are an odd permutation of } 1, 2, 3, \\ 0 & \text{if } i, j, k \text{ are not a permutation of } 1, 2, 3. \end{cases} \tag{1.46}$$

1.5.2   Examples

a) $\delta_{11} = 1$, $\delta_{12} = 0$, $\delta_{31} = 0$, $\delta_{22} = 1$,

b) $\epsilon_{123} = 1, \epsilon_{321} = -1, \epsilon_{112} = 0, \epsilon_{213} = -1$,

c) $\delta_{jj} = \delta_{11} + \delta_{22} + \delta_{33} = 3$,

d) Verify that $\epsilon_{ijk}\epsilon_{ijm} = 2\delta_{km}$.

Note first that the number of free indices on the left and on the right are the same, as they should be. Let us now expand $\epsilon_{ijk}\epsilon_{ijm}$ for a range of $i, j, k, m = 1, 2, 3$:

$$\epsilon_{ijk}\epsilon_{ijm} = \epsilon_{12k}\epsilon_{12m} + \epsilon_{13k}\epsilon_{13m} + \epsilon_{21k}\epsilon_{21m} + \epsilon_{23k}\epsilon_{23m} + \epsilon_{31k}\epsilon_{31m} + \epsilon_{32k}\epsilon_{32m},$$

since if any two indices are the same, then $\epsilon_{ijk} = 0$, e.g., $\epsilon_{11k} = 0$, and therefore,

$$\text{for} \quad k = 1, \quad \epsilon_{ijk}\epsilon_{ijm} = \epsilon_{231}\epsilon_{23m} + \epsilon_{321}\epsilon_{32m} = \begin{cases} 2 & \text{if } m = 1 \\ 0 & \text{if } m \neq 1 \end{cases}$$

$$\text{for} \quad k = 2, \quad \epsilon_{ijk}\epsilon_{ijm} = \epsilon_{132}\epsilon_{13m} + \epsilon_{312}\epsilon_{31m} = \begin{cases} 2 & \text{if } m = 2 \\ 0 & \text{if } m \neq 2 \end{cases}$$

$$\text{for} \quad k = 3, \quad \epsilon_{ijk}\epsilon_{ijm} = \epsilon_{123}\epsilon_{12m} + \epsilon_{213}\epsilon_{21m} = \begin{cases} 2 & \text{if } m = 3 \\ 0 & \text{if } m \neq 3 \end{cases}$$

As a result,

$$\epsilon_{ijk}\epsilon_{ijm} = \begin{cases} 2 & \text{if } k = m \\ 0 & \text{if } k \neq m \end{cases}$$

and thus,

$$\epsilon_{ijk}\epsilon_{ijm} = 2\delta_{km}, \tag{1.47}$$

as we were asked to verify.

The Kronecker delta is generally used in cases such as an orthogonal transformation of a coordinate system to another, or as a substitution operator to obtain compact forms of some equations as we shall see later on. δ_{ij} is a second-order tensor (see Section 2.6).

The permutation symbol (or alternator) is generally used for the cross product of vectors. For instance, if $a = a_1 e_1 + a_2 e_2 + a_3 e_3$ and $b = b_1 e_1 + b_2 e_2 + b_3 e_3$ then $a \times b = c$ can be written as

$$\epsilon_{ijk}a_j b_k = c_i. \tag{1.48}$$

Equation (1.48) can be expanded to see that this is indeed the case. ϵ_{ijk} is a third-order tensor (see Section 2.6).

Note also that if two indices of the permutation symbol are exchanged, then the value of the permutation symbol will change the sign. For example,

$$\epsilon_{ijk} = -\epsilon_{jik}, \tag{1.49}$$

and this can be shown by simply expanding the indices for their range. Therefore, ϵ_{ijk} is a skew-symmetric tensor. Equation (1.49) is frequently used in expanding equations written in indicial notation or to prove identities. Also note that the product of a skew symmetric tensor and a symmetric tensor is zero.

Another identity commonly used is the $\epsilon - \delta$ identity given by

$$\epsilon_{ijk}\epsilon_{klm} = \delta_{il}\delta_{jm} - \delta_{im}\delta_{jl}, \qquad (1.50)$$

Note also that

$$\delta_{ij}x_j = x_i, \qquad \delta_{kl}v_l = v_k, \qquad (1.51)$$

1.6   Concluding Remarks

In this chapter, we reviewed some of the basic descriptions and relationships important to the use of vectors. Part of our focus here was on vector algebra, specifically, the different ways in which vectors can be multiplied with each other, to give either scalar or vector results, namely, the inner product, and the cross product, respectively. We considered some examples representative of situations often encountered in marine applications, and the fluid mechanics basics that one needs to be familiar with when working with bodies in water and fluid flows. Part of our attention here was on the differential calculus relations that are important to solving the problems of interest to marine situations. Such problems may require analysis of the combined effects of multiple vector quantities, particularly, how their interaction determines behavior, of a body in water or the water itself. In Chapter 2, we will review the integral calculus of vectors, with a special emphasis on some of the classical relationships and theorems of vector integral calculus that are frequently used in problem-solving.

1.7   Self-Assessment

1.7.1

Consider the vectors

$$\mathbf{c} = [3,2,-1] \text{ and } \mathbf{d} = [-2,3,1]. \qquad (1.52)$$

Evaluate: $\mathbf{c} \times \mathbf{d}$ and $\mathbf{d} \times \mathbf{c}$.

1.7.2

Given the two vectors $\mathbf{c}$ and $\mathbf{d}$ above, and a third vector $\mathbf{e} = [1,0,10]$, consider the three-dimensional volume defined by the three vectors.

(a) What is the area of the parallelogram formed by $\mathbf{c}$ and $\mathbf{d}$ (given by $\|\mathbf{c} \times \mathbf{d}\|$)?

(b) What is the volume defined by the three vectors, given by $|\mathbf{c} \cdot \mathbf{d} \times \mathbf{e}|$? What would the volume be if the three vectors are coplanar? Thus, suggest a test for coplanarity of three vectors.

1.7.3

For $\phi = x^2 y^3 z^3$, obtain

(a) $\nabla\phi$,

(b) $\nabla\phi$ at $(2,1,0)$ along the y axis (i.e., in the $\mathbf{j}$ direction).

1.7.4

Verify the following relationships listed in Section 1.4:

$$\nabla^2 = \nabla \cdot \nabla = \frac{\partial^2}{\partial x^2} + \frac{\partial^2}{\partial y^2} + \frac{\partial^2}{\partial z^2}$$

$$\nabla \cdot (f\mathbf{v}) = \nabla f \cdot \mathbf{v} + f \nabla \cdot \mathbf{v}$$

$$\nabla \times (f\mathbf{v}) = \nabla f \times \mathbf{v} + f \nabla \times \mathbf{v}$$

$$\nabla \cdot (\mathbf{u} \times \mathbf{v}) = \mathbf{v} \cdot (\nabla \times \mathbf{u}) - \mathbf{u} \cdot (\nabla \times \mathbf{v}) \tag{1.53}$$

1.7.5

Show that

$$\nabla \times (\nabla \times \mathbf{a}) = \nabla(\nabla \cdot \mathbf{a}) - \nabla^2 \mathbf{a}, \quad \nabla^2 = (\nabla \cdot \nabla)\mathbf{a}. \tag{1.54}$$

1.7.6

If $\mathbf{a} = (2,1,3)$, what is $\nabla \times (\nabla \times \mathbf{a})$?

2 Vector Calculus-II

In this chapter, we continue our journey into the mathematics of vectors. Having reviewed the different ways in which vectors may be multiplied together to lead to vector or scalar results, the del operator, and its various operations on vectors, we now broaden our treatment somewhat and consider how a vector quantity or operations on vector quantities behave under integration. For field variables such as density, pressure, velocity, and so forth, integration may be required over a line or a curve, over a surface, or over a volume. Integrals defined over specified lines or curves are useful in computing the work done by a force, for instance, the work done on a ship by the net force on it due to surface drag as the ship travels over a particular trajectory. Integration over surfaces typically arises when one needs to compute the net force acting on the immersed surface of a floating or submerged body. Sometimes when the center of mass or the center of buoyancy of ships or floating bodies in oscillation are to be evaluated in real time, volume integrals need to be evaluated. Interesting relationships exist between volume integrals within regions enclosed by closed surfaces and surface integrals evaluated over the enclosing surfaces. These can provide valuable, sometimes profound, insight into the behavior of the field variables appearing within these integrals. As before, a more extensive discussion of the topics of this chapter can be found in Kreyszig, Kreyszig, and Norminton (2011) and Greenberg (1978, 1998).

We begin our study with integrals to be evaluated over specified lines or curves in space.

2.1 Line Integrals

Suppose we want to evaluate the total energy spent by a ship or its scaled-down model in moving at a specified speed through a certain "test trajectory" in different wave conditions, including calm water. The trajectory or path may be curvilinear, and an open curve, i.e., simply take the vessel from point A to point B, or it may be a closed curve in that the vessel traces a specified trajectory that brings it back to its point of origin. In any event, it is important to evaluate the work done by the vessel in overcoming the resistance forces acting on it as it completes its trajectory. Calculation of work done over a trajectory requires evaluation of a line integral. Line integrals are important, and as in our exploration of the calculus of variations in Chapter 16, we will ask and answer questions such as follows: What trajectory leads to the minimum

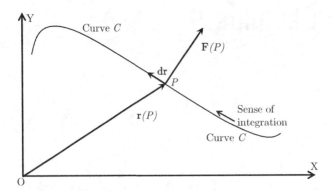

Figure 2.1 Figure illustrating a line integral of a vector (e.g., force) along a prescribed path (trajectory) designated as C. See equation (2.2). P is an arbitrary point on the path C.

energy expended, what trajectory leads to the minimum travel time, what trajectory leads to both, and so forth. Here, if the work done by the ship in overcoming a resistance force $\mathbf{F}$ in order to travel over a path element $\mathbf{dr}$ is written as a the inner product (see Figure 2.1).

$$dW = \mathbf{F} \cdot \mathbf{dr}. \tag{2.1}$$

If the path or trajectory to be followed by the vessel is denoted by C, then the total work done can be found using an integral evaluated over the path C,

$$W = \int_C \mathbf{F} \cdot \mathbf{dr}. \tag{2.2}$$

If the trajectory is a closed curve, then

$$W = \oint_C \mathbf{F} \cdot \mathbf{dr}. \tag{2.3}$$

We may also define the line integral in equation (2.2) more directly as integral over the incremental length of a path element dS, as

$$W = \int_C \left(\mathbf{F} \cdot \frac{\mathbf{dr}}{ds} \right) ds \tag{2.4}$$

where the quantity in parentheses is seen to be a scalar. Other approaches defining line integrals can also be used. For instance, if the trajectory is semicircular (see Figure 2.2), one may describe the radius vector to any point on the trajectory, for unit radius, as

$$\mathbf{r} = \cos\theta\mathbf{i} + \sin\theta\mathbf{j}. \tag{2.5}$$

This implies that

$$\frac{\mathbf{dr}}{d\theta} = -\sin\theta\mathbf{i} + \cos\theta\mathbf{j}. \tag{2.6}$$

If the force were specified in suitable units as

$$\mathbf{F} = 3\mathbf{r} - 4\frac{\mathbf{dr}}{ds}, \tag{2.7}$$

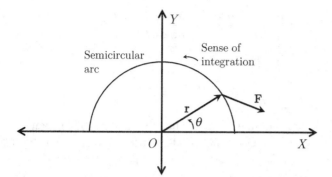

Figure 2.2 Line integral expressed in terms of a scalar variable (θ here). See equations (2.5–2.10). The trajectory is a semicircular arc.

then in terms of the variable θ,

$$\mathbf{F} = (3\cos\theta + 4\sin\theta)\,\mathbf{i} + (3\sin\theta - 4\cos\theta)\,. \tag{2.8}$$

so that

$$\mathbf{F} \cdot \frac{d\mathbf{r}}{d\theta}\,d\theta = \left(-3\cos\theta\sin\theta - 4\sin^2\theta + 3\cos\theta\sin\theta - 4\cos^2\theta\right)d\theta. \tag{2.9}$$

The expression in equation (2.9) simplifies to

$$W = -4\int_0^\pi d\theta = -4\pi. \tag{2.10}$$

As expected, in general, the value of a line integral depends on the vector function being integrated, and on the curve or path along which the integration is carried out. For a large category of functions, however, the value of a line integral only depends on the end points regardless of the integration path. We recall that under some conditions, vector functions may be defined by means of a potential function. Thus, for instance, a vector field $\mathbf{v}(x,y,z)$, if a potential function can be found such as

$$\mathbf{v} = \nabla\phi(x,y,z) = \frac{\partial\phi}{\partial x}\mathbf{i} + \frac{\partial\phi}{\partial y}\mathbf{j} + \frac{\partial\phi}{\partial z}\mathbf{k}. \tag{2.11}$$

Realizing that $d\mathbf{r} = dx\mathbf{i}, dy\mathbf{j}, dz\mathbf{k}$, and writing $\mathbf{v} = u\mathbf{i} + v\mathbf{j} + w\mathbf{k}$,

$$\int_C \mathbf{v}\cdot d\mathbf{r} = \int_C (u\,dx + v\,dy + w\,dz) = \int_a^b \left(u\frac{dx}{ds} + v\frac{dy}{ds} + w\frac{dw}{ds}\right)ds. \tag{2.12}$$

The integral is thus evaluated between the limits $s = a$ and $s = b$. Inserting the right-hand side of equation (2.11) into equation (2.12),

$$\int_C \mathbf{v}\cdot d\mathbf{r} = \int_a^b \left(\frac{\partial\phi}{\partial x}\frac{dx}{ds} + \frac{\partial\phi}{\partial y}\frac{dy}{ds} + \frac{\partial\phi}{\partial z}\frac{dz}{ds}\right)ds. \tag{2.13}$$

The expression in parentheses is simply

$$\left(\frac{\partial\phi}{\partial x}\frac{dx}{ds} + \frac{\partial\phi}{\partial y}\frac{dy}{ds} + \frac{\partial\phi}{\partial z}\frac{dz}{ds}\right) = \frac{d\phi}{ds}ds = d\phi. \tag{2.14}$$

so that

$$\int_C \mathbf{v}\cdot d\mathbf{r} = \int_a^b d\phi = \phi(b) - \phi(a). \tag{2.15}$$

This is a remarkable result in that we find that the value of the integral only depends on the end points a and b, and is independent of the actual path or the curve along which the integral is to be evaluated. All paths in this case lead to the same value as long as the limits are the same, and paths lie within a single, connected spatial domain (such as the inside of a cube or a sphere, but not the inside of a torus). What would happen if the integral were defined over a closed curve, i.e., $a = b$? The integral in equation (2.15) would then evaluate to zero.

We conclude then that as long as the function to be integrated can be defined as a gradient of a potential function, the line integral would be path independent. The work done by a force, as expressed in equation (2.9), would be path independent and evaluate to zero over a closed curve if the force $\mathbf{F}$ could be expressed as a gradient of a potential function. This is certainly the case for gravity or inertial forces, or indeed for an ideal spring with zero dissipation. The slightest presence of friction or energy dissipation or energy radiation (as in the case of floating bodies that make outward traveling waves with their oscillations) would change everything, however. In the ideal case when no frictional forces exist, and only gravity and inertial forces act on the body, the work done along a closed curve in a connected domain is zero (with the curve never leaving the domain). The forces for which this is true are referred to as "conservative forces."

Work done by frictional forces, such as ship resistance forces (e.g., frictional drag), wavemaking resistance, and so forth, would clearly depend on the path followed, longer paths leading to greater work, requiring greater energy. Resistive or dissipative forces are referred to as "nonconservative forces." Conservative forces can be defined as gradients of some potential field, such as potential energy (as in the case of gravity forces) or kinetic energy (as in the case of inertial forces). We summarize by saying that for a line integral of a vector to be path independent, $\mathbf{v} = \nabla\phi$. We recall that, in this case, $\nabla \times \mathbf{v} = 0$.

We add an important note to this discussion. The domain referred to earlier is a "simply connected domain." A domain is said to be simply connected when a closed curve in that domain can progressively be shrunk down to a point without leaving the domain. We now move our discussion to surface integrals.

2.2   Surface Integrals

Surface integrals are integrals of functions evaluated over surfaces in three-dimensional space. Often it is helpful to define surfaces using a generalization of the parametric representation of curves, with two parameters rather than one. Thus, we may define a surface using parameters (μ, β), where a surface may be defined by all points represented by a position vector $\mathbf{r}(\mu, \beta)$ such that

$$\mathbf{r}(\mu, \beta) = x(\mu, \beta)\mathbf{i} + y(\mu, \beta)\mathbf{j} + z(\mu, \beta)\mathbf{k}. \tag{2.16}$$

This representation is useful in that it allows us to define planes that are tangential to the surface at a chosen point on the surface, as well as vectors that are normal to the surface at that point. "Surface normals" are important in fluid-flow problems, helping us determine the net inflow or outflow (flux) of fluid (or for that matter, heat, or electric charge) across a surface. Surface normals are extensively used in fluid mechanics discussions (as we will see in Chapters 10, 11, etc.). The partial derivatives of the position vector $\mathbf{r}(\mu, \beta)$ mentioned above with respect to μ and β define two directions lying in a plane that is tangential to the surface at the point $\mathbf{r}(\mu, \beta)$. Figure 2.3 helps to illustrate this discussion. We have

$$\mathbf{r}_\mu = \frac{\partial \mathbf{r}}{\partial \mu} = \frac{\partial x}{\partial \mu}\mathbf{i} + \frac{\partial y}{\partial \mu}\mathbf{j} + \frac{\partial z}{\partial \mu}\mathbf{k},$$

$$\mathbf{r}_\beta = \frac{\partial \mathbf{r}}{\partial \beta} = \frac{\partial x}{\partial \beta}\mathbf{i} + \frac{\partial y}{\partial \beta}\mathbf{j} + \frac{\partial z}{\partial \beta}\mathbf{k}. \tag{2.17}$$

With $\mathbf{r}(\mu, \beta)$ defining our surface, now if we want to integrate a scalar function h of (μ, β) over our surface, we can use

$$\iint_D h(\mu, \beta) dS = \iint_D h(\mu, \beta) \left\| (\mathbf{r}_\mu \times \mathbf{r}_\beta) \right\| d\mu d\beta. \tag{2.18}$$

Here, $\| \cdot \|$ denotes magnitude. We return to this cross product momentarily. If $\mathbf{h}(\mu, \beta)$ is a vector, then its surface integral is defined as

$$\iint_D \mathbf{h}(\mu, \beta) \cdot \mathbf{n} dS, \tag{2.19}$$

where $\mathbf{n}$ is the surface normal. The surface normal can be defined as a vector perpendicular to the tangent plane at $\mathbf{r}$. From what we know of vector or cross products, we can argue that any vector lying along the cross product of the two vectors $\partial \mathbf{r}/\partial \mu$ and $\partial \mathbf{r}/\partial \beta$ must be perpendicular to the plane containing $\partial \mathbf{r}/\partial \mu$ and $\partial \mathbf{r}/\partial \beta$. Thus, we

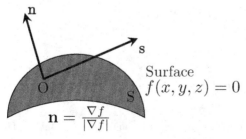

Figure 2.3 Schematic showing a surface $f(x, y, z) = 0$ and the surface normal $\mathbf{n}$. Also shown is a direction vector $\mathbf{s}$.

can define the unit vector normal to the tangent plane (and to the surface) at the point represented by $\mathbf{r}$ as

$$\mathbf{n} = \frac{\mathbf{r}_n}{|\mathbf{r}_n|}, \tag{2.20}$$

where $\mathbf{r}_n$ is the cross product of the two vectors $\partial\mathbf{r}/\partial\mu$ and $\partial\mathbf{r}/\partial\beta$ given by $\partial\mathbf{r}/\partial\mu \times \partial\mathbf{r}/\partial\beta$, or $\mathbf{r}_\mu \times \mathbf{r}_\beta$.

Oftentimes, a surface is represented in the form

$$f(x, y, z) = 0. \tag{2.21}$$

In such a representation, the vector ∇f is normal to the surface f, and the unit normal to the surface is given by (see Figure 2.3)

$$\mathbf{n} = \frac{\nabla f}{|\nabla f|}. \tag{2.22}$$

Suppose that we need to evaluate the surface integral for a vector $\mathbf{h} = x\mathbf{i} + y\mathbf{j} + z\mathbf{k}$ over a surface specified as $f = z - (x + y) = 0$, or alternatively, as $z = g(x, y) = x + y$. $\mathbf{h}$ may represent a flow field, and we want to evaluate the flux of $\mathbf{h}$ across the surface f. Our region is bounded within the limits $0 \leq x \leq 2$, $0 \leq y \leq 3$. Denoting the flow flux as q, we thus need

$$q = \iint_S \mathbf{h} \cdot \mathbf{n} \, dS. \tag{2.23}$$

Now, we can convert this surface integral into a double integral over the xy plane using the relation, $z = g(x, y)$ as

$$dS = \left[\sqrt{\left(\frac{\partial g}{\partial x}\right)^2 + \left(\frac{\partial g}{\partial y}\right)^2 + 1} \right] dxdy. \tag{2.24}$$

This is a case of the method in equation (2.23), where $\mathbf{r} = x\mathbf{i} + y\mathbf{j} + g(x, y)\mathbf{k}$. Here we find,

$$\mathbf{n} = -\mathbf{i} - \mathbf{j} + \mathbf{k}, \tag{2.25}$$

and

$$dS = \sqrt{3} \, dxdy. \tag{2.26}$$

Thus,

$$q = \int_0^2 \int_0^3 (-x - y + z) \, \sqrt{3} \, dxdy. \tag{2.27}$$

This integral can now be evaluated as a straightforward double integral. This example illustrates the general procedure with the help of quantities given by simple algebraic relations.

The directional derivative of the surface f along a direction denoted by a vector $\mathbf{s}$ can be found as

$$\frac{df}{ds} = \nabla f \cdot \mathbf{s}. \tag{2.28}$$

If the vector $\mathbf{s}$ is tangential to the surface f, then the right-hand side of equation (2.28) becomes zero. Therefore, the vector $\mathbf{n}$ is normal to the surface.

For instance, the representation $f(x, y, z) = 0$ for a spherical surface of radius R and centered at the origin would take the form

$$f(x, y, z) = (x^2 + y^2 + z^2) - R^2 = 0. \tag{2.29}$$

For the case of the sphere in equation (2.29),

$$\nabla f = 2x\mathbf{i} + 2y\mathbf{j} + 2z\mathbf{k}. \tag{2.30}$$

Therefore, the unit normal to the surface at a point (x, y, z) on the sphere will be

$$\mathbf{n} = \frac{\nabla f}{|\nabla f|} = \frac{x}{R}\mathbf{i} + \frac{y}{R}\mathbf{j} + \frac{z}{R}\mathbf{k}, \tag{2.31}$$

where we have used

$$|\nabla f| = 2\sqrt{x^2 + y^2 + z^2} = 2R. \tag{2.32}$$

2.3   Gauss's Divergence Theorem

This is an important relationship between integrals of scalar or vector quantities evaluated over volumes and integrals evaluated on surfaces. To develop an understanding of the divergence theorem, we start by considering the following surface integral (Greenberg 1978).

$$\int_\sigma \mathbf{n} \cdot \mathbf{v} \, dS, \tag{2.33}$$

where $\mathbf{n}$ represents a unit vector normal to the surface, and $\mathbf{v}$ represents a vector field, though it is easier to visualize it as a fluid velocity vector $\mathbf{v}$. dS is a surface element that forms part of a small surface σ. σ encloses a small volume v. It is interesting to consider the quantity

$$\frac{\int_\sigma \mathbf{n} \cdot \mathbf{v} dS}{v}. \tag{2.34}$$

If, following Greenberg (1978), we let σ enclose a cubical volume $v = dxdydz$, the integral evaluated in the x direction becomes

$$\int_{xdir} = v_x\left[\left(x_0 + \frac{\Delta x}{2}, y_1, z_1\right) - v_x\left(x_0 - \frac{\Delta x}{2}, y_2, z_2\right)\right] \Delta y \Delta z. \tag{2.35}$$

This follows from the mean value theorem, whereby $\int_a^b f(x)dx = f(\xi)(b-a)$, where ξ is a point in the interval $[a, b]$. Now dividing the quantity on the right of equation (2.35) by v, and letting $v \to 0$, the right-hand side of equation 2.35 becomes

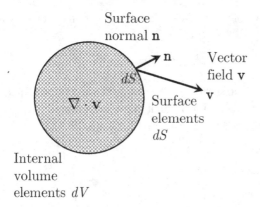

Figure 2.4 Schematic illustrating the reasoning behind Gauss' divergence theorem. dV are volume elements within the sphere, while dS are surface elements on the bounding surface of the sphere.

$$\lim_{v \to 0} = v_x \frac{\left[\left(x_0 + \frac{\Delta x}{2}, y_1, z_1\right) - v_x \left(x_0 - \frac{\Delta x}{2}, y_2, z_2\right)\right] \Delta y \Delta z}{\Delta x \Delta y \Delta z}$$

$$\lim_{\Delta x \to 0} \frac{v_x \left(x_0 + \frac{\Delta x}{2}, y_0, z_0\right) - v_x \left(x_0 - \frac{\Delta x}{2}, y_0, z_0\right)}{\Delta x}$$

$$\frac{\partial v_x}{\partial x}(x_0, y_0, z_0). \tag{2.36}$$

Performing these operations on the faces in the y and z directions (letting (x_0, y_0, z_0) be any arbitrarily chosen point) and adding the results,

$$\frac{\int_\sigma \mathbf{n} \cdot \mathbf{v} dS}{v} = \frac{\partial v_x}{\partial x} + \frac{\partial v_y}{\partial y} + \frac{\partial v_z}{\partial z} = \nabla \cdot \mathbf{v}, \tag{2.37}$$

where we recall that $\nabla \cdot \mathbf{v} = \text{div}\mathbf{v}$. Now if we argue that the surface σ encloses a small element within a larger region, to be represented as σ_i, with $dV_i \to 0$ being the volume occupied by the element (see Figure 2.4).

Then, if we rewrite equation (2.37),

$$\text{div}\mathbf{v}dV_i \approx \int_{\sigma_i} \mathbf{n} \cdot \mathbf{v} d\sigma_i. \tag{2.38}$$

Now if we add together similar expressions to equation (2.38) over the entire region and evaluate the limit $dV_i \to 0$,

$$\lim_{dV_i \to 0} \sum_i^{\infty} (\text{div}\mathbf{v}dV_i) = \int_V \text{div}\mathbf{v}dV. \tag{2.39}$$

The summation of the term on the right is interesting. Thus, consider

$$\sum_{i=1}^{\infty} \int_{\sigma_i} \mathbf{n} \cdot \mathbf{v} \, d\sigma_i. \tag{2.40}$$

We argue next that adjacent elements share sides over which the surface normal $\mathbf{n}_i$ has opposite directions. The contributions along all such shared sides must cancel. In fact, the only contributions that do not cancel are those along the outer bounding surface S of the region (i.e., the surface enclosing the region). Denoting the elemental surface on the bounding surface S as dS_i, we have

$$\sum_{i=1}^{\infty} \int_{\sigma_i} \mathbf{n} \cdot \mathbf{v} d\sigma_i = \sum_{i=1}^{\infty} \mathbf{n} \cdot \mathbf{v} dS_i \rightarrow \int_{S} \mathbf{n} \cdot \mathbf{v} dS, \tag{2.41}$$

where $\mathbf{n}$ now denotes the outer normal to the surface at a point on the surface. Based on equations (2.38)–(2.41), we are able to say

$$\int_{V} \mathrm{div}\, \mathbf{v}\, dV = \int_{V} \nabla \cdot \mathbf{v} dV = \int_{S} \mathbf{n} \cdot \mathbf{v} dS. \tag{2.42}$$

Equation (2.42) is in fact Gauss' divergence theorem, which allows us to convert surface integrals into volume integrals, and vice versa, depending on the need of the application. Interestingly, if $\mathbf{v}$ in equation (2.42) were replaced by a scalar field f, then the integrand in the volume integral in equation (2.42) would be ∇f, which would be a vector field. By the same token, the integrand in the surface integral in equation (2.42) would become $f\mathbf{n}$, which would also be a vector field. One application arises in the analysis of fluid forces on an immersed body (see, for instance, Newman (1978)). Thus, if the pressure at a point in the fluid is p, then the force vector acting on the body can be expressed as

$$\mathbf{F} = - \int_{S_B} p\mathbf{n} dS. \tag{2.43}$$

The negative sign is in recognition of the fact that we choose $\mathbf{n}$ to be an outward normal at a point on the body surface. An application of the Gauss theorem to this problem would lead to an expression such as

$$\mathbf{F} = - \int_{V} \nabla p\, dV. \tag{2.44}$$

The interpretation of the volume V needs to be discussed. Since p exists exterior to the body and is the cause of the force on it, it would make no sense for V to be the volume enclosed by the body, as no fluid pressure exists therein. V must therefore be exterior to the body. As discussed in Newman (1978), another surface enclosing the body is introduced, typically at a distance far enough away from the influence of the body. Together, S_B and S_∞ should define a closed surface, and V is then the volume enclosed by the closed surface $S_B + S_\infty$.

We will use Gauss' divergence theorem in Chapter 11. Several interesting mathematical relations can be derived using the Gauss divergence theorem. Some of these have profound significance in the study of floating body dynamics. We refer in particular to the two Green's identities. We again follow the treatment of Greenberg (1978) in developing an understanding of Green's identities.

If we suppose that ϕ_1 and ϕ_2 are two continuous scalar fields with continuous first and second partial derivatives in a fluid region, then $\nabla \phi_2$ is a vector field, as is the product $\phi_1 \nabla \phi_2$. If we set up the product $\nabla \cdot (\phi_1 \nabla \phi_2)$, we have

$$\nabla \cdot (\phi_1 \nabla \phi_2) = \nabla \phi_1 \cdot \nabla \phi_2 + \phi_1 \nabla^2 \phi_2, \tag{2.45}$$

which is essentially an application of the chain rule. Similarly, if we consider the relation $\mathbf{n} \cdot (\phi_1 \nabla \phi_2)$, we find

$$\mathbf{n} \cdot (\phi_1 \nabla \phi_2) = \phi_1 \mathbf{n} \cdot \nabla \phi_2 = \phi_1 \frac{\partial \phi_2}{\partial n}, \tag{2.46}$$

where we have applied the realization that the second term in equation (2.46) is simply the normal derivative of ϕ_2, i.e., a component of the derivative or gradient vector in the direction of the surface normal.

Now if we decide to apply the divergence theorem to this vector field within a fluid volume V that is bounded by a surface S, then

$$\int_V \nabla \cdot (\phi_1 \nabla \phi_2) \, dV = \int_S \mathbf{n} \cdot (\phi_1 \nabla \phi_2) \, dS. \tag{2.47}$$

Using what we learn from equations (2.45) and (2.46), we have

$$\int_V \left(\nabla \phi_1 \cdot \nabla \phi_2 + \phi_1 \nabla^2 \phi_2 \right) dV = \int_S \phi_1 \frac{\partial \phi_2}{\partial n} \, dS. \tag{2.48}$$

Equation (2.48) is in fact Green's first identity. If we switch the two scalar fields ϕ_1 and ϕ_2 in equation (2.48), we have

$$\int_V \left(\nabla \phi_2 \cdot \nabla \phi_1 + \phi_2 \nabla^2 \phi_1 \right) dV = \int_S \phi_2 \frac{\partial \phi_1}{\partial n} \, dS. \tag{2.49}$$

Next, if we subtract equation (2.49) from equation (2.48) and realize that the first terms on the left in both equations are equal to each other,

$$\int_V \left(\phi_1 \nabla^2 \phi_2 - \phi_2 \nabla^2 \phi_1 \right) dV = \int_S \left(\phi_1 \frac{\partial \phi_2}{\partial n} - \phi_2 \frac{\partial \phi_1}{\partial n} \right) dS. \tag{2.50}$$

This in fact is Green's second identity, which can be used to obtain several powerful results in determining fluid forces mainly dominated by nonviscous effects. Note that these relations were derived with the help of Gauss' divergence theorem. We will return to Green' identities in Chapter 11. We consider next a theorem more widely applicable to flows where viscous effects are present and flows cannot be considered irrotational.

2.4   Stokes's Theorem

Just as the divergence theorem allows a back-and-forth transformation from volume integrals to surface integrals and vice versa, Stokes' theorem provides a way to express surface integrals in terms of a line integral along a closed curve bounding the surface. It essentially relates the inner product of a unit surface normal and the curl of a vector integrated over the surface to the line integral of that vector along the bounding curve. Stokes' theorem forms the basis of some of our discussions in Chapters 10 and 11. We follow the treatment of Greenberg (1978) once again. Referring to the surface S shown in Figure 2.5, we see that the bounding curve C is shown with a direction.

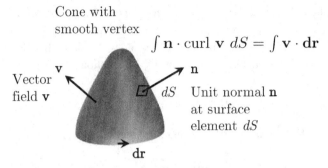

Figure 2.5 Schematic illustrating the application of Stokes' theorem for a cone with a smooth vertex. dS are surface (area) elements on the cone surface, while **dr** are line elements along the bounding circle forming the base of the cone.

The normal vector to the surface **n** is understood to have positive sense based on the right-hand rule. Thus, with the fingers of the right-hand curling along the direction of the bounding curve, the direction in which the thumb stretches out defines the positive direction in this case. To begin, we recall that the curl of a vector **v**, where **v** defines some vector field, is expressed as

$$\mathrm{curl}\,\mathbf{v} = \nabla \times \mathbf{v}, \tag{2.51}$$

where $\nabla = \frac{\partial}{\partial x}\mathbf{i} + \frac{\partial}{\partial y}\mathbf{j} + \frac{\partial}{\partial z}\mathbf{k}$. Sometimes, the curl operation is also defined as (Greenberg 1978),

$$\mathrm{curl}\,\mathbf{v} = \lim_{\upsilon \to 0}\left(\frac{1}{\upsilon}\int_{\sigma} v \times \mathbf{v}\,d\sigma\right). \tag{2.52}$$

If we suppose that υ represents the volume of a very small cylinder placed on the surface S, with σ denoting the small surface area of the cylinder, equation (2.52) can approximately be interpreted to write

$$\mathbf{n} \cdot \mathrm{curl}\,\mathbf{v} \approx \frac{1}{\upsilon}\int_{\sigma} \mathbf{n} \cdot v \times \mathbf{v}\,d\sigma, \tag{2.53}$$

where we point out the distinction between surface normal to the small cylinder element placed on S and the surface normal to the surface S itself. Realizing that the surface of the cylinder is comprised of the top and bottom circular areas and the cylinder side, the overall integral is the sum of the individuals performed over each of these three surfaces. Now, since the cylinder is infinitesimally small, we use **v** as defined at the center of the cylinder, to find that the integrals over the top and bottom faces must cancel because their respective normals are in opposite directions (such that $v = \mathbf{n}$ on the top surface, and $v = -\mathbf{n}$ on the bottom surface). The normal to the curved side surface of the cylinder can be represented as $\mathbf{e}_n$. We see that $\mathbf{e}_n$ is a unit vector along the cylinder radius vector **r**. We can use the unit vector $\mathbf{e}_t$ to define the tangential direction (or the direction of the unit tangent) to the curved surface. Then, recalling that, for the scalar triple product,

$$\mathbf{n} \cdot \mathbf{e}_n \times \mathbf{v} = \mathbf{n} \times \mathbf{e}_n \cdot \mathbf{v} = \mathbf{e}_t \cdot \mathbf{v}. \tag{2.54}$$

For our small cylinder, the volume $v = s_f l$, where l is the length of the cylinder and s_f is the area of the circular top and bottom faces. The elemental area for the side surface is then $d\sigma = h ds$, where s is the length measured along the circle. It follows from this and from equation (2.54) that the right-hand side of equation (2.53) would devolve to

$$\frac{1}{s_f l} \int_C \mathbf{v} \cdot \mathbf{e}_t l ds = \frac{1}{s_f} \int_C \mathbf{v} \cdot d\mathbf{r} \qquad (2.55)$$

since $\mathbf{e}_t ds = d\mathbf{r}$. Thus, equation (2.52) can now be expressed as

$$\mathbf{n} \cdot \text{curl} \mathbf{v} s_f \approx \int_C \mathbf{v} \cdot d\mathbf{r}. \qquad (2.56)$$

We need to sum this relation for N such small elements, where $N \to \infty$.

$$\sum_{n=1}^{N} \mathbf{n} \cdot \text{curl} \mathbf{v} s_{fn} \approx \sum_{n=1}^{N} \int_{C_n} \mathbf{v} \cdot d\mathbf{r}. \qquad (2.57)$$

Realizing that the bounding curve directions on abutting elements are opposite (leading to cancellations), only the bounding curve of the entire surface S survives the summation. In the limit, the area $s_f \to ds$ and the summations approach integrals over the entire surface S and the entire bounding curve C. Thus, using dS to represent an element of the entire surface S,

$$\int_S \mathbf{n} \cdot \text{curl } \mathbf{v} \, dS = \int_C \mathbf{v} \cdot d\mathbf{r} \qquad (2.58)$$

Equation (2.58) in fact is a statement of the Stokes' theorem (see Figure 2.5 for an illustration). The presence of the curl or ($\nabla\times$) operator on the left side takes us back to Section 1.4, where the curl was associated with rotation and vorticity. Thus, if we apply Stokes' theorem to the vector field defined by the fluid velocity $\mathbf{v}$ (i.e., if $\mathbf{v}$ in equation (2.57) were the fluid velocity vector), then $\nabla \times \mathbf{v}$ represents the vorticity in the fluid flow past the surface bounded by the curve C. Stokes' theorem then provides a way to define a quantity known as "circulation," which is sometimes denoted as Γ and expressed as

$$\Gamma = \int_C \mathbf{v} \cdot d\mathbf{r} = \int_C \mathbf{v} \cdot \left(\frac{\partial \mathbf{r}}{ds}\right) ds. \qquad (2.59)$$

Let us verify the Stokes' theorem with an example. Suppose that a fluid contained within a cylindrical region of radius R is given a uniform angular velocity ω about the vertical axis, with which the axis of the cylindrical region coincides (Lamb 1932). Any point within the cylinder can be represented via a position vector $\mathbf{r} = x\mathbf{i} + y\mathbf{j} + z\mathbf{k}$. The angular velocity ω in vector form can be written as $\omega\mathbf{k}$, with the fluid rotating such that a corkscrew turning with the fluid will advance in the positive z direction (i.e., the right-hand rule). The linear velocity of a fluid particle within the cylindrical region can be found as

$$\mathbf{v} = \omega \times \mathbf{r} = -\omega y\mathbf{i} + \omega x\mathbf{j}. \qquad (2.60)$$

Now focusing on the top surface capping the cylindrical region, the left-hand side of equation (2.58) can be expressed as

$$\int_{S_T} \mathbf{n} \cdot \text{curl} \mathbf{v} dS = \int_{S_T} \mathbf{n} \cdot \nabla \times \mathbf{v} dS = \int_{S_T} \mathbf{k} \cdot \nabla \times \mathbf{v} dS. \qquad (2.61)$$

It can be verified that $\nabla \times \mathbf{v} = 2\omega\mathbf{k}$, uniformly over the entire top surface S_T. The second right-hand side of equation (2.61) thus becomes

$$\int_{S_T} \mathbf{k} \cdot \nabla \times \mathbf{v} dS = 2\pi\omega R^2. \qquad (2.62)$$

Now proceeding to the right-hand side of equation (2.58), and realizing that the contour C is just the circle of radius R outlining the top surface S_T,

$$d\mathbf{r} = dx\mathbf{i} + dy\mathbf{j} = -R \sin\theta\mathbf{i} + R\cos\theta\mathbf{j}, \qquad (2.63)$$

where we have used $x = R\cos\theta, y = R\sin\theta$. Since on the circular contour C, $\mathbf{v} = -\omega R \sin\theta\mathbf{i} + \omega R \cos\theta\mathbf{j}$,

$$\int_C \mathbf{v} \cdot d\mathbf{r} = \int_0^{2\pi} \omega R^2 \left(\sin^2\theta + \cos^2\theta\right) d\theta = 2\pi\omega R^2. \qquad (2.64)$$

The end results for both the left and the right sides of Stokes' theorem are thus found to be equal for the example we have considered.

2.5   Helmholtz Theorem

We consider finally what is referred to as the Helmholtz theorem, which essentially states that any velocity field in a fluid can be expressed as a sum of an irrotational velocity field and a rotational velocity field. Thus,

$$\mathbf{v} = \mathbf{v}_{\text{irr}} + \mathbf{v}_{\text{rot}}, \qquad (2.65)$$

where

$$\mathbf{v}_{\text{irr}} = \nabla\phi \qquad (2.66)$$

is the irrotational part expressed as a gradient of a scalar potential field. The rotational part is expressed as

$$\mathbf{v}_{\text{rot}} = \nabla \times \mathbf{A}. \qquad (2.67)$$

The irrotational part satisfies the condition,

$$\nabla \times \mathbf{v}_{\text{irr}} = \nabla \times \nabla\phi = 0, \qquad (2.68)$$

which essentially recalls what we learned before, namely that the curl of an irrotational velocity field is zero, i.e., the vorticity at any point in such a field is zero. Next, if we set up an inner product of the gradient operator with the rotational part, we find

$$\nabla \cdot \nabla \times \mathbf{A} = 0. \qquad (2.69)$$

This finding implies that $\nabla \cdot \mathbf{v}_{\mathrm{rot}} = \mathrm{div}\mathbf{v}_{\mathrm{rot}} = 0$. Vector fields that satisfy this condition are referred to as "solenoidal" fields. Examples of such fields include the magnetic field around a magnetic dipole (i.e., a standard magnet), or around a moving electric charge. Indeed, for any fluid flow of an incompressible fluid, the divergence of fluid particle velocity is zero. Such fields are solenoidal. When flows of incompressible fluids are also irrotational, the velocity vector satisfies both the zero divergence and the zero curl conditions. Fluid flows that are both solenoidal and irrotational are important in marine applications. Water can be thought to be incompressible for most applications.[1] Hence, we must have

$$\nabla \cdot \mathbf{v} = 0, \quad \Rightarrow \mathbf{v} = \nabla \times \mathbf{\Psi}, \tag{2.70}$$

where $\mathbf{\Psi}$ represents the vector *stream function*. In addition, if the flow is also irrotational, then,

$$\nabla \times \mathbf{v} = 0, \quad \Rightarrow \mathbf{v} = \nabla \phi, \tag{2.71}$$

where we recall that ϕ is the scalar velocity potential function. Both $\mathbf{\Psi}$ and ϕ are, in general, space and time dependent, i.e., functions of the spatial coordinates (x, y, z) and t. Equations (2.70) and (2.71) imply that $\mathbf{\Psi}$ and ϕ are related to each other according to

$$\nabla \phi = \nabla \times \mathbf{\Psi}. \tag{2.72}$$

Broken down into components, equation (2.72) takes the form,

$$\frac{\partial \phi}{\partial x} = \left(\frac{\partial \psi_z}{\partial y} - \frac{\partial \psi_y}{\partial y} \right),$$

$$\frac{\partial \phi}{\partial y} = \left(\frac{\partial \psi_x}{\partial z} - \frac{\partial \psi_x}{\partial z} \right),$$

$$\frac{\partial \phi}{\partial z} = \left(\frac{\partial \psi_x}{\partial y} - \frac{\partial \psi_y}{\partial x} \right). \tag{2.73}$$

We will return to these relationships in Chapter 11, noting here that they are of particular interest in two-dimensional fluid flows that are incompressible (i.e., solenoidal) and irrotational.

2.6   Cartesian Tensors

Here we continue our study of indicial notation and discuss the use of tensors which provide an alternative, compact way to express the relationships we have discussed in Sections 2.2–2.5. Briefly, the motivation for the use of tensors is as follows.

---

[1] Sound propagation in water is represented as pressure waves with alternating peaks and troughs, or localized regions of compression and expansion. Outside of these small, localized regions of pressure and density changes, the flow field can be assumed to be incompressible.

Physical laws should remain valid independent of any coordinate system that they are referred to. In other words, a mathematical equation that describes a certain physical phenomenon must be invariant under a coordinate transformation or change of fundamental units. Such physical laws can be represented by tensors that are mathematical entities that obey certain laws.

Tensors are classified by rank or order. A tensor of rank zero is called a *scalar*. A tensor of rank one is called *vector*. A second-rank tensor is called a *dyadic*. These tensors have one, three, and nine components, respectively, in the three-dimensional Euclidean space, i.e., the number of components of a tensor is 3^K, where K is the rank or order of the tensor. However, the number of dimensions in general can be referred to a space of functions, sequence of numbers, and so on. For example, the second-order added-mass tensor that will be discussed in Chapter 14 can be described in a response or motion mode space, and since a rigid body can have six degrees of freedom, it has $6^2 = 36$ components in the three-dimensional Euclidean space. An M-dimensional vector, e.g., can have $M^{K=1} = M$ components in a one-dimensional Euclidean space.

The fundamental requirement for a physical quantity to be a tensor is that the components of a tensor must transform from one coordinate system to another by obeying the following law:

$$T'_{ijk} = a_{ip} a_{jq} a_{km} \cdots T_{pqm} \cdots .$$ (2.74)

Here, a_{ip}, a_{jq}, and so on, are the components of the transformation or rotation matrix.

A well-known example of a second-order tensor is the stress tensor, and it has three normal stress components and six shear stress components, as shown in Figure 2.6:

$$\begin{bmatrix} \tau_{ij} \end{bmatrix} = \begin{bmatrix} \tau_{11} & \tau_{12} & \tau_{13} \\ \tau_{21} & \tau_{22} & \tau_{23} \\ \tau_{31} & \tau_{32} & \tau_{33} \end{bmatrix}.$$ (2.75)

The arrows actually indicate the direction of the force components associated with each stress component.

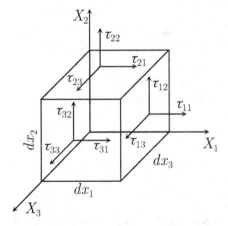

Figure 2.6 The nine components of the stress tensor on a cubic fluid element.

2.7   Concluding Remarks

The focus of this chapter was vector integral calculus. We reviewed integrals of vector quantities over paths in space defined by lines, straight or curved. In particular, we saw how line integrals could be reduced to integrals defined with respect to a parameter such as the length coordinate measured along a path. We also saw how they could be evaluated using three integrals, using component representations. We discussed surface integrals using different ways to represent surfaces in space, such as using two parameters, in an extension of the approach used with line integrals. We discussed the relevance of surface normals, i.e., unit vectors defined to be normal to the surface at chosen points. This led to a review of important relationships between surface integrals and volume integrals, and surface integrals and line integrals, namely the theorems of Gauss and Stokes. Via the Gauss divergence theorem, we could express integrals defined over surfaces enclosing volumes in terms of volume integrals of the divergence of the quantities to be integrated. The Gauss divergence theorem is widely used in the interpretation of forces and moments applied by fluid flows on bodies within fluids or bodies immersed in fluids. Some extensions of the divergence theorem, such as the two identities of Green (assuming certain properties for the functions involved), were also reviewed, particularly in view of their relevance to diffraction theory of wave forces on floating or submerged bodies. Stokes' theorem was also reviewed, relating as it does integrals involving the curl of a vector field over surfaces to line integrals evaluated around the curves bounding the surfaces. Finally, we briefly reviewed the Helmholtz theorem, which states that any general velocity field can be represented as a sum of an irrotational velocity field and a rotational velocity field.

2.8   Self-Assessment

2.8.1

The velocity field of a flow is expressed as

$$u_1 = \frac{(x_1^2 - x_2^2)}{r^4}; \quad u_2 = \frac{2x_1 x_2}{r^4}; \quad r = \sqrt{x_1^2 + x_2^2}. \qquad (2.76)$$

(a) Does this flow field represent an irrotational flow?
(b) Does this flow field represent an incompressible flow?
(c) Calculate the circulation Γ for this flow,

$$\Gamma = \oint \mathbf{v} \cdot \mathbf{dx} = \oint (u_1 dx_1 + u_2 dx_2), \qquad (2.77)$$

around the rectangle $ABCD$, where the vertices are $A(1,2)$, $B(3,2)$, $C(3,6)$, and $D(1,6)$.

2.8.2

Consider the box enclosed within the region $0 \leq x \leq 2, 0 \leq y \leq 2$, and $0 \leq z \leq 2$. Evaluate the following integral over this box,

$$\iint_S \mathbf{F} \cdot \mathbf{ds}, \tag{2.78}$$

using Gauss' divergence theorem. $\mathbf{F}$ is given by

$$\mathbf{F} = \left[2x^2 + 3y^3, \ 3y^2 - xz, \ 5x - \cos y \right]. \tag{2.79}$$

2.8.3

For a vector $\mathbf{F} = \left[x - z, \ x + y, \ x \right]$, determine

$$\oint \mathbf{F} \cdot \mathbf{dr} \tag{2.80}$$

over an ellipse defined as

$$\frac{x^2}{16} + \frac{y^2}{4} = 1. \tag{2.81}$$

2.8.4

Suppose a flow is described as

$$\mathbf{v} = \left[y^2, \ -x^3, \ 0 \right]. \tag{2.82}$$

(a) Is the flow solenoidal?
(b) Is the flow irrotational?

2.8.5

The hydrostatic pressure in a fluid is given by $p = \rho g z$, where z is the depth below the surface. If z is defined as

$$z = x^2 + y, \tag{2.83}$$

evaluate the force acting on the surface, using the definition of the surface normal $\mathbf{n}$.

3 Complex Variables

The algebra and calculus of complex variables are important in mathematical modeling, particularly when it comes to marine applications, since many ocean phenomena
are dependent on where they occur and how they evolve over scales ranging from a
few centimeters or a few seconds to several kilometers or several days. The response
of structures in waves, for instance, is a function of the spatial distribution of the
wave field over the immersed portion of the structure, and the time dependence of
not only the wave field but also, in general, of the structure's response itself. Many
time-dependent phenomena can be represented as sums of purely oscillatory functions
of time. Many spatially varying phenomena can be considered as sums of oscillatory
functions of the spatial coordinates. We can understand oscillatory behavior in a consistent manner using quantities such as temporal or spatial frequencies, amplitudes,
and phases. Complex numbers allow us to combine amplitude and phase into a single "number," one which contains a real part and an imaginary part, or an amplitude
and a phase angle. The amplitude is the square root of the sum of the squares of the
real and imaginary parts, while phase is the inverse tangent of the ratio of the imaginary and real parts. Functions of complex variables are generally complex valued, and
under some conditions, also satisfy relationships defined in the calculus of real variables. Some features of the calculus of complex variables and functions are unique,
however, and an understanding of such features often provides fundamental insights.
Complex variables feature prominently in hydrodynamics, vibrations and dynamics,
control, signal processing and analysis, and structural mechanics. Much of our work
in later chapters in marine applications (e.g., Chapters 11–15) will make use of complex variables and complex functions. Contour integrals and their "easy" evaluation
via residues are particularly important in this respect. Our goal in this chapter is to
provide the basic building blocks leading up to contour integration, which, as we will
see, have applications that range far and wide. Once again, treatments that go deeper
and wider into complex analysis can be found in Greenberg (1978) and Kreyszig,
Kreyszig, and Norminton (2011).

3.1   Complex Variables

Complex numbers seem to accomplish what two-dimensional vectors do, although
many vector functions may in turn be complex valued. In essence, they allow us

to carry two numbers in a single package through standard arithmetic, algebraic, and calculus operations (under some conditions), within a consistent and generalizable framework. What enables such a framework is the notion of imaginary operator $i = \sqrt{-1}$ (recall that square roots of negative numbers are not defined without the assistance of the imaginary operator i). A complex number is just a particular value of a complex variable z, which is represented as

$$z = x + iy, \tag{3.1}$$

where x and y are both real variables that take on particular values to define a particular value of the complex variable z. x is the real part of z, represented as $x = \Re z$, and y is the imaginary part, with $y = \Im z$. A complex variable is frequently paired with its conjugate, referred to as "complex conjugate," defined as

$$z^* = x - iy. \tag{3.2}$$

Two complex variables may be added according to

$$z_1 + z_2 = (x_1 + x_2) + i(y_1 + y_2). \tag{3.3}$$

Subtraction is equivalent to

$$z_1 - z_2 = (x_1 - x_2) + i(y_1 - y_2). \tag{3.4}$$

Product of two complex variables can be evaluated as

$$z_1 z_2 = (x_1 + iy_1)(x_2 + iy_2) = (x_1 x_2 - y_1 y_2) + i(x_1 y_2 + y_1 x_2). \tag{3.5}$$

It is interesting to see that the addition, subtraction, and multiplication of a complex number with its conjugate are

$$
\begin{aligned}
z + z^* &= 2x = 2\Re(z), \\
z - z^* &= 2iy = 2i\Im(z), \\
z z^* &= x^2 + y^2.
\end{aligned}
\tag{3.6}
$$

Defining the "amplitude" or "modulus" of a complex number z as $|z|$, we have

$$|z| = \sqrt{x^2 + y^2} = \sqrt{(\Re z)^2 + (\Im z)^2} \tag{3.7}$$

so that

$$z z^* = |z|^2. \tag{3.8}$$

It is interesting to note that

$$z^2 = x^2 + y^2 + 2ixy. \tag{3.9}$$

Thus, $z z^*$ is real, whereas z^2 itself is complex.

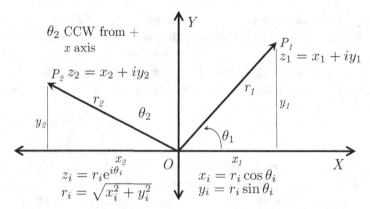

Figure 3.1 Representation of complex variables on a complex z-plane. Note that θ_i are measured counterclockwise from the positive x direction.

Division of two complex numbers can be performed using

$$
\begin{aligned}
\frac{z_1}{z_2} &= \frac{x_1 + iy_1}{x_2 + iy_2}, \\
&= \frac{(x_1 + iy_1)(x_2 - iy_2)}{(x_2 + iy_2)(x_2 - iy_2)}, \\
&= \frac{(x_1 x_2 + y_1 y_2)}{x_2^2 + y_2^2} + i\frac{(x_1 y_2 + x_2 y_1)}{x_2^2 + y_2^2}.
\end{aligned}
\tag{3.10}
$$

As indicated in Figure 3.1, a complex variable $z = x + iy$ can be represented as a point on the Cartesian xy plane, where r can be used to denote the radius vector to that point (from the origin), and θ can define the angle between the positive x axis and the radius vector so that

$$
r = \sqrt{x^2 + y^2}, \quad \theta = \tan^{-1}\left(\frac{y}{x}\right).
\tag{3.11}
$$

With $\tan\theta$ defined as in equation (3.11), it can be verified that

$$
\cos\theta = \frac{x}{\sqrt{x^2 + y^2}}, \quad \text{and} \quad \sin\theta = \frac{y}{\sqrt{x^2 + y^2}}.
\tag{3.12}
$$

An alternative representation of a complex variable z is frequently more convenient,

$$
z = re^{i\theta}.
\tag{3.13}
$$

Here, r denotes the modulus or amplitude of the complex variable z, while θ, the argument over the exponential, is the "phase" of the variable z. The representation of equation (3.13) can be verified as follows:

$$
\begin{aligned}
z = x + iy &= \sqrt{x^2 + y^2}\left(\frac{x}{\sqrt{x^2 + y^2}} + i\frac{y}{\sqrt{x^2 + y^2}}\right), \\
&= |z|(\cos\theta + i\sin\theta).
\end{aligned}
\tag{3.14}
$$

But, by equation (3.11), $z = r$. As we know from Euler's formula,

$$
e^{i\theta} = \cos\theta + i\sin\theta.
\tag{3.15}
$$

Equations (3.14) and (3.15) enable us to verify that

$$z = re^{i\theta}. \tag{3.16}$$

Using equation (3.16), we have, for the complex conjugate,

$$z^* = re^{-i\theta} \tag{3.17}$$

The form (equation (3.16)) is particularly convenient for products and divisions, since

$$z_1 z_2 = r_1 r_2 e^{i(\theta_1 + \theta_2)},$$

$$\frac{z_1}{z_2} = \frac{r_1}{r_2} e^{i(\theta_1 - \theta_2)}. \tag{3.18}$$

It is now easy to see that

$$zz^* = r^2, \text{ and } z^2 = r^2 e^{i2\theta}. \tag{3.19}$$

Complex variables prove particularly useful in describing phenomena with oscillatory behavior in time or space, or both. Floating body oscillations in response to incoming waves are an important example. Other examples include wave propagation, whether on the ocean surface, or within the ocean (e.g., acoustic pressure waves, internal waves at density stratifications, etc.), tidal oscillations, position vector of a point on the earth as the earth rotates about an axis and revolves about the sun, etc. The translational oscillations of a floating body (i.e., surge, sway, and heave) can often be represented as

$$\mathbf{x}(t) = \Re\left[\mathbf{X}e^{i\omega t}\right], \tag{3.20}$$

where $\mathbf{x}(t) = [x_1(t) \; x_2(t) \; x_3(t)]^T$ is the vector of instantaneous oscillations in surge, sway, and heave. ω denotes the angular frequency of oscillation, in radians per second. When the body response to waves can be supposed to be linear (i.e., for most of the problems studied in this text), ω equals the angular frequency of the waves exciting the oscillations. $\mathbf{x}(t)$ represents the instantaneous real-valued displacements of the body in the surge, sway, and heave modes. We expect that these oscillations typically will be at different phases relative to each other. The amplitudes $\mathbf{X}(i\omega)$ are thus complex valued, expressed as

$$\mathbf{X}_j(i\omega) = \left|\mathbf{X}_j\right| e^{i\theta_j}, \text{ where, } j = 1, 2, 3. \tag{3.21}$$

X_j are the so-called complex amplitudes of the oscillations, containing not only amplitude but also phase information, as given by θ_j. Therefore, the instantaneous displacements in the three modes can be expressed as

$$x_j(t) = \Re\left[X_j e^{i\omega t}\right] = \Re\left[\left|X_j\right| e^{i(\omega t + \theta_j)}\right], \text{ where } j = 1, 2, 3. \tag{3.22}$$

3.2   Functions of Complex Variables

Many of the function definitions for real variables extend to complex variables, though a little effort may be required in establishing them as such. Some functions require interpretation in order to satisfy the basic requirement for a function, namely that a function cannot be a multivalued map from one variable to another. Recall that a function is a mapping from one variable space to another under some prescribed rule. A function could be a relationship that maps one variable in a given space uniquely to one variable in a different space (one-to-one map), or it may map multiple variable values in one space into a single variable value in a different space (many to one). One example of the former is a function $f(x) = Ax + B$, where A and B are known constants. An example of the latter is the cosine function, $f(x) = \cos x$, which is a periodic function with values repeating themselves periodically as x increases (or decreases). A one-to-many map is inherently ambiguous, and hence does not qualify as a function. These notions become important in the use of complex functions, where "branch cuts" need to be defined in order to remove "multi-valuedness" that arises when we carry over real-variable functions into the complex variable space.

More generally, a function $f(z)$ that maps, in a single-valued manner, a variable $z = x + iy$ in one space into a variable $w = u + iv$ in another (see Figure 3.2) can be described as

$$w = f(z) = g(x, y) + ih(x, y). \tag{3.23}$$

For example, a function $f(z) = Az + Bz^2$ can be expressed as

$$f(z) = A(x + iy) + B(x + iy)^2. \tag{3.24}$$

Evaluating further,

$$f(z) = \left[Ax + B\left(x^2 - y^2\right)\right] + i\left(Ay + 2Bxy\right). \tag{3.25}$$

Or, in terms of the definition of equation (3.23), i.e., $f(z) = g(x, y) + ih(x, y)$,

$$g(x, y) = Ax + B(x^2 - y^2), \text{ and } h(x, y) = Ay + 2Bxy. \tag{3.26}$$

Some other functions such as $f(z) = \cos z$ and $f(z) = \sin z$ need more thought. We could start by recalling Euler's formula for real variables, e.g.,

$$e^{ix} = \cos x + i \sin x. \tag{3.27}$$

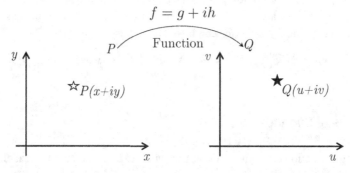

Figure 3.2 Representation of a complex function mapping a point $P(x + iy)$ into a point $Q(u + iv)$.

This allows us to define

$$\cos x = \frac{1}{2} \left(e^{ix} + e^{-ix} \right) \text{ and} \tag{3.28}$$

$$\sin x = \frac{1}{2i} \left(e^{ix} - e^{ix} \right). \tag{3.29}$$

We consider next a function defined as $f(z) = e^z$, which could be expanded as

$$e^z = e^{(x+iy)} = e^x e^{iy}. \tag{3.30}$$

Recalling equation (3.27),

$$e^z = e^x \left(\cos y + i \sin y \right). \tag{3.31}$$

Similarly, a function $f(z) = e^{-z}$ can be expressed as

$$e^{-z} = e^{-x} \left(\cos y - i \sin y \right). \tag{3.32}$$

Further, functions $f(z) = e^{iz}$ and $f(z) = e^{-iz}$ can also be written as

$$e^{iz} = e^{-y} \left(\cos x + i \sin x \right),$$
$$e^{-iz} = e^{y} \left(\cos x - i \sin x \right). \tag{3.33}$$

Equations (3.31)–(3.33) can be used to describe

$$\cosh z = \frac{1}{2} \left(e^z + e^{-z} \right),$$
$$\sinh z = \frac{1}{2} \left(e^z - e^{-z} \right),$$
$$\cos z = \frac{1}{2} \left(e^{iz} + e^{-iz} \right),$$
$$\sin z = \frac{1}{2i} \left(e^{iz} - e^{-iz} \right). \tag{3.34}$$

Using equations (3.31) and (3.32), we can verify that

$$\cosh z = (\cosh x \cosh iy - \sinh x \sinh iy). \tag{3.35}$$

Here, the relationships $\cosh iy = \cos y$ and $\sinh iy = -i \sin y$ have been used. Note that substituting $z = x + iy$ in equation (3.35), we find

$$\cosh z = \cosh(x + iy) = \cosh x \cos y + i \sinh x \sin y. \tag{3.36}$$

In a similar manner,

$$\cos z = \cos(x + iy) = \cos x \cos iy - \sin x \sin iy = \cos x \cos y - i \sin x \sinh y,$$
$$\sin z = \sin(x + iy) = \sin x \cos iy + \cos x \sin iy = \sin x \cosh y + i \cos x \sinh y. \tag{3.37}$$

Complex functions (i.e., functions of complex variables) are thus a little more complicated than functions of purely real variables. The question whether or not a complex function is analytic or not often becomes important in applications, particularly those where complex integration is involved. To understand analyticity, we need to understand differentiability, assuming a complex function is continuous in the "$\delta - \epsilon$

sense" (see, e.g. Greenberg (1978)). Derivatives of complex functions are important in applications. These are only defined if a complex function satisfies the so-called Cauchy–Riemann conditions, which essentially require that, for a function $f(z)$ to be considered differentiable, the following must hold, where

$$f(z) = f(x + iy) \equiv g(x,y) + ih(x,y), \tag{3.38}$$

$$\frac{\partial g}{\partial x} = \frac{\partial h}{\partial y}; \quad \frac{\partial h}{\partial x} = -\frac{\partial g}{\partial y}, \tag{3.39}$$

where the partial derivatives g_x, g_y, h_x, and h_y are all also continuous. Equation (3.39) is a statement of the Cauchy–Riemann conditions. If a function $f(z)$ is continuous and differentiable at some point and differentiable in some neighborhood of that point, then it is said to be *analytic* at that point. If the function is analytic at any arbitrary point (i.e., at all points) in a particular domain, such as the upper xy half-plane, then the function is analytic over that entire domain.

There is one other property that becomes important in applications. This is the property of "single-valuedness" or lack thereof. Recall that, for real variables, a function is a map from a domain to a range, and it does not have to be a one-to-one map. It can be a many-to-one map, such as in the case of the sine function, or any other periodic function, where $\sin x = \sin(x + 2n\pi)$, $n = 1, 2, \ldots$, and so forth. However, a function cannot be a one-to-many map, as that would lead to ambiguity, unless some additional information guiding the choice of a particular value over any other value is available. The need for such additional information does arise in the case of some complex functions. Consider, for instance, the natural log function,

$$f(z) = \ln z. \tag{3.40}$$

It is easiest to proceed further if we use $z = re^{i\theta}$, noting that θ and $\theta + 2n\pi$, $n = 1, 2, \ldots$, all lead to the same z. However,

$$\ln z = \ln(re^{i\theta}) = \ln r + i\theta. \tag{3.41}$$

Here, we see a potential problem, in that, $\ln z$ is also $\ln r + i(\theta + 2\pi)$, $\ln r + i(\theta + 4\pi)$, and so forth. We thus have a one-to-many map. This is where the notion of *branch cuts* comes in. A branch cut is simply a way to preclude multivaluedness, by introducing an artificial cut or wall in the range of values defining a function. Figure 3.3 shows an example. The branch cut along the line $\theta = 2\pi$ just makes the function definition unique as long as we understand that $[0, 2\pi)$ is the region we are to consider if we expect $\ln z$ to be unique. If we cross the cut or wall, we accrue an additional 2π in the θ value (i.e., the argument) going into $\ln z$. Other choices for branch cuts are also possible, though branch cuts are just a way to limit the region into which a function definition takes us so as to keep the function single-valued. It is the additional information needed to eliminate ambiguity that would otherwise come from multivaluedness. Other functions requiring branch cuts include $f(z) = \sqrt{\mu^2 - k^2}$, $f(z) = 1/\sqrt{\mu^2 - k^2}$, and so forth, which sometimes arise in wave propagation problems. The argument changes from 0 to $\pi/2$ so that the function goes from a real value

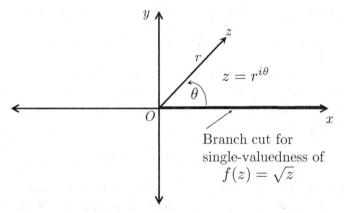

Figure 3.3 Figure showing a branch cut to prevent $f(z) = \sqrt{z}$ from becoming multivalued. The branch cut precludes θ from reaching $\theta + 2\pi$ and over.

R to an imaginary value $Re^{i\pi/2}$ as k, increasing from, say, 0 to ∞, crosses the real value μ.

It is interesting to see that in the case of the function $f(z) = \sqrt{z}$, where $z = re^{i\theta}$, the function $\sqrt{z}$ leads to two values even if $e^{i\theta} = e^{i(\theta+2\pi)}$. The two values are $\sqrt{r}(\cos\theta/2 + i\sin\theta/2)$ and $-\sqrt{r}(\cos\theta/2 + i\sin\theta/2)$. To prevent this happening, we need to eliminate the possibility $\theta + 2\pi$ by introducing a branch cut such as shown in Figure 3.3.

3.3   An Example of Cauchy–Riemann Conditions

An interesting connection between the discussion of Chapter 1 and the present complex functions can be seen below. It was pointed out in Chapter 1 that, for incompressible flows, the divergence of the velocity field is zero (which follows from mass conservation or continuity). Then we can represent the local velocity vector using a vector function such that

$$\mathbf{v} = \nabla \times \Psi, \tag{3.42}$$

where Ψ is the so-called stream function in vector form. It becomes a scalar function in two-dimensional flows, i.e., flows where the flow field is essentially planar, with negligible changes in the perpendicular direction. Long-crested ocean surface waves or flows perpendicular to the long axis of a cylinder, or flows past long hydrofoils (resembling long aircraft wings), are some examples. Similarly, the stream function is a scalar in flow fields that are three-dimensional but axisymmetric, for instance, flows produced by the motion of a circular disk in a direction along its out-of-plane axis. In two-dimensional flows, for instance, flows in the xy plane where the velocity component along the z direction is zero, the velocity vector can be represented using

$$\mathbf{v} = u\mathbf{i} + v\mathbf{j}, \quad w = 0. \tag{3.43}$$

Here, only a scalar stream function ψ suffices, such that

$$u = \frac{\partial \psi}{\partial y}, \quad v = -\frac{\partial \psi}{\partial x}. \tag{3.44}$$

The flow fields only need to be two-dimensional and incompressible, for equation (3.44) to hold, and they can be viscous and rotational. However, now in addition, if the flow is also irrotational, such that the vector velocity is a gradient of the scalar potential function, then we also have

$$\mathbf{v} = \nabla \phi = \frac{\partial \phi}{\partial x}\mathbf{i} + \frac{\partial \phi}{\partial y}\mathbf{j} + \frac{\partial \phi}{\partial z}\mathbf{k}. \tag{3.45}$$

For two-dimensional irrotational flows in the xy plane, the component in the z direction is zero. Clearly, for two-dimensional compressible and irrotational flows,

$$\mathbf{v} = \nabla \phi = \nabla \times \psi, \tag{3.46}$$

where both ϕ and ψ are scalar quantities.

$$u = \frac{\partial \phi}{\partial x}; \quad v = \frac{\partial \phi}{\partial y}. \tag{3.47}$$

It follows, from equations (3.44) and (3.45), that, for two-dimensional incompressible, irrotational flows,

$$\frac{\partial \phi}{\partial x} = \frac{\partial \psi}{\partial y},$$

$$\frac{\partial \phi}{\partial y} = -\frac{\partial \psi}{\partial x}. \tag{3.48}$$

The two equations in equation (3.48) are in fact just a statement of the Cauchy–Riemann conditions for a complex function χ that can be defined as

$$\chi = \phi + i\psi, \tag{3.49}$$

where χ is the complex potential function. Because the physical situation ensures that the Cauchy–Riemann conditions are satisfied, the function χ is analytic in the xy plane.

3.4   Complex Integration Overview

As alluded to in the introductory paragraph of this chapter, integration of complex functions sometimes turns out to be a lot easier than would seem possible when one first confronts the function to be integrated. Indeed, it is frequently easier to evaluate an integral of a function of a real variable by converting it into a complex integral. Important to complex integration are (1) Cauchy–Goursat theorem, (2) Cauchy's integral formula, and (3) the residue theorem, which leads to the method of residues.

Briefly, the Cauchy–Goursat theorem states that, if a function $f(z)$ is analytic within and on a closed curve C, then its integral over that closed curve is zero. In other words,

$$\int_C f(z)dz = 0. \tag{3.50}$$

For example, consider a function $f(z) = z^2$. This function is continuous and differentiable around the origin $z = 0$. For our closed curve over which to integrate $f(z)$, we choose the circle $|z| = 1$. $f(z)$ is thus analytic inside and on the curve C. We can express, $z = re^{i\theta}$, so that, performing the integration counterclockwise,

$$\int_C f(z)dz = \int_C r^2 e^{2i\theta} d\theta = \int_0^{2\pi} e^{2i\theta} d\theta. \tag{3.51}$$

It can be seen that

$$\int_C f(z)dz = \frac{1}{2i}\left(e^{2\pi i} - e^{i0}\right) = 0. \tag{3.52}$$

Cauchy's integral formula allows us to say that, if a function $f(z)$ is analytic inside and on a closed curve C, and further, if a point $z = a$ is inside the curve but not on it, then,

$$\frac{1}{2\pi i}\int_C \frac{f(z)}{z - a}dz = f(a). \tag{3.53}$$

As an example, consider a function $f(z) = z^2 - 3z + 6$. Letting the closed, counterclockwise curve be $z = 3$, we see that the function is analytic inside and on the curve $|z| = 3$. For a point $z = 2$ inside the curve,

$$\int_C \frac{z^2 - 2z + 6}{z - 2}dz = 2\pi i \left(z^2 - 3z + 6\right)\bigg|_{z=2} = 10\pi i. \tag{3.54}$$

Before moving on to the residue theorem, it would be well to probe a little deeper into the notion of singularities of complex functions, i.e., points at which a complex function approaches infinity or becomes undefined. Consider first a function such as

$$f(z) = \frac{1}{z - 1}; \quad f'(z) = -\frac{1}{(z - 1)^2}. \tag{3.55}$$

As $z \to 1$, $f(z) \to \infty$, and $f'(z) \to \infty$. Elsewhere, however, the function is well defined. In fact, one can visualize a region around $z = 1$ defined by the open interval, $1 < |z| < r$, where $r > 0$, over which $f(z)$ is analytic. The singularity here is said to be an isolated singularity. Further, $z = 1$ defines a "pole" of the function $f(z)$. It is moreover a first-order pole. If the function $f(z)$ were $1/(z - 1)^2$, $z = 1$ would be a second-order pole. An isolated singularity can define an mth-order pole, as long as m is finite. When $m \to \infty$, such as for a function represented by a series,

$$f(z) = \frac{1}{z - 3} + \frac{1}{(z - 3)^2} + \ldots + \frac{1}{(z - 3)^m} + \ldots, m \to \infty. \tag{3.56}$$

The function $f(z)$ in equation (3.56) has a singularity at $z = 3$, but it is not an isolated singularity. Consider next the function $f(z) = \ln z$. We have seen that

$$f(z) = \ln r + i(\theta + 2n\pi). \tag{3.57}$$

By defining a branch cut along the positive x axis, we can choose the branch $0 \leq \theta \leq 2\pi$ as our "principal branch." It should be noted that $r > 0$ or always positive. For real variables R, $\ln(R)$ is not defined for $R < 0$. For complex variables, however, we can define $\ln(-z)$ as $\ln\left(re^{i\theta+i\pi}\right) = \ln(r) + i\,(\theta + \pi)$. In other words, $\ln(-z)$ lies on a different branch. Hence, operations defined on the principal branch are not valid for the branch required for negative z.

We recall that $\ln z$ is not defined at $z \to 0$. However, the Cauchy–Riemann conditions in polar coordinates (Kreyszig, Kreyszig, and Norminton 2011) state that, for a function $f(z) = g(r,\theta) + ih(r,\theta)$,

$$\frac{\partial g}{\partial r} = \frac{1}{r}\frac{\partial h}{\partial \theta}; \quad \frac{\partial h}{\partial r} = -\frac{1}{r}\frac{\partial g}{\partial \theta}. \tag{3.58}$$

For $z \neq 0$ on the branches that do not allow z values on the negative real axis,

$$\ln z = \ln r + i\,(\theta + 2n\pi) \equiv g(r,\theta) + ih(r,\theta), \tag{3.59}$$

where $g(r,\theta) = \ln(r)$ and $h(r,\theta) = \theta + 2n\pi$;

$$\frac{\partial g}{\partial r} = \frac{1}{r} = \frac{1}{r}\frac{\partial h}{\partial \theta} = \frac{1}{r},$$

$$\frac{\partial g}{\partial \theta} = 0 = -\frac{1}{r}\frac{\partial h}{\partial r} = 0. \tag{3.60}$$

Under the conditions that $z > 0$ and z not be on the negative real axis, $\ln z$ is differentiable. However, the two conditions that make it so also ensure that $\ln z$ does not have an isolated singularity at $z = 0$.

The residue theorem only applies to functions that have isolated singularities (see Figure 3.4). It essentially asserts, first of all, that, if a complex function $f(z)$ has isolated singularities inside a closed curve C, then the integral of that function around the closed curve is not zero, and second, that, the value of the integral is related to the "residues" of that function at the singularities included within the closed curve. In fact, if there are J isolated singularities enclosed within a closed curve, then

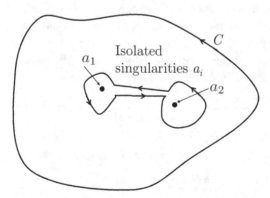

Figure 3.4 Closed curve C with isolated singularities at a_1 and a_2. Curve can be deformed to the smaller shape shown.

$$I = \int_C f(z)dz = 2\pi i \sum_{j=1}^{J} R_j. \tag{3.61}$$

R_j denotes the jth residue, at point $z = a_j$, where a_j represents the jth isolated singularity of $f(z)$. At each a_j, R_j is evaluated according to

$$R_j = \lim_{z \to a_j} \left[(z - a_j)f(z) \right]. \tag{3.62}$$

As an example, we consider a function such as

$$f(z) = \frac{z^2 - 2}{z^2 + 7z + 12}. \tag{3.63}$$

We may also express $f(z)$ as

$$f(z) = \frac{(z+2)(z-2)}{(z+3)(z+4)}. \tag{3.64}$$

$f(z)$ has isolated singularities at $z = -3$ and $z = -4$. Assuming a closed curve or contour of integration that includes both singularities, we proceed to find the residues at both. Thus,

$$R_1 = \lim z \to -3 (z+3)f(z) = \lim_{z \to -3} \frac{(z+3)(z+2)(z-2)}{(z+3)(z+4)} = \frac{(-1)(-5)}{(1)} = -5. \tag{3.65}$$

Similarly,

$$R_2 = \lim z \to -4 (z+4)f(z) = \lim_{z \to -4} \frac{(z+4)(z+2)(z-2)}{(z+3)(z+4)} = \frac{(-2)(-6)}{(-1)} = 12. \tag{3.66}$$

The integral then evaluates to

$$\int_C f(z)dz = 2\pi i(-5 + 12) = -7\pi i. \tag{3.67}$$

3.5   Contour Integration

As we mentioned previously, the residue theorem leads to the elegant technique of Contour integration, which enables analytical solutions to several problems related to wave motion and floating-body dynamics that at first seem overwhelming. We will return to contour integration in our treatment of Fourier and Laplace transforms as ways to solve dynamics problems relevant to marine applications. Here, we will review contour integration in mostly general terms. For illustration, we begin with an integral,

$$I = \int_0^{2\pi} \sin^2 \theta d\theta. \tag{3.68}$$

Using the substitution, $\sin^2 \theta = \frac{1}{2}(1 - \cos^2 \theta)$, it can be seen that this integral evaluates to a value equal to π. If we were to use contour integration, we would begin with the substitution,

$$\sin\theta = \frac{e^{i\theta} - e^{-i\theta}}{2i}. \tag{3.69}$$

In equation (3.69), we see that the magnitude of the two parts making up the function is unity. Integration over θ from 0 to 2π is thus equivalent to a contour integration with respect to z over the closed circle given by the unit circle $|z| = 1$. Then,

$$z = e^{i\theta} \Rightarrow dz = ie^{i\theta}d\theta, \Rightarrow d\theta = \frac{dz}{iz}. \tag{3.70}$$

Thus, the integral I in equation (3.68) can be rewritten as

$$I = \int_C \frac{i(e^{2iz} + e^{-2iz} - 2)}{-4(z-0)} dz. \tag{3.71}$$

Since the only enclosed singularity is at $z \to 0$,

$$I = 2\pi i R_1, \tag{3.72}$$

where R_1 is the residue.

$$R_1 = \lim_{z\to 0}(z-0)\frac{i(e^{2iz} + e^{-2iz} - 2)}{-4(z-0)} = -\frac{i}{2}. \tag{3.73}$$

By equation (3.72),

$$I = 2\pi i R_1 = \pi. \tag{3.74}$$

This result agrees with the known result stated under equation (3.68).

As another example, consider an integral defined for the real variable x,

$$I_x = \int_{-\infty}^{\infty} \frac{dx}{x^2 + 2x + 4}. \tag{3.75}$$

The integral in equation (3.75) is an "improper integral," because there is an ∞ involved in the integration limits, in this case, both limits. The functions do converge to 0 at either ∞, however, so that the integral as a whole also converges (i.e., the function is integrable). In order to evaluate this integral, we may begin with a complex integral over a closed contour C expressed as

$$I_z = \int_C \frac{dz}{z^2 + 2z + 4}. \tag{3.76}$$

The integrand has poles at $z = p_1, p_2$, where

$$p_1, p_2 = -\frac{1}{4} \pm i\frac{\sqrt{3}}{4}. \tag{3.77}$$

It follows that

$$I_z = \int_C \frac{dz}{(z - p_1)(z - p_2)}. \tag{3.78}$$

The two singularities here are in the form of a complex conjugate pole pair, and are indicated as in Figure 3.5. For contour integration, we select the closed contour shown in Figure 3.5. Note that the sense of integration is counterclockwise. This contour

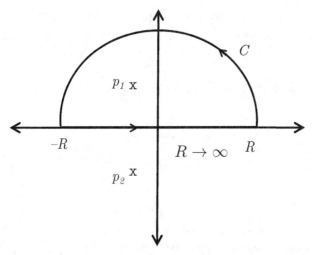

Figure 3.5 Illustration of contour integration, with $R \to \infty$. The closed curve C encloses the pole p_1, and integration along it is counterclockwise. Equivalent result would be obtained with a reflected version of the curve that includes p_2 instead with integration along it proceeding clockwise.

encloses the pole p_1, and the pole p_2 is excluded. The integral over the contour can be written as

$$I_z = \int_{-\infty}^{\infty} \frac{dz}{(z - p_1)(z - p_2)} + \int_{C_1} \frac{dz}{(z - p_1)(z - p_2)}, \qquad (3.79)$$

where C_1 represents the counterclockwise semicircular curve with radius $R \to \infty$. On C_1, $z = Re^{i\theta}$ so that $dz = iRe^{i\theta} d\theta$. Thus,

$$\int_{C_1} \frac{dz}{(z - p_1)(z - p_2)} \to \lim_{R \to \infty} \int_0^{\pi} \frac{iRe^{i\theta} d\theta}{R^2 e^{i2\theta}} \to 0. \qquad (3.80)$$

Next, by residue theorem, since there is one singularity within the closed contour C,

$$I_z = 2\pi i R_1, \qquad (3.81)$$

where the residue R_1 can be found as

$$R_1 = \lim z \to p_1 (z - p_1) \frac{1}{(z - p_1)(z - p_2)} = \frac{2}{i\sqrt{3}}. \qquad (3.82)$$

Hence,

$$I_z = \int_{-\infty}^{\infty} \frac{dz}{(z - p_1)(z - p_2)} = \int_{-\infty}^{\infty} \frac{dx}{(x - p_1)(x - p_2)} = \frac{4\pi}{\sqrt{3}}. \qquad (3.83)$$

Closing the contour through the left-half plane below the x axis will mean that the pole p_2 on the negative imaginary axis is now included and the pole p_1 is excluded so that

$$I_z = -2\pi R_2. \tag{3.84}$$

Note the negative sign, given that the contour is in the clockwise sense. Here, R_2 is given by

$$R_2 = \lim_{z \to p_2} (z - p_2) \frac{1}{(z - p_1)(z - p_2)} = -\frac{2}{i\sqrt{3}}. \tag{3.85}$$

Thus, once again,

$$I_z = \frac{4\pi}{\sqrt{3}}. \tag{3.86}$$

In applications, how does one decide whether to use the contour closing through the top half-plane or the contour closing through the bottom half plane? In wave propagation problems, where, for instance, the space–time dependence is expressed as $e^{-i(kx-\omega t)}$, the choice is driven by (i) the so-called "radiation condition," i.e., whether or not the wave propagates to the right or to the left, and whether (ii) time dependence is expressed as $e^{i\omega t}$, or as $e^{-i\omega t}$. A particularly interesting application of contour integration arises in time-domain representation of floating body oscillations, as discussed in texts such as Korde and Ringwood (2016).

3.6   Principal Value Integrals

Oftentimes in marine applications, it is meaningful to evaluate integrals such that the singular points are excluded. As an example, we consider a real-valued function such as

$$f(x) = \frac{1}{x - 4}. \tag{3.87}$$

This function becomes singular as $x \to 4$. Hence, an integral such as

$$I = \int_0^6 \frac{dx}{x - 4} \tag{3.88}$$

does not really exist, because the left limit of $f(x)$ as $x \to 4$ from the left does not equal the right limit, as $x \to 4$ from the right, the former being negative, with the latter being positive. However, the integral in equation (3.88) could be evaluated in the "principal value sense." If we define

$$PV \int_0^6 \frac{dx}{x - 4} = \lim_{\epsilon \to 0} \left[\int_0^{4-\epsilon} \frac{dx}{x - 4} + \int_{4+\epsilon}^6 \frac{dx}{x - 4} \right], \tag{3.89}$$

one may be able to find its "Cauchy principal value." This could be done numerically with reasonable accuracy in cases where analytical evaluation of the integral is not practical. In the case of the integral in equation (3.89), the principal value integral becomes

$$PV \int_0^6 \frac{dx}{x-4} = \lim_{\epsilon \to 0} \left[\ln |\epsilon| - \ln |4| + \ln |2| - \ln |\epsilon| \right]. \qquad (3.90)$$

Since the two terms involving ϵ cancel, the principal value of the integral can be seen to be

$$PV \int_0^6 \frac{dx}{x-4} = \ln \left| \frac{1}{2} \right|. \qquad (3.91)$$

We may consider another example where contour integration can be applied to evaluate the principal value of an integral. Consider the function

$$f(z) = \frac{1}{z(z-3)}. \qquad (3.92)$$

We wish to evaluate the principal value of the integral of $f(z)$ in equation (3.92) about the singularity at $z = 0$. For illustration, we may define a counterclockwise contour C with vertices at points $1, 1+i$, and i, completing it with three quarters of a small circle of radius ϵ. We wish to find the principal value integral

$$I_{PV} = PV \int \frac{dz}{z(z-3)}. \qquad (3.93)$$

$f(z)$ has two poles: one at $z = 0$ and the other at $z = 3$. Only the $z = 0$ pole is included in the chosen contour C. Therefore, the integral I around the closed contour can be found as

$$I = \int_C \frac{dz}{z(z-3)} = 2\pi i R_1, \qquad (3.94)$$

where

$$R_1 = \lim_{z \to 0} z \frac{1}{z(z-3)} = -\frac{1}{3}. \qquad (3.95)$$

Thus,

$$I = \int_C \frac{dz}{z(z-3)} = -\frac{2\pi i}{3}. \qquad (3.96)$$

Now, we can express the integral I as

$$I = I_{PV} + \lim_{\epsilon \to 0} \int_{\epsilon i}^{\epsilon} \frac{dz}{z(z-3)}. \qquad (3.97)$$

The second integral in equation (3.97) is defined along the three-fourths circle with radius ϵ. For $\epsilon \to 0$, the denominator $z^2 - 3z$ in the integrand can be approximated as $\approx -3z$. Then,

$$\lim_{\epsilon \to 0} \int_{\epsilon i}^{\epsilon} \frac{dz}{z(z-3)} \to -\lim_{\epsilon \to 0} \int_{\epsilon i}^{\epsilon} \frac{dz}{3z}. \qquad (3.98)$$

The right-hand side of equation (3.98) can be evaluated as

$$-\lim_{\epsilon \to 0} \left[\ln \left(|\epsilon| e^{2\pi i} \right) - \ln \left(|\epsilon| e^{i\pi/2} \right) \right] = -(2\pi - \pi/2)i = -\frac{3\pi i}{2}. \qquad (3.99)$$

Given equation (3.97),

$$I_{PV} = -\frac{2\pi i}{3} + \frac{\pi i}{2} = \frac{\pi i}{6}.$$ (3.100)

It can be verified that redefining the closed contour so that the singularity is avoided by means of a small quarter-circle arc leads to the same result.

3.7   Complex Velocity Potentials

As mentioned in Chapter 1, for flows that are irrotational and incompressible, the velocity potential function ϕ and the stream function Ψ can be used to gain much insight into such flows, particularly in situations that can be approximated as two-dimensional, for then the stream function is a scalar, denoted as ψ. Examples of such flow situations include long-crested waves, tidal currents through channels, ship roll in beam seas, and so on. We will have a chance to use complex potentials in Chapter 11. In two-dimensional flows, the velocity potential ϕ and the stream function ψ can be combined into a single complex function $F(z)$ such that

$$F(z) = \phi + i\psi.$$ (3.101)

Realizing that, for irrotational flows at velocity $\mathbf{v}$, we have

$$\mathbf{v} = \nabla \phi = \nabla \times \Psi.$$ (3.102)

For two-dimensional irrotational flows, equation (3.102) implies that the velocity $\mathbf{v} = u\mathbf{i} + v\mathbf{j}$. Therefore,

$$u = \frac{\partial \phi}{\partial x} = \frac{\partial \psi}{\partial y},$$
$$v = \frac{\partial \phi}{\partial y} = -\frac{\partial \psi}{\partial x}.$$ (3.103)

Incompressibility further implies that $\nabla \cdot \mathbf{v} = (\partial u/\partial x) + (\partial v/\partial y) = 0$. Then, in terms of ϕ, we have

$$\frac{\partial^2 \phi}{\partial x^2} + \frac{\partial^2 \phi}{\partial y^2} = \nabla^2 \phi = 0.$$

Equation (3.104) is the Laplace equation describing the spatial dependence of the velocity potential function ϕ. The surfaces in space on which ϕ is a constant ($\phi = c$) are the so-called equipotential surfaces. In two-dimensional space, $\phi = c$ represents equipotential lines. The velocity vector defined by $\nabla \phi$ is then normal to the equipotential surfaces. It can be verified that the directional derivative along the surface is zero, as it should be for surfaces on which ϕ is supposed to be constant. Are there analogous surfaces or lines defined by the stream function Ψ? In two dimensions, the lines on which the scalar $\psi = 0$ are the stream lines, representing the lines along which flow takes place. If we are observing flow around an object or over a boundary (e.g., over a wall), then the stream lines $\psi = 0$ represent the boundaries of the object or the physical boundary such as a wall.

We observe that, for irrotational flows, as expressed by equation (3.104), the velocity potential ϕ can be described by the Laplace equation. As discussed earlier, for flows that are also incompressible, we can also use the stream function Ψ to describe the flow. Narrowing our discussion to two-dimensional flows, it is reasonable to ask then if the stream function might also satisfy the Laplace equation? We can see that it does so, by arguing that, for irrotational flows, $\nabla \times \mathbf{v} = 0$. In two-dimensional flows, this implies that

$$\frac{\partial u}{\partial y} - \frac{\partial v}{\partial x} = 0. \tag{3.104}$$

Since $u = \partial\phi/\partial x$ and $v = \partial\phi/\partial y$, we have, by substitution,

$$\frac{\partial^2 \phi}{\partial x \partial y} - \frac{\partial^2 \phi}{\partial y \partial x} = 0. \tag{3.105}$$

Given that ϕ and ψ are related by equation (3.103), another substitution leads to

$$\frac{\partial^2 \psi}{\partial x^2} + \frac{\partial^2 \psi}{\partial y^2} = 0. \tag{3.106}$$

We have used equation (3.103) to arrive at this conclusion. As already seen in Section 3.3, equations (3.103) represent the Cauchy-Riemann conditions. Since these are satisfied here, $F(z)$ as defined by equation (3.101) is an analytic function in the flow domain. Thus, it can be seen that

$$\frac{dF(z)}{dz} = u - iv. \tag{3.107}$$

Finally, as seen above, the real and imaginary parts of the complex potential function $F(z)$ both satisfy the Lapace equation. The condition $F(z) = C$, where C is a complex constant, thus describes both equipotential lines and stream lines for two-dimensional incompressible, irrotational flows.

3.8 Conformal Mapping

For two-dimensional irrotational, incompressible flows that are described by the complex potential $F(z)$, sometimes it may be easier to understand and derive flow fields by means of a variable transformation, i.e., by mapping the z domain to another two-dimensional complex domain ζ via a function $f(z)$ such that $\zeta = f(z)$. Such an approach is advantageous when the flow field in the mapped-to domain is well known and has a compact representation in that domain. For instance, for unit constant velocity $U = 1$, the flow past a straight wall or a plate in the mapped-to domain ζ can be described by a complex potential

$$F(\zeta) = \zeta. \tag{3.108}$$

This straightforward expression can be used to solve for the flow fields around a corner, flow around an ellipse, or a flow around a circle. We remind ourselves that we are

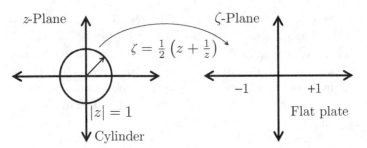

Figure 3.6 Illustration showing how a complex mapping can be used to solve flow problems. In this case, a unit-circle in the z-plane maps into a straight line in the ζ-plane.

discussing irrotational, incompressible flow, i.e., an ideal flow. Following Newman (1978), if we consider a mapping such as

$$\zeta = z^{\pi/\theta}, \tag{3.109}$$

for $\theta = \pi$, we simply map the domain onto itself, and the flow fields in both, the mapped-from and mapped-to, domains are identical, being described as $F(z) = z$. If on the other hand, $\theta = \pi/2$, then a 90° corner in the z domain maps to a flat plate in the ζ domain via

$$\zeta = z^2. \tag{3.110}$$

The flow in the mapped-to domain is then described by equation (3.108) and is simply the flow past a straight wall or flat plate, i.e., $F(\zeta) = \zeta$. The flow in our z domain can be found by substituting $z = \zeta^{1/2}$ in equation (3.108) so that $F(\zeta) = \zeta$ changes to

$$F(z) = z^2. \tag{3.111}$$

More generally, for an angle θ, the complex potential $F(z)$ would be

$$F(z) = z^{\pi/\theta}. \tag{3.112}$$

On the other hand, let us consider a map such as (Newman 1976)

$$\zeta = \frac{1}{2}\left(z + \frac{1}{z}\right). \tag{3.113}$$

We can see that $z = e^{i\theta}$ represents a circle with radius $r = 1$, with θ ranging from 0 to 2π. The figure represented by $|z| = 1$ is thus a circle with unit radius in the z domain. For $\theta = 0$, $\zeta = \cos\theta = 1$, while for $\theta = \pi$, $\zeta = -1$, with $\zeta = 2\pi$ again leading to $\zeta = 1$. Thus, the unit circle maps to a straight line from -1 to 1 in the ζ domain (see Figure 3.6). The line representing a flat plate in two dimensions, we know that the flow past this plate is simply given by

$$F(\zeta) = \zeta. \tag{3.114}$$

The inverse map form ζ to z basically describes the flow past a unit circle in two dimensions (which essentially is a cylinder that is infinitely long in the direction perpendicular to the plane of the paper), as

$$F(z) = \frac{1}{2}\left(z + \frac{1}{z}\right). \tag{3.115}$$

If we denote the velocity potential and stream functions in the ζ plane as $\overline{\phi}$ and $\overline{\psi}$, respectively, then equation (3.114) can be rewritten as

$$F(\zeta) = \overline{\phi} + i\overline{\psi}. \tag{3.116}$$

Letting $\zeta = \xi + i\eta$, for uniform flow U over a horizontal flat plate, the vertical velocity component is zero, so that

$$\frac{\partial \overline{\phi}}{\partial \xi} = U; \quad \frac{\partial \overline{\phi}}{\partial \eta} = 0. \tag{3.117}$$

The Cauchy-Riemann conditions tell us that

$$\frac{\partial \overline{\psi}}{\partial y} = \frac{\partial \overline{\phi}}{\partial x} = U; \quad \frac{\partial \overline{\psi}}{\partial x} = -\frac{\partial \overline{\phi}}{\partial y} = 0. \tag{3.118}$$

Equations (3.117) and (3.118) can be integrated to give

$$\overline{\phi} = U\xi; \quad \overline{\psi} = U\eta. \tag{3.119}$$

The complex potential $F(\zeta)$ therefore becomes, from equation (3.116),

$$F(\zeta) = U\zeta. \tag{3.120}$$

Now, using the transformation described by equation (3.115) and substituting for ζ in equation (3.120),

$$F(z) = \frac{1}{2}U\left(z + \frac{1}{z}\right). \tag{3.121}$$

$F(z)$ describes the complex velocity potential for flow around a cylinder of unit radius. We can see that, if we were to observe the flow far downstream of the cylinder (i.e., as $x \to \infty$), the effect of the cylinder is negligible, so that the flow velocity is restored to the original free stream velocity far up stream of the cylinder (i.e., as $x \to -\infty$), the flow velocity is $U/2$. The more general case of flow around an ellipse (i.e., a cylinder with an elliptical cross section), the following map can be used (see, for instance, Newman (1978)).

$$\zeta = Az + \frac{B}{z}. \tag{3.122}$$

In general, the functional maps involved in conformal mapping satisfy the following conditions: (1) the function is analytical, and (2) the map preserves angles between two lines in going from one domain to another (both magnitude and sense). Condition (2) only needs to be satisfied in a very local sense, however, as discussed further in Greenberg (1978). It is through a series of conformal maps that one obtains the flow field around an airfoil or a hydrofoil. Thus, conformal mapping has an important place in fluid flow analysis problems.

3.9   Concluding Remarks

In this chapter, we reviewed complex variables and complex functions. We discussed how multivalued maps can be treated as single-valued functions by introducing branch cuts and specifying the branches of interest to the operation at hand. We outlined the conditions under which a complex function is differentiable, and under what conditions a function is considered analytical. We discussed integration of complex functions, and defined singularities, particularly, how one might actually use singularities to compute complicated integrals of complex functions via the residue theorem. We studied contour integration next, and worked through examples. We followed that up with a brief consideration of principal value integrals, since these too arise frequently in marine applications. We concluded the chapter with an introduction to conformal mapping, application of which is vital in analytical solutions to several fluid flow problems, including flows around submerged bodies in the ocean.

3.10   Self-Assessment

3.10.1

If $z = 2/(3 = 4i)$, what is the real part of z? What is the imaginary part of z?

3.10.2

Evaluate the complex integral,

$$\int_C z\,dz, \tag{3.123}$$

where C goes from $z = 2 + 3i$ to $4 + 3i$.

3.10.3

If a flow field is described using a complex potential $F(z)$ defined as $F(z) = z^2$, map the stagnation points of the flow. Use contour equation to evaluate the integral,

$$\int_0^\infty \frac{1}{1 + x^2}\,dx. \tag{3.124}$$

3.10.4

Use contour equation to evaluate the integral,

$$\int_0^\infty \frac{1}{1 + x^2}\,dx. \tag{3.125}$$

3.10.5

Use contour integration to evaluate the integral,

$$\int_{-\infty}^{\infty} \frac{e^{-ikx}}{k^2 - k_o^2} dk. \tag{3.126}$$

(a) Include the singularity k_o and exclude $-k_o$ in the contour.
(b) Include $-k_o$ and exclude k_o.
(c) Discuss the result in each case.

4   Fourier Analysis

One of the most frequently used techniques in the analysis of fluid flow and bodies interacting with fluids is that of Fourier analysis, which in some ways is just a layer of abstraction, though one that can often provide much needed insights. The foundational idea behind Fourier analysis is that some classes of functions (specifically, periodic functions) can be represented as summations of cosine and sine functions, though sometimes summations of other functions such as Bessel functions and Legendre polynomials could also be used. The original functions may be formulated in the spatial domain or in the temporal domain. Often, in practical situations, the functions one encounters seem to contain a large number of periods (frequencies), making it hard to determine where or at near which frequencies the most consequential part of the function resides. Use of just this "distilled out" dominant part of the given function in subsequent steps can often provide a more meaningful assessment and a better intuitive understanding of the practical situation. The process of "breaking down" possibly complicated functions into series of "everyday" functions (basis functions) is referred to as Fourier decomposition. The reverse process can be called "Fourier synthesis."

As alluded to earlier, by observing the magnitudes of the "coefficients" of the basis functions, especially the largest or the dominant ones, we can understand the dominant traits of the complicated functions. In filtering type applications, therefore, oftentimes we just keep the dominant basis functions, discard the least dominant basis functions, and reverse the decomposition process; i.e., simply add the selected basis functions together to generate a "filtered" version of the original complicated functions. The approach still works when the complicated functions are not periodic, as then the series of basis functions ("Fourier series") turn into integrals where often the cosine and sine functions are combined into complex exponential functions ("Fourier transforms").

In the process of Fourier transformation also lies a powerful technique for solving differential equations, partial and ordinary. Broadly speaking, Fourier transformation of both sides of a partial differential equation leads to an ordinary differential equation, with a further transformation turning the ordinary differential equation into an algebraic equation. With each step, we exchange the resulting ease of obtaining solutions for the need to think about variables expressed in a more abstract domain. Once the final results are obtained in the transformed domains, a series of inverse transformations (which are not always straightforward) can provide the solutions to the original equations in the original domain. Fourier techniques are valuable in understanding the spatial behavior of marine phenomena, as well as temporal behavior of bodies interacting with such phenomena. Wave motion provides particularly interesting opportunities, in that both spatial and temporal transforms can be used to provide clarity in how waves and their energy travel. In this chapter, we review Fourier series and Fourier transforms, and their use in either direction. The techniques summarized here form the basis for many of the developments of Chapters 7, 8, and 15. As before, a more detailed treatment of these topics can be found in Greenberg (1978), Brown and Churchill (2008), and so on.

4.1 Requirements for Fourier Analysis

One of the requirements for Fourier analysis to be valid is that the function to be decomposed be integrable. This is because the two-way process from the original function to cosine and sine functions and back requires integration. In informal terms, the integration operation on a function only leads to sensible results if the function is piecewise continuous. Recall that a function is continuous at a point x_0 if, for every $\epsilon > 0$, however small, there exists a $\delta(x_0, \epsilon) > 0$ such that $|f(x) - f(x_0)| < \epsilon$, whenever $|x - x_0| < \delta(x_0, \epsilon)$. In other words, $f(x)$ is very close to $f(x_0)$, whenever x is very close to x_0. This means that a function such as $f(x) = x^2$ is continuous everywhere, but consider a function defined such that

$$f(x) = x^2, \quad x < 5;$$
$$f(x) = 10, \quad x = 5;$$
$$f(x) = x^2, \quad x > 5. \tag{4.1}$$

The value of $f(x)$ suddenly drops to 10 from very nearly 25 for a point very close to 5 on either side of 5. The function $f(x)$ is therefore not continuous at $x = 5$. It can be seen that the $\epsilon - \delta$ criterion does not quite work out for this function near $x = 5$. If we define a limit L of a function $f(x)$ at a point x_0 as follows,

$$\lim_{x_0} f(x) = L, \tag{4.2}$$

if for every $\epsilon > 0$, however small, there exists a $\delta(x_0, \epsilon) > 0$ such that $|x - x_0| < \delta(x_0, \epsilon)$. In other words, the closer x gets to x_0, the closer $f(x)$ gets to L. $f(x)$ is continuous at x_0 when its value $f(x_0) = L$. Consider a function defined as

$$f(x) = 2, \quad x < 5,$$
$$f(x) = 8, \quad x \geq 5. \tag{4.3}$$

This function is interesting in that it has two different limits at 5, depending on whether it is approached from the left or from the right. Thus, the left limit at 5 is 2, while the right limit is 8. The left limit does not equal the value of $f(x)$ at 5, though the right limit equals the value of $f(x)$ at 5. There is thus a jump discontinuity at 5. However, over the interval $-\infty < x < 5$, the function value equals its limit, meaning the function is continuous, while over the interval $x \geq 5$ also the function is

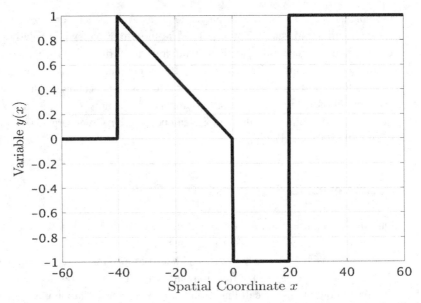

Figure 4.1 A plot showing an example of piecewise continuous functions. Note that the jumps across the discontinuities are finite.

continuous. Equation (4.3) is an example of a piecewise continuous function. Functions are piecewise continuous over a given interval if this interval can be divided into a finite number of subintervals and the functions are continuous within each subinterval and have finite limits at each end point (left or right). Thus, any "jumps" between two adjacent pieces are finite. Figure 4.1 shows an example piecewise continuous function. Such functions can be integrated. Piecewise continuity is thus a requirement for Fourier analysis to be meaningful for a given function. It should be noted, however, that some functions may be continuous but are not integrable in the sense that the integrals diverge (i.e., approach infinity). The function $xy = c$, where c is a constant, which represents a hyperbola, is an example of such functions. The two parts of the function (one in the first quadrant and the other in the third quadrant) are continuous in that the function values approach their limits at all points, but the jumps between the two parts are not finite. Hence, the function $xy = c$ is not piecewise continuous. Another familiar example is $y = \tan x$, which approaches $\pm\infty$ as x approaches $\pm\pi/2$. Only two branches of this periodic function are shown in Figure 4.2. Once again, the function values approach their limits as x gets close to $\pm\pi/2$, $\pm3\pi/2$, and so on. However, the jumps between values where one branch "hands off" to the next branch are not finite.

4.2   More about Functions

It is probably best to begin by understanding periodicity. A function $y = f(x)$ of a real variable is periodic if for some real valued ξ, $f(x + \xi) = f(x)$. ξ defines the period.

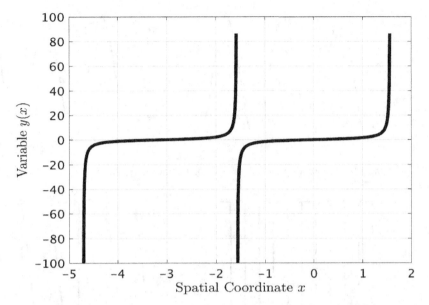

Figure 4.2 A plot showing an example of functions that are not piecewise continuous. Note that the jumps across the discontinuities are infinite.

For functions such as $\sin\theta$ and $\cos\theta$, $\xi = 2\pi$ (see Figure 4.3). These two functions also satisfy the condition

$$\sin\theta = \sin(\theta + 2n\pi), \; n = 1,2,3,\ldots,\infty$$
$$\cos\theta = \cos(\theta + 2n\pi), \; n = 1,2,3,\ldots,\infty. \qquad (4.4)$$

For a more general variable x with a more general period ξ, we can define θ such that

$$\theta = \frac{2\pi x}{\xi}, \qquad (4.5)$$

so that

$$\sin\theta = \sin(\theta + 2n\pi) \Rightarrow \sin x = \sin(x + n\xi), \; n = 1,2,3,\ldots,\infty \text{ and}$$
$$\cos\theta = \cos(\theta + 2n\pi) \Rightarrow \cos x = \cos(x + n\xi), \; n = 1,2,3,\ldots,\infty. \qquad (4.6)$$

Figure 4.4 shows some more examples of periodic functions.

It is interesting to consider the so-called orthogonality property of sine and cosine functions. Thus,

$$\int_0^{2\pi} \sin n\theta \sin m\theta\, d\theta = 0, n \neq m,$$

$$\int_0^{2\pi} \sin n\theta \sin m\theta\, d\theta = \pi, n = m, n \geq 1, \qquad (4.7)$$

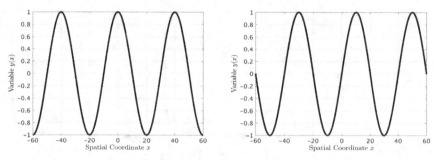

Figure 4.3 Plots showing the cosine and sine functions.

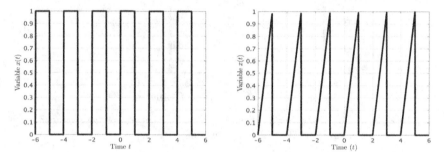

Figure 4.4 Plots showing more examples of periodic functions, namely the square wave and sawtooth wave functions.

and

$$\int_0^{2\pi} \cos n\theta \cos m\theta \, d\theta = 0, n \neq m,$$

$$\int_0^{2\pi} \cos n\theta \cos m\theta \, d\theta = \pi, n = m, n \geq 1,$$

$$\int_0^{2\pi} \cos n\theta \cos m\theta \, d\theta = 2\pi, m = n = 0. \tag{4.8}$$

Further,

$$\int_0^{2\pi} \sin n\theta \cos m\theta \, d\theta = 0, \text{for all } n, m. \tag{4.9}$$

Equation (4.7) asserts that two sine functions with arguments $m\theta$ and $n\theta$ represent two mutually orthogonal "directions" in a space of functions, just like two unit vectors in Cartesian space do. When $n = m$, the same direction is represented, and hence the integral is nonzero. The statement applies to two cosine functions with arguments $m\theta$ and $n\theta$. Interestingly, a sine function of $m\theta$ is always orthogonal to a cosine function of $n\theta$, regardless of the values of n and m. Summations of sine and cosine functions, with m and n allowed to be as large as needed, are thus particularly well suited to being used to approximate other functions or unknown solutions to physical problems. It is also interesting to note that $\sin m\theta$, $\cos m\theta$, $\sin n\theta$, and $\cos n\theta$, $(m \neq n)$, are all, besides being mutually orthogonal, also linearly independent, meaning that one cannot be expressed as a linear combination of others (just like functions such as $x, x^2, x^3, \ldots$).

It is interesting to consider that just like real variables, some functions of real variables can be described as odd functions and some as even functions. An arbitrary function can be expressed as a sum of an odd function and an even function. Briefly, a function $y = f(x)$ is even if $f(x) = f(-x)$. The function plot on the negative x axis is thus a mirror image of the function plot on the positive real axis. Functions $y = \cos x$, $y = x^2$, $y = |x|$, and $y = \cosh x$ are examples of even functions. On the other hand, if a function satisfies the relationship $f(-x) = -f(x)$, it is an odd function. Thus, $y = \sin x$ is an odd function, as are the functions $y = x$, $y = x^3$, $y = \sinh x$, and so on.

However, the function $y = e^{-2x}$ is neither odd nor even. Can we represent it as an odd + even combination? We recall that

$$\cosh 2x = \frac{e^{2x} + e^{-2x}}{2}, \quad \text{and} \quad \sinh 2x = \frac{e^{2x} - e^{-2x}}{2}. \tag{4.10}$$

Subtracting the right part of equation (4.10) from the left part, we find

$$f(x) = e^{-2x} = f_e(x) + f_o(x) = \cosh 2x - \sinh 2x, \tag{4.11}$$

where the even part $f_e(x) = \cosh 2x$, and the odd part $f_o(x) = -\sinh 2x$.

4.3   Fourier Trigonometric Series

The basis for Fourier series decomposition and synthesis is simply that, if a function $f(x)$ is periodic with a period 2π and is piecewise continuous, then it can be represented using an infinite series comprised of sin and cos functions. Although x is often used as a spatial variable associated with position, with $f(x)$ being considered a function representing the spatial behavior of some quantity, Fourier series decomposition of course works equally well for functions of time, and finds particular value in time-varying functions of interest to marine applications (e.g., waves, body motions in water, current flows, etc.), to say nothing of its ubiquitous use in the processing of signals and data from ongoing phenomena. Specifically,

$$f(x) = \frac{1}{2}a_0 + \sum_{n=1}^{\infty} (a_n \cos nx + b_n \sin nx). \tag{4.12}$$

It is often the case that $f(x)$ is a known function (perhaps piecewise continuous) and lacks a single analytical form, or it may be that one is particularly interested in the "frequency" composition of the function (spatial or temporal), which would be highlighted by the relative magnitudes of the "Fourier coefficients" a_n and b_n for various n. If the function is unknown but we have some measured data on its behavior, we might make a judicious assumption that it can be represented as in equation (4.12), and use some type of "best fit" criterion to estimate the coefficients a_n and b_n up to a chosen N. This at any rate would give us a single analytical expression that approximately captures physical behavior inherent in the unknown function. Assuming that $f(x)$ is known, we can find the Fourier coefficients via a few integrations as follows:

$$\int_{-\pi}^{\pi} f(x)dx = \frac{1}{2}a_0(2\pi),$$ (4.13)

and, multiplying both sides of equation (4.12) by $\cos nx$ ($\sin nx$) and integrating both sides from $-\pi$ to π,

$$\int_{-\pi}^{\pi} f(x) \cos nxdx = a_n\pi,$$

$$\int_{-\pi}^{\pi} f(x) \sin nxdx = b_n\pi.$$ (4.14)

It will be noted that we have used the orthogonality property of the sine and cosine functions as summarized in equations (4.7)–(4.9) in evaluating the integrals mentioned earlier. The relations can be summarized in a more compact form as

$$a_n = \frac{1}{\pi} \int_{-\pi}^{\pi} f(x) \cos nxdx, \quad (n = 0,1,2,\ldots),$$

$$b_n = \frac{1}{\pi} \int_{-\pi}^{\pi} f(x) \sin nxdx, \quad (n = 1,2,3,\ldots).$$ (4.15)

Equations (4.13)–(4.15) should help to clarify why integrability of $f(x)$ is an important requirement for a Fourier decomposition to be valid. Consider an example, of a square-wave defined, over one period as

$$f(t) = 0, \ -\pi \le t < 0,$$

$$f(t) = 1, \ 0 \le t < \pi.$$ (4.16)

This is a periodic function. In order to express this function using a Fourier series, we apply the first of equations (4.15), with $n = 0$, to find

$$a_0 = \frac{1}{\pi} \int_0^{\pi} (1)dt = 1.$$ (4.17)

Next, applying the first of equations (4.15) with $n \ne 0$, we find

$$a_n = \frac{1}{\pi} \int_0^{\pi} \cos ntdt = \frac{1}{n\pi}(\sin n\pi - \sin 0) = 0.$$ (4.18)

On the other hand,

$$b_n = \frac{1}{\pi} \int_0^{\pi} \sin ntdt = -\frac{1}{n\pi}(\cos n\pi - \cos 0).$$ (4.19)

Realizing that $\cos n\pi = (-1)^n$,

$$b_n = \frac{1}{n\pi}\left[(-1)^n - 1\right].$$ (4.20)

The Fourier decomposition for the periodic square wave of equation (4.16) can be written as

$$f(t) = \frac{1}{2} + \sum_{n=1}^{\infty} \frac{1}{n\pi}\left[1 - (-1)^n\right] \sin nt$$ (4.21)

To consider another example,

$$f(x) = |x|, -\pi < x \le \pi.$$ (4.22)

This is an even function. Since the product of an even function and an odd function is an odd function, all integrals involving $\sin n\theta$ from $-\pi$ to π must equal zero. Therefore, all coefficients $b_n = 0$. Since $\cos n\theta$ is an even function, however, the coefficients a_n must be nonzero.

$$a_0 = \frac{1}{\pi} \int_{-\pi}^{\pi} |x| \, dx = \frac{2}{\pi} \int_{0}^{\pi} x \, dx = \pi. \tag{4.23}$$

Further,

$$a_n = \frac{2}{\pi} \int_{0}^{\pi} x \cos nx \, dx = \frac{2}{n^2 \pi} \left[(-1)^n - 1 \right], \; n \text{ odd}. \tag{4.24}$$

Note that

$$a_n = 0, \; n \text{ even}. \tag{4.25}$$

The Fourier decomposition for the function $f(x) = |x|$ is thus

$$|x| = \frac{\pi}{2} + \sum_{n=1}^{\infty} \frac{2}{n^2 \pi} \left[(-1)^n - 1 \right] \cos nx, \; n \text{ odd}. \tag{4.26}$$

We generalize the function period here, from $-\pi \leq \theta \leq \pi$, for instance, to $-T_0 \leq t \leq T_0$, where t may be time t, or $-L \leq x \leq L$, where x defines the spatial coordinate. The Fourier decomposition in such a case can be written as

$$f(x) = \frac{a_0}{2} + \sum_{n=1}^{\infty} \left[a_n \cos \left(\frac{n\pi x}{L} \right) + b_n \sin \left(\frac{n\pi x}{L} \right) \right], \tag{4.27}$$

where

$$a_0 = \frac{1}{L} \int_{-L}^{L} f(x) \, dx,$$

$$a_n = \frac{1}{L} \int_{-L}^{L} f(x) \cos \left(\frac{n\pi x}{L} \right) dx,$$

$$b_n = \frac{1}{L} \int_{-L}^{L} f(x) \sin \left(\frac{n\pi x}{L} \right) dx. \tag{4.28}$$

Using t as the independent variable, we consider another function that is periodic in t and defined as

$$f(t) = \frac{2t}{T}, \; -T/2 \leq t \leq T/2. \tag{4.29}$$

Because $f(t)$ is an odd function, only the sin terms in the expansion have nonzero coefficients. Therefore, the expansion takes the form

$$f(t) = \sum_{n}^{\infty} b_n \sin \left(\frac{2\pi nt}{T} \right), \tag{4.30}$$

where

$$b_n = \frac{2}{T} \int_{0}^{T/2} (2) \frac{2t}{T} \sin \left(\frac{2\pi nt}{T} \right) dt. \tag{4.31}$$

Upon integration, we find

$$b_n = -\frac{4}{nT}(-1)^n .$$

(4.32)

In addition, $a_0 = 0$, and $a_n = 0$, $n = 1, 2, \ldots, \infty$.

4.4   Fourier Exponential Series

A more compact Fourier series expansion can be obtained by using the relationships

$$\cos\left(\frac{2n\pi x}{L}\right) = \frac{1}{2}\left[\exp\left(\frac{2in\pi x}{L}\right) + \exp\left(\frac{2in\pi x}{L}\right)\right]$$

(4.33)

and

$$\sin\left(\frac{2n\pi x}{L}\right) = \frac{1}{2i}\left[\exp\left(\frac{2in\pi x}{L}\right) - \exp\left(\frac{2in\pi x}{L}\right)\right].$$

(4.34)

Thus, we can expand a piecewise continuous function $f(x)$ as

$$f(x) = \sum_{\infty}^{\infty} c_n \exp\left(\frac{2in\pi x}{L}\right).$$

(4.35)

Note that the summation is now from $n \to -\infty$ to $n \to \infty$. It can be verified (realizing that cos and sin functions are even and odd functions, respectively) that, equation (4.35) is equivalent to

$$f(x) = \sum_{n=0}^{\infty}\left[a_n \cos\left(\frac{2n\pi x}{L}\right) + b_n \sin\left(\frac{2n\pi x}{L}\right)\right],$$

(4.36)

where

$$a_n = c_n + c_{-n}; \text{ and } b_n = i(c_n - c_{-n}).$$

(4.37)

It is then seen that

$$c_n = \frac{1}{L}\int_{-L/2}^{L/2} f(x)\exp\left(-\frac{2in\pi x}{L}\right) dx.$$

(4.38)

Just like the trigonometric functions in the expansion with differing indices n and m, the exponential functions are also mutually orthogonal, such that,

$$\int_{-L/2}^{L/2} \exp\left(\frac{2in\pi x}{L}\right)\exp\left(\frac{2im\pi x}{L}\right) dx \qquad = L, \quad n = m,$$

$$= 0, \quad n \neq m.$$

(4.39)

Consider an example of a periodic function $f(x)$, expressed as

$$f(x) = 1 + x, \quad -1 \leq x \leq 1.$$

(4.40)

$f(x)$ in equation (4.40) is periodic, with a period $L = 2$. To expand $f(x)$ in a Fourier exponential series,

$$f(x) = \sum_{-\infty}^{\infty} c_n \exp\left(\frac{2in\pi x}{L}\right), \tag{4.41}$$

where

$$c_n = \frac{1}{L} \int_{L/2}^{L/2} f(x) \exp\left(\frac{-2in\pi x}{L}\right) dx. \tag{4.42}$$

Here, for $n = 0$, we have

$$c_0 = \frac{1}{2} \int_{-1}^{1} (1 + x)\, dx, \tag{4.43}$$

which evaluates to $c_0 = 1$. For all other n,

$$c_n = \frac{1}{2} \int_{-1}^{1} (1 + x)\, e^{-in\pi x} dx. \tag{4.44}$$

Working through the integration, we find, for $n \neq 0$,

$$c_n = \frac{1}{n\pi} (-1)^n \left(1 - \frac{1}{n\pi}\right). \tag{4.45}$$

4.5 Fourier Transforms

It is possible to carry Fourier expansions to the extreme, where a function to be expanded may not be periodic at all. Examples of such functions are

$$f(x) = 1; -1 \leq x \leq 1, \quad f(x) = 0, \text{ elsewhere,}$$
$$f(x) = |x|; -2 \leq x \leq 1, \quad f(x) = 0, \text{ elsewhere,}$$
$$f(x) = e^{-2x}; -1 \leq x \leq 1, \quad f(x) = 0, \text{ elsewhere,}$$
$$f(x) = \sin x; -1 \leq x \leq 1; \quad f(x) = 0, \text{ elsewhere.} \tag{4.46}$$

Such functions can arise frequently in marine applications, when used to approximate loads on structures, or when expressed as time-dependent signals, in underwater acoustics, for instance. An example of some significance is ocean surface wave patterns when expressed between intervals of negligible wave activity or surface areas of localized wave activity. Another example may be the action on a ship's hull of a rotary machine such as a compressor that only works intermittently on an as-needed basis. The oscillatory force on the ship's hull is therefore finite length and nonperiodic. The use of Fourier transforms may be appropriate in such a situation, and for the compressor example, it would transform a function expressed in the time domain into its transform or map the frequency domain. Transformation from the time domain to the frequency domain is referred to as the "forward transformation," while a reverse transformation from the frequency domain to the time domain is referred to as the inverse Fourier transformation. For the purpose of obtaining a Fourier transform, the summation in equation (4.41) is extended into an integral, with, broadly speaking, the complex coefficients turning into functions.

In brief, the forward Fourier transform from time domain to frequency domain is defined as

$$F(i\omega) = \int_{-\infty}^{\infty} f(x)e^{-i\omega t} \, dt. \tag{4.47}$$

The argument $i\omega$ indicates that the function F is in general complex valued. The reverse operation is defined as

$$f(t) = \frac{1}{2\pi} \int_{-\infty}^{\infty} F(\omega)e^{i\omega t} \, d\omega. \tag{4.48}$$

The presence of the factor $(1/2\pi)$ can be attributed to the fact that the transform is from t to $\omega = 2\pi v$, rather than from t to v, where $v = 1/t$ is the frequency in Hertz (Hz) or cycles per second, while ω as defined earlier, is the angular frequency in radians per second. Other definitions can also be used, for instance, the factor $(1/2\pi)$ is sometimes split up into $\sqrt{(1/2\pi)}$ and applied to both operations, forward and backward (Gradshteyn and Ryzhik 1994). It is necessary just to use mutually consistent definitions at all steps so that unnecessary calculation errors can be avoided.

The Fourier transform pair in equations (4.47) and (4.48) extends the Fourier series expansions to nonperiodic functions in the sense that, whereas functions with a periodic behavior require infinite series of cosines and sines (or exponentials) to approximate them, functions with a nonperiodic behavior require integrals stretching between $-\infty$ and ∞. Thus, while periodic functions need countably infinite summations sweeping over the space of integers with integer increments, nonperiodic functions require the summations to sweep over the space of real numbers with real increments that can get infinitesimally small so that they are in fact integrals over the real-number space. Thus, whereas a Fourier expansion of a periodic function comprises lines of different lengths spaced at integer-frequency intervals along the frequency axis, the Fourier transform of a nonperiodic function is a continuous "spectrum" of values along the frequency axis. Typically, of course, one would have both amplitude and phase associated with the Fourier transform, and each may be plotted separately. The functions have to be piecewise continuous, as before. In addition, for the integrals in the two-way transforms to be defined (i.e., converge), we need

$$f(t) \to 0, \text{ and } f(-t) \to 0; \quad t \to \infty, \tag{4.49}$$

$$|F(i\omega)| \to 0, \text{ and, } |F(-i\omega)| \to 0, \quad \omega \to \infty. \tag{4.50}$$

In other words, $f(t)$ or $F(i\omega)$ can be nonperiodic, but they need to become vanishingly small as the integration limits in each case are approached.

Some examples may be in order at this point. Suppose we want the forward Fourier transform of a function defined as

$$f(t) = 1; \quad -t_0 \le t \le t_0; f(t) = 0, \text{ elsewhere.} \tag{4.51}$$

The Fourier transform of this function can be written as

$$F(i\omega) = \int_{-\infty}^{\infty} f(t) e^{-i\omega t} dt = \int_{-t_0}^{t_0} e^{-i\omega t} dt. \tag{4.52}$$

Integration, followed by a few algebraic steps, leads to

$$F(i\omega) = \frac{\sin \omega t_0}{\omega t_0}. \tag{4.53}$$

Similar transformations apply to the spatial domain, where for instance, functions $f(x)$ transform into functions $F(ik)$, where k has the significance of a spatial frequency, which for problems involving surface waves and/or underwater pressure waves or internal waves is more generally referred to as wave number. Thus,

$$F(ik) = \int_{-\infty}^{\infty} f(x) e^{-ikx} dx \tag{4.54}$$

and

$$f(x) = \frac{1}{2\pi} \int_{-\infty}^{\infty} F(ik) e^{ikx} dk. \tag{4.55}$$

As an example, let

$$f(x) = x; \quad -x_0 \le x \le x_0; \quad f(x) = 0, \text{ elsewhere.} \tag{4.56}$$

The spatial Fourier transform of $f(x)$ in equation (4.56) can be found using

$$F(ik) = \int_{-\infty}^{\infty} x e^{-ikx} dx = \int_{-x_0}^{x_0} x e^{ikx} dx. \tag{4.57}$$

The integral in equation (4.57) can be evaluated via integration by parts. The result is found to be

$$F(ik) = \frac{2x_0^2 \sin kx_0}{kx_0} - \frac{2ix_0}{k} \frac{\sin kx_0}{kx_0}. \tag{4.58}$$

The Fourier transform here is seen to be complex valued. Functions of the form $\sin ax/ax$ are referred to as "sinc" functions, and arise frequently in signal processing and applications where a rectangular distribution in the time or space is to be Fourier transformed for analysis. As seen in the following text, the sinc function also arises when applying the inverse Fourier transform (see Figure 4.5).

$$F(i\omega) = 1, \quad -\omega_0 \le \omega \le \omega_0. \tag{4.59}$$

The inverse Fourier transform of this function is

$$f(t) = \frac{1}{2\pi} \int_{-\infty}^{\infty} F(i\omega) e^{i\omega t} d\omega = \frac{1}{2\pi} \int_{-\omega_0}^{\omega_0} e^{i\omega t} d\omega. \tag{4.60}$$

Working through the algebra, we arrive at

$$f(t) = \frac{\sin \omega_0 t}{\pi \omega_0 t}. \tag{4.61}$$

This result is shown in Figure 4.5 in a normalized form (so the maximum value is unity).

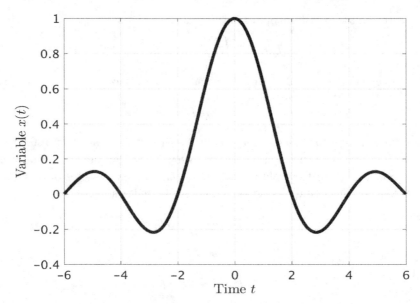

Figure 4.5 A plot showing the normalized sinc function. Note that the peak narrows as ω_0 increases (i.e., as the frequency distribution in equation (4.59) gets wider).

It is interesting to note how a rectangular distribution in one domain appears to get transformed as a function concentrated around one point in the other domain. Indeed, the longer x_0 gets in the x domain, the larger the magnitude becomes, and the narrower its spread gets in the k domain. If we take this reasoning further, intuitively, it seems as though a rectangular distribution that almost stretches from negative infinity to positive infinity, but becomes zero before actually reaching either infinity, would have a transform that has a very large peak (almost approaching infinity) that is focused around zero in the transformed domain. In other words, a wide-band process in the frequency domain should look like an "impulse" at zero in the time domain, and a wide-band constant function in the time domain should transform into an "impulse" around zero in the frequency domain. This makes sense, since a constant "DC" signal in the time domain (stretching from almost negative infinity to positive infinity) has only one frequency, and that is zero. Conversely, a constant in the frequency domain represents a very wide range of frequencies (from almost negative infinite frequency to positive infinite frequency), but appears as an impulse with an almost zero width in the time domain. Thus, one of the standard tests for oscillatory behavior and vibration response is application of an impact or impulse force to mechanical systems and structures. The response of a ship hull model to such a load contains a range of frequencies (with values close to the various natural frequencies of the beam-like hull structure) deriving their energy from the impulsive excitation.

This is perhaps a good point at which to define the Dirac delta function, as outlined in Section 4.6.

4.6   The Delta Function

Despite the argument mentioned earlier involving sinc functions, it is easier to understand the Dirac delta function following the approach of Greenberg (1978). We consider a sequence of functions that defined as (see Figure 4.6),

$$h_a(t) = \frac{a}{2}, \quad -\frac{1}{a} \le t \le \frac{1}{a}; \quad h_a(t) = 0, \quad |t| > \frac{1}{a}. \tag{4.62}$$

It is interesting to note that

$$\lim_{a \to \infty} \int_{-\infty}^{\infty} h_a(t) dt = \lim_{a \to \infty} \frac{a}{2} \cdot \frac{2}{a} = 1. \tag{4.63}$$

Thus, as a approaches ∞, the height of the pulse increases as its width decreases, such that the area within the pulse remains constant at 1. This fact makes it easier to see how

$$\lim_{a \to \infty} \int_{-\infty}^{\infty} h_a(t) f(t) dt = \lim_{a \to \infty} \frac{a}{2} f(0) \frac{2}{a} = f(0). \tag{4.64}$$

In words, the function $h_a(t)$ mentioned earlier has allowed us to pick the value of another function $f(t)$ at $t = 0$. Indeed, equation (4.63) provides the definition for the Dirac delta function $\delta(t)$. Thus, $\delta(t)$ is defined by its "sifting" ability, as described by

$$\int_{-\infty}^{\infty} \delta(t) f(t) dt = f(0). \tag{4.65}$$

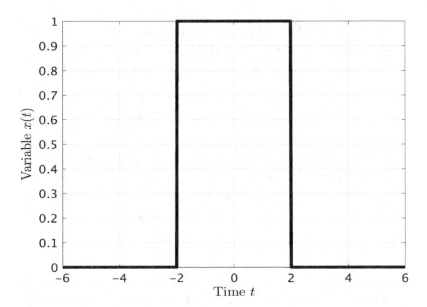

Figure 4.6 A plot showing a rectangular pulse of unit height.

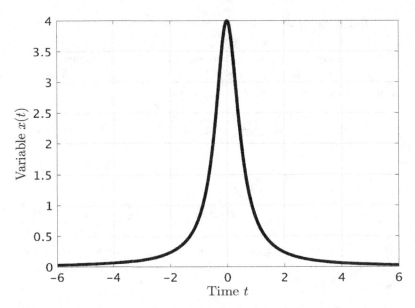

Figure 4.7 A plot showing a pulse defined as in equation (4.68), which becomes narrower as $a \to 0$, and resembles $\delta(t)$.

Similarly, $\delta(x)$ can be defined via

$$\int_{-\infty}^{\infty} \delta(x) f(x) dx = f(0). \tag{4.66}$$

Just as $\delta(t)$ allows us to pick the value of $f(t)$ at $t = 0$, a function $\delta(t - \xi)$ provides $f(\xi)$ at $t = \xi$.

Physically, the delta function represents an infinitesimally spread impulse of infinite magnitude centered at 0, but such that the area under its zone of definition remains unity. Returning to our discussion of Fourier transforms, it follows from equation (4.66) that, the Fourier transform of $\delta(t)$ is

$$\int_{-\infty}^{\infty} \delta(t) e^{-i\omega t}\, dt = e^{-i\omega 0} = 1. \tag{4.67}$$

Equation (4.67) illustrates how widening the rectangular distribution as alluded to near the end of Section 4.5 leads to an impulse at $t = 0$.

It is often useful to think of a pulse function shaped as shown in Figure 4.7 and defined as

$$f(t) = \frac{2a}{a^2 + t^2}. \tag{4.68}$$

Note that this pulse becomes narrower around $t = 0$ as $a \to 0$.

Before we illustrate the inverse Fourier transform of the frequency domain function $F(i\omega) = 1$, we need to consider the following.

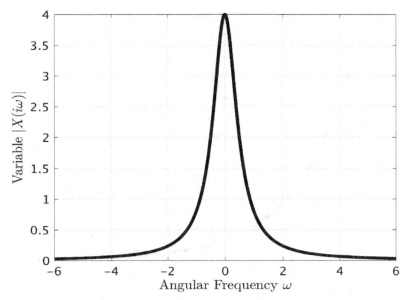

Figure 4.8 A plot showing a pulse defined as in equation (4.71), which becomes narrower as $a \to 0$. As $a \to 0$, this function resembles a delta function in the frequency domain.

Let $f(t)$ be a function defined over $(-\infty, \infty)$ as

$$f(t) = e^{-at}, a > 0, t \geq 0,$$
$$f(t) = e^{at}, a > 0, t < 0. \tag{4.69}$$

This function is definitely Fourier transformable, and its Fourier transform can be computed using

$$F(i\omega) = \int_{-\infty}^{0} e^{(a-i\omega)t} \, dt + \int_{0}^{\infty} e^{-(a+i\omega)t} \, dt. \tag{4.70}$$

Some algebra shows that equation (4.70) results in

$$F(i\omega) = \frac{a}{a - i\omega} + \frac{a}{a + i\omega} = \frac{2a}{a^2 + \omega^2}. \tag{4.71}$$

The result on the right-hand side of equation (4.71) resembles the expression (4.68) except that t in that equation is here replaced by ω (see Figure 4.8). An inspection of the end result in equation (4.71) also shows that as $a \to 0$, $F(i\omega) \to 2/a$, and for $|\omega| \to 0$, the smaller the a becomes, the steeper the function on the right sides gets. The two expressions in equation (4.69) suggest that as $a \to 0$, $f(t)$ on both sides of $t = 0$ approaches a constant value $= 1$. Consider a function that approaches $2\pi\delta(\omega)$, where $\delta(\omega)$ is the delta function in the frequency domain. In this sense, the following inverse Fourier transformation produces

$$f(t) = \frac{1}{2\pi} \int_{-\infty}^{\infty} 2\pi\delta(\omega)e^{i\omega t} \, d\omega = e^{i0t} = 1. \tag{4.72}$$

In other words, the Fourier transform of a constant function $f(t) = 1$ is $2\pi\delta(\omega)$.

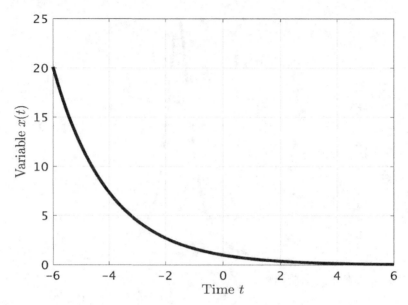

Figure 4.9 A plot showing the function defined in equation (4.73).

Now consider just the right half of the function in equation (4.70), defined as

$$f(t) = e^{-at}, a > 0, t \geq 0,$$
$$f(t) = 0, t < 0. \tag{4.73}$$

This function is shown in Figure 4.9.

The Fourier transform of $f(t)$ in equation (4.73) is

$$F(i\omega) = \int_0^\infty e^{-(a+i\omega)t} dt. \tag{4.74}$$

Following the integration, we have

$$F(i\omega) = \frac{1}{a + i\omega} = \frac{a - i\omega}{a^2 + \omega^2}. \tag{4.75}$$

We note that as $a \to 0$, the function in equation (4.73) approaches a one-sided step function, starting at $t = 0$ but with a jump at $t = 0$. If we try to find the limiting expression for $F(i\omega)$ as $a \to 0$,

$$\lim_{a \to 0} F(i\omega) = \lim_{a \to 0} \frac{a}{a^2 + \omega^2} + \lim_{a \to 0} \frac{-i\omega}{a^2 + \omega^2},$$
$$= \pi\delta(\omega) + \frac{1}{i\omega}. \tag{4.76}$$

The first term is one-half of the end result $2\pi\delta(\omega)$ following equation (4.71). The one-sided step function mentioned earlier (the expression in (4.73) as $a \to 0$) is referred to as the Heaviside step function, with the symbol $H(t)$ being commonly used and with the function often defined as (see Figure 4.10)

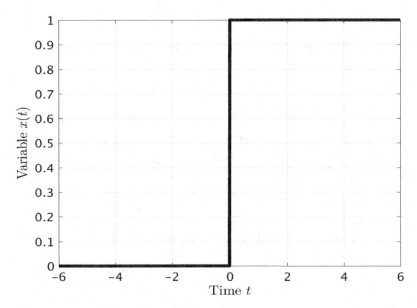

Figure 4.10 A plot showing a part of the Heaviside step function. Note that the function value remains at unity all the way through $t \to \infty$.

$$H(t) = 1, t > 0,$$
$$= 0, t < 0,$$
$$= \frac{1}{2}, t = 0. \tag{4.77}$$

The Fourier transform of $H(t)$ is thus

$$\int_{-\infty}^{\infty} H(t) e^{-i\omega t}\, dt = \pi\delta(\omega) + \frac{1}{i\omega}. \tag{4.78}$$

Frequently, the Dirac delta function is seen as the derivative of the Heaviside step function, as in

$$\delta(t) = \frac{dH(t)}{dt}. \tag{4.79}$$

The functions $\delta(t)$ and $H(t)$ are often referred to as generalized functions. These functions will appear frequently in our subsequent discussions.

4.7   Fourier Transforms of Derivatives and Integrals

Of particular value in solutions of dynamic problems involving variables and functions defined over time and space is the way the Fourier transformation operation works on derivatives of functions. These relationships can often be used to great advantage, in converting ordinary differential equations, for instance, into algebraic equations, partial differential equations into ordinary differential equations, and so forth. Thus, with $F(i\omega)$ representing the Fourier transform of a function $f(t)$, if we are looking for the Fourier transform of $\dot{f}(t) = df/dt$, then

$$\int_{-\infty}^{\infty} \dot{f}(t)e^{-i\omega t} dt = \left[f(t)e^{-i\omega t}\right]_{t\to\infty} - \left[f(t)e^{-i\omega t}\right]_{t\to-\infty}$$

$$- \int_{-\infty}^{\infty}(-i\omega)f(t)e^{-i\omega t} dt, \tag{4.80}$$

where we have used integration by parts. We realize that the function $e^{-i\omega t}$ is bounded at both $\pm\infty$. Furthermore, $f(t) \to 0, t \to \pm\infty$, because that was what we required of it in order to be able to define its Fourier transform. Thus, the boundary terms in equation (4.80) vanishe, and we are left with

$$\int_{-\infty}^{\infty} \dot{f}(t)e^{-i\omega t} dt = i\omega F(i\omega). \tag{4.81}$$

It is convenient here to designate a symbol $\mathcal{F}$ for Fourier transformation so that equation (4.81) can be rewritten as

$$\mathcal{F}\left(\dot{f}(t)\right) = i\omega \mathcal{F}\left(f(t)\right). \tag{4.82}$$

For the second, third, through nth derivatives of $f(t)$, through an analysis similar to that followed in equation (4.80), we see that

$$\mathcal{F}\left(\ddot{f}(t)\right) = (i\omega)^2 \mathcal{F}\left(f(t)\right),$$
$$\mathcal{F}\left(f^{(n)}(t)\right) = (i\omega)^n \mathcal{F}\left(f(t)\right). \tag{4.83}$$

The reverse relationships are also true. In other words, integration in the variable's own domain is equivalent to a division $i\omega$ in the transformed domain. Thus,

$$\mathcal{F}\left(\int f(t)dt\right) = \frac{1}{i\omega}\mathcal{F}\left(f(t)\right),$$
$$\mathcal{F}\left(\int g(t)dt\right) = \frac{1}{(i\omega)^2}\mathcal{F}\left(f(t)\right). \tag{4.84}$$

Here, $g(t) \equiv \int f(t)dt$.

4.8   Concluding Remarks

In this chapter, we reviewed one of the very fundamental and important analysis techniques used in science and engineering contexts, including in the marine world. The Fourier series expansion allows us to represent as series of sines and cosines any piecewise continuous functions, as long as they are periodic. This type of analysis allows us to understand the "frequency content" in functions that are not well known or functions that can only be observed through measurement, experiment, or computations. This general approach can be extended to nonperiodic functions through the Fourier transform, once again, providing a way to break unknown functions, variables, or signals down into their frequency content, whereby we can tell the dominant frequencies over which amplitudes are large or frequencies where significant phase changes occur.

The techniques of Fourier transformation and its close relative, the inverse Fourier transformation are often the pillars upon which many solution techniques for linear systems are built. We also discussed generalized functions like the Dirac delta and Heaviside step functions, and concluded the chapter by defining the Fourier transforms of derivatives and integrals of Fourier transformable functions.

4.9   Self-Assessment

4.9.1

Sketch the following functions and find their Fourier series expansions, where applicable, using the property that they may be odd or even functions.

(a) With a period 6 units,

$$f(x) = 6, \quad 0 \le x \le 3,$$
$$= -6, \quad 3 \le x \le 6. \tag{4.85}$$

(b) $f(x) = 4x$, $0 \le x \le 8$, with a period 8 units.

4.9.2

Expand $f(x) = \cos x$, $0 \le x \le \pi$ and its periodic extensions into a Fourier sine series.

4.9.3

Find the Fourier exponential series expansion for a rectangular wave defined as

$$f(x) = 6, \quad 0 < x < 4,$$
$$= -6, \quad 4 < x < 8. \tag{4.86}$$

4.9.4

Find the Fourier transform of the function ($H(t)$ represents the Heaviside step function),

$$f(t) = \cos \omega t \, [H(t) - H(t - 2)]. \tag{4.87}$$

4.9.5

Find the Fourier transform of the function,

$$f(t) = \sin \omega_0 t. \tag{4.88}$$

4.9.6

Show that the Fourier transform of $f'(t)$ is $i\omega F(i\omega)$, where $F(i\omega)$ is the Fourier transform of $f(t)$.

4.9.7

Consider the following integral equation,

$$\int_{-\infty}^{\infty} f(u)g(x-u)du = e^{-x^2}. \tag{4.89}$$

If $g(x) = e^{-2x^2}$, then use the convolution theorem to find the solution $f(x)$.

Part II

Understanding Dynamic Systems

Wave motion, floating body oscillations, unsteady heat flow and diffusion, and unsteady fluid flow are modeled using time-dependent quantities and relationships that involve rates of change relative to both time and space. Thus, here we review differential equations for discrete systems such as a small buoy responding to waves. In this context, we describe the technique of Laplace transformation, which often makes it convenient to develop solutions when linearity and linear superposition hold and parameters (i.e., coefficients on the variables and their derivatives) remain constant. We consider systems that need to be described using linear differential equations for which the parameters are not constant. Thus, we consider Bessel equations and their solutions, and Legendre equations and their solutions.

We next discuss systems and processes that need more than one independent variable in order to describe them meaningfully. Thus, we study partial differential equations of commonly encountered types such as parabolic, hyperbolic, and elliptic. We consider both finite and infinite domains. For finite domains, we study eigenvalues and eigenfunctions and the important insights they provide. We also discuss how we can express arbitrary functions and solutions in terms of summations of eigenfunctions. We next study flows and waves in infinite domains. For wave problems, we illustrate dispersion relations and the insights they can provide.

5   Ordinary Differential Equations-I

In this chapter, we turn our attention to the mathematics of bodies in water. Here we are primarily concerned with rigid bodies engineered by humans (e.g., buoys, boats and ships, offshore platforms, etc.), although in a subsequent chapter we will turn our attention to flexible bodies such as cables, drill strings, and flexible rafts. Our interest is in analyzing the motion response of such bodies when excited by forces common to the marine environment, for instance, forces due to waves, currents, wind, and so on. Such an analysis may be required for effective performance in normal conditions. In addition, if no simplifying restrictions are made on the magnitudes of the forces and body response, such analysis could also be used to inform design calculations to assure safety of structure and operator in extreme conditions. Recall that our treatment of forces and bodies in Chapter 1 assumed that all bodies were stationary. Our forces and moments all summed to zero so that there was no net force vector, nor net moment vector present. Indeed, the same situation would occur if the all entities we considered were in uniform translation, i.e., moving at constant velocity. In this chapter, we consider objects that move, gather speed and slow down, oscillate, rotate about their own axes as well as other axes and other bodies. Such motions occur on ship-borne objects or tools and equipment on offshore platforms, as well as underwater vehicles. Motions of floating bodies are more complicated to work with than bodies in air or bodies deeply submerged.

Our focus here is not on understanding the actions of the fluid (seawater in many cases) through which a body moves, but rather on the mathematical relations that help us describe and understand the body motions once the forces on it are known. Thus, here our interest is in ordinary differential equations, assuming that all of the objects we deal with here are rigid (i.e., suffer only negligible structural deformation during the motions of interest). The forces, moments, and motions of rigid bodies mostly are just, or can be reduced to, functions of time. The forces and moments we encounter in fluid flow are rarely just functions of time, depending as they do on the spatial position of the fluid portion being considered. The difference arises because all particles that are part of a rigid body always stay with the rigid body. In other words, their motion is "constrained" such that the particles remain bound to, and maintain their relative place within, a body while the body itself may be undergoing arbitrary motion. No such constraint generally exists in the case of fluids, and hence we need to track the motion of fluid particles in both time and space, for which we need partial differential equations, treated in Chapter 6.

5.1   Simple Examples: First-Order Systems

Dynamic models for which the highest-order derivative is an nth-order derivative are said to model an nth-order system. Thus, when $n = 1$, we have a first-order system described by first-order equations. It is interesting to consider one or two very simple situations where differential equations make an appearance. This section draws on the foundational Dynamics text (Meriam and Kraige 1997). For example, suppose a ship is traveling at a constant speed v, with the propellers providing thrust that is just enough to overcome the resistance force R the ship encounters in traveling through water. Suppose next that a decision is now made to turn off the ship's engines. If we ask questions such as, all other factors remaining the same, (i) how much time will it take for the ship's speed to drop down to a tenth of the value it had when the propeller thrust almost became negligible, and (ii) how much farther will the ship have traveled by then; we are suddenly asking questions that involve more than a simple force balance where all external forces and moments sum to zero. Thus, while the ship was traveling at constant speed, we had, $\sum_n \mathbf{F}_n = 0$, or

$$B - W = 0; \text{ and } T - R = 0. \tag{5.1}$$

Here B denotes the vertically upward buoyancy force due to the ship's submerged volume (displacement), and W is the ship's weight in air. T denotes the steady thrust applied by the propellers, while R denotes the resistance to be overcome by the ship. Equation (5.1) describes the static equilibrium condition at constant velocity. Once the thrust force ceases to act, the situation changes in the horizontal direction. Since R will act as long as the ship has a velocity, and since R will act to oppose the ship's travel, the ship must decelerate at a rate a such that

$$ma = -R, \tag{5.2}$$

where a represents the ship's acceleration, negative when speed is decreasing, and positive when it is increasing, with m denoting the ship's in-air mass. Equation (5.2) is essentially a statement of Newton's second law in our context. It we realize, however, that $a = dv/dt$, and $v = dx/dt$, where v and x denote the velocity and displacement, respectively, then, equation (5.2) can be rewritten as

$$m\frac{dv}{dt} = m\frac{d^2x}{dt^2} = -R. \tag{5.3}$$

Typically, we have some idea of the dependence of R on the ship's speed v. In practice, this is rarely a closed-form relationship, but if for the sake of discussion we express R as $R = cv^2$, where c is a constant, then (see Figure 5.1)

$$m\frac{dv}{dt} = -cv^2; \text{ or } m\frac{d^2x}{dt^2} = -cv^2. \tag{5.4}$$

Equations (5.3) and (5.4) are ordinary differential equations, relating derivatives (in theory, up to nth order) and variables to other variables and/or other functions. Time t is here the independent variable, while x and its derivatives are dependent variables. The time dependence of the ship's position and velocity can fully be expressed by

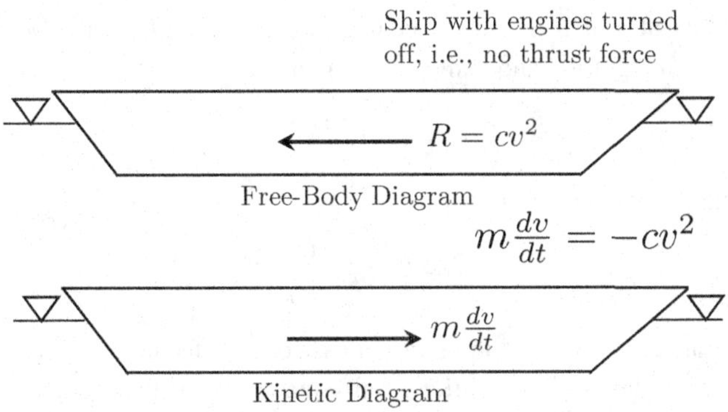

Ship with engines turned
off, i.e., no thrust force

Free-Body Diagram

$$m\frac{dv}{dt} = -cv^2$$

Kinetic Diagram

Figure 5.1 Figure illustrating the situation in the example of equations (5.4-sol2). Forces on the free-body and kinetic diagram are positive rightward. dv/dt is seen to be a deceleration as a result of the dynamic equilibrium equation (5.4).

solving the differential equation (5.3) or (5.4). To that end, we observe that the first of equation (5.4) can be written as

$$\frac{dv}{v^2} = -\frac{c}{m}dt. \tag{5.5}$$

Integrating both sides from $t = 0$ to the present time t,

$$\frac{1}{v} = \frac{c}{m}t + \frac{1}{v_0} \Rightarrow v(t) = \frac{m}{ct + m/v_0}. \tag{5.6}$$

We see from the solution $v(t) \to 0$, as $t \to 0$. The first of the questions posed earlier can now be answered, using equation (5.6). Thus, v drops down to ϵv_0, when

$$t_\epsilon = \frac{m}{cv_0}\left(\frac{1}{\epsilon} - 1\right). \tag{5.7}$$

The time taken for the velocity to reach one-tenth of its initial value v_0 can be found by substituting $\epsilon = 0.1$. It is easy to verify that the dimensions (units of the quantities) on the two sides of the equalities (5.6) and (5.7) are consistent with each other. To find the displacement when the velocity drops to ϵv_0, we argue that

$$\frac{dv}{dt} = \frac{dv}{dx}\frac{dx}{dt} = v\frac{dv}{dx}. \tag{5.8}$$

Substituting the right-hand side of equation (5.8) for dv/dt in the first of equation (5.4),

$$mv\frac{dv}{dx} = -cv^2 \Rightarrow \frac{dv}{v} = -\frac{c}{m}dx, \tag{5.9}$$

which can be rewritten as

$$\int_{v_0}^{v}\frac{dv}{v} = -\frac{c}{m}\int_{0}^{x}dx, \tag{5.10}$$

where the instant at which the thrust ceases is defined as the zero for the distance traveled x. Upon integration and some rearrangement,

$$x = \frac{m}{c} \, ln \left(\frac{v_0}{v} \right). \tag{5.11}$$

By substituting $v = \epsilon v_0$, we find

$$x_\epsilon = \frac{m}{c} \, ln \left(\frac{1}{\epsilon} \right). \tag{5.12}$$

We have illustrated the "separation of variables" solution technique, which is straightforward for first-order ordinary differential equations when the equations are in a form amenable to variable separation such as used earlier. Note that this differential equation was nonlinear, due to the term v^2, which means that a sum of two or more solutions is not necessarily a solution to this equation, nor is some constant α times a solution such as found in equations (5.7) and (5.12) necessarily a solution to equation (5.4).

It may be useful to consider another example to illustrate some interesting situations sometimes encountered in marine applications, where the mass of the object of interest is changing with time. It is important to consider that Newton's second law is defined for systems with constant mass/inertia. Therefore, for systems with variable mass, an additional force caused by the entering or exiting mass must be included. Examples include firefighting vessels spraying out water, vessels paying out mooring chain and anchor, and jet-propelled surface or underwater vehicles. Thus, suppose that a vessel is moving to the right at a velocity v, and as it moves, it releases mass at the rate m' such that the released mass has a velocity v_0, also to the right, but such that $v_0 < v$. Then, the force that the body applies on the released mass in order to decelerate it is R, which acts leftward, such that

$$R = m'(v - v_0), \tag{5.13}$$

where it is understood that $v - v_0 = u$ is the rightward velocity of the released mass relative to the body velocity, and the force R acts leftward. An equal and opposite force must therefore act on the body. Thus, if F denotes the existing external rightward force acting on the body, then R must add to it such that

$$m\frac{dv}{dt} = F + \frac{dm}{dt}u. \tag{5.14}$$

In case the body is initially moving at constant velocity v (so that $F = 0$), and mass is released at relative velocity u at a rate dm/dt, then, the mass release causes a thrust acting to the right, which will accelerate the body such that

$$m\frac{dv}{dt} = \frac{dm}{dt}u \equiv T. \tag{5.15}$$

T is the rightward thrust produced on the body by releasing mass at a rate dm/dt, which causes the body to accelerate in the rightward direction. Equation (5.15) is, of course, a first-order differential equation, and unless the rate of mass release is intended to depend on the instantaneous body velocity v, equation (5.15) is linear.

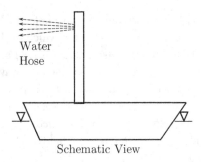

Schematic View

(a) Schematic View

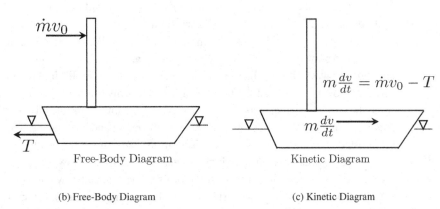

Free-Body Diagram

Kinetic Diagram

(b) Free-Body Diagram   (c) Kinetic Diagram

Figure 5.2 Figures showing the dynamics as modeled in equation (5.16) in the horizontal direction. The moment equation is not considered here.

Note that it is the relative velocity $u = v - v_0$ on the right side. It is also worth noting further that the argument that the rate of change of momentum equals the external force on a body only applies in case the body has constant mass.

We consider a very simple example, wherein a firefighting vessel of mass M is at rest spraying water at a mass flow rate $\dot{m}$ at a leftward velocity $-v_0$ (see Figure 5.2). We will find that, based on equation (5.15),

$$M\frac{dv}{dt} = \dot{m}\left[v - (-v_0)\right] = \dot{m}v_0, \tag{5.16}$$

realizing that the vessel is at rest so that $v = 0$. In order to maintain position, therefore, the propellers of the vessel will need to apply a leftward thrust T given by

$$T = \dot{m}v_0. \tag{5.17}$$

5.2 Simple Examples: Second-Order Systems

If we turn our attention next to oscillatory motion of bodies in water, particularly floating bodies in waves, we find an interesting interplay between the external forces applied on the body by the surrounding wave field, the hydrostatic restoring force that acts on the body, and the inertial force with which the body responds. The wave field surrounding the body comprises the ambient wave field that exists without the body being immersed in water (modeled as the "incident wave field"), the distortions introduced by the body (modeled as another wave field called the "scattered wave field" superimposed on top of the incident wave field), and the wave field created by the body as it oscillates on the water surface (the so-called "radiation wave field" superimposed on top of the incident and scattered wave fields). This interpretation is based on linear theory, which assumes that all waves are small enough in slope (height/length) to allow any higher-order effects to be neglected. We will study these components and the forces due to them in Chapters 12–14. Here, we consider differential equations that can be used to describe the oscillations of our floating body without going too far into the hydrodynamics. If we consider a cylindrical buoy heaving in response to approaching waves, we may approximate its oscillations as follows. We use m to denote the in-air mass of the buoy and A_{wp} to denote the water-plane area, with ρ and g, respectively, standing for water density and gravity acceleration. Static equilibrium requires that the upward buoyancy force on the body balance the downward weight force due to gravity at the still water line. This is the so-called "hydrostatic equilibrium" condition, and forces arising due to the interplay between body weight and buoyancy are often termed "hydrostatic forces." When the body oscillations cause it sink to below the still water line, it feels excess buoyancy, which pushes the body back upward. When the body rises above still water line, it loses some of its buoyancy, and now the weight force on it pulls it back down. The hydrostatic force thus forms a restoring force on the body, much like a spring force in spring-mass systems in air. If the heave displacement of body is denoted as x, the hydrostatic restoring force on it can be expressed as

$$F_h = \rho g A_{wp} x, \tag{5.18}$$

which holds true as long as the buoy is strictly cylindrical. Associated with oscillations induced by the radiated wave field referred to earlier is the so-called "infinite-frequency added mass" to be denoted as $\overline{a}(\infty)$. This term thus adds to the inertia or mass term m. We will denote the remainder of the radiation force as F_R. Consider in addition, a situation where there is a rotary generator/motor on the buoy, and there is a slight eccentricity in this machine's rotor (see Figure 5.3). Consequently, there is an oscillatory vertical force F_e due to this effect, which could influence the dynamics if this machine's angular frequency (related to revolutions per minute) is comparable to the heave natural frequency of the buoy in water. The force due to the incident wave field and the scattered wave together can be denoted as F_f. Additionally, if for small oscillation amplitudes we use a linear approximation (represented by a coefficient $c > 0$) for the viscous skin-friction damping acting on the buoy, the relationship expressing dynamic equilibrium can be written as

$$[m + \overline{a}(\infty)]\ddot{x} = -kx - c\dot{x} + F_R + F_f + F_e,$$
$$\Rightarrow [m + \overline{a}(\infty)]\ddot{x} + c\dot{x} + kx = F_f + F_R + F_e. \tag{5.19}$$

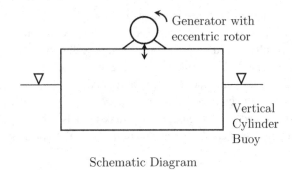

(a) Schematic Diagram

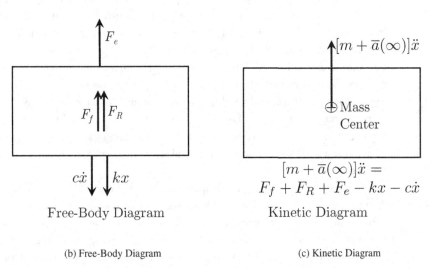

(b) Free-Body Diagram

(c) Kinetic Diagram

Figure 5.3 Figures showing the dynamics as modeled in equation (5.19) in the vertical direction. Incoming/radiated waves are not shown in the schematic diagram. The moment equation for pitch mode and the equation for surge mode are not considered here.

Systems for which inertial effects are comparable to any other effects such as spring and damping, a second-order system results, requiring a second-order differential equation. For equation (5.19) to make sense physically, the mass, spring, and damping (i.e., the coefficients attached to the inertia, stiffness, and dissipation terms) must all be positive when nonzero. So long as the waves and body oscillations in response are assumed to be small enough for nonlinearities to be ignored. Even so, a large amount of interesting complexity is embedded in equation (5.19), which we will discuss in Chapters 13 and 14. Absent any specific representation for F_R, or any assumptions as to the nature of the wave input, with $\overline{a}(\infty)$ set to zero, equation (5.19) is simply a second-order ordinary differential equation with constant coefficients. Given the nature of the wave forces (due to incident + scattered waves and radiated waves), oscillations of floating bodies cannot be captured by such equations. Further, typically, floating body oscillations in response to waves do not have a distinct "initial moment" that would provide an initial condition. Recall that all second-order differential equations require two initial conditions, before a unique solution can be found. Hence, since waves can be said to have continued from a long time into the past, and are expected to go on a long time into the future, floating body oscillations caused by approaching wave fields are typically not amenable to treatment under the initial value problem framework (forced oscillations of wave tank models driven by external actuators in the absence of waves can be treated as initial value problems, however).

5.2.1   Natural or Unforced Response

Here we consider the response of the system with all external forces assumed to be zero. Our system will have a nonzero response only if it was initially displaced from its equilibrium position and/or if it had an initial non-zero velocity. The fact that the system and, therefore, our model of its oscillation are linear means that linear super-position applies. Thus, the combined response of the system to two or more sources of excitation can be found simply by summing together its responses to each source of excitation on its own, with the excitation from all other sources set at zero. Thus, when a system has nonzero initial conditions (position and velocity) at the instant when the external force/excitation starts to act, its response to the external force in the presence of nonzero initial conditions can be found by evaluating the system's response to initial conditions and its response to external excitation separately and then adding the two together.

Returning to the spring-mass-damper system in equation (5.19), consider a situation when the mass is released from some nonzero and released at a nonzero velocity. For the moment, we suppose that there are no forces acting on the mass, so its displacement is simply a response to its initial conditions in the presence of the spring and the damper attached to it. This response defines the inherent properties of our system, and is therefore considered its natural response. Given that we have a mass and a spring here (and thereby meeting the prerequisites for oscillatory motion), we expect to see oscillations in the response of our system. Further, because we also have a damper taking energy away from the system, we expect these oscillations gradually to decrease in magnitude. Absent the damper, our spring-mass system will remain oscillating, with the energy input into it by us via the nonzero displacement against the spring force (potential energy) and nonzero velocity (kinetic energy) will go on cycling back and forth between the mass and the spring. The frequency at which it would oscillate is defined as the system's natural frequency, given by

$$\omega_n = \sqrt{\frac{k}{m + \overline{a}(\infty)}}. \tag{5.20}$$

To return to the question of determining the response of our spring-mass-damper system, we rewrite equation (5.19) without any forcing terms on the right-hand side of equation (5.21), and with the initial conditions indicated explicitly.

$$[m + \overline{a}(\infty)]\ddot{x} + c\dot{x} + kx = 0; \quad x(0) = x_0, \dot{x}(0) = v(0) = v_0. \tag{5.21}$$

A solution to equation (5.21) would give us the response we are looking for. We use a trial solution of the form $x(t) = C \exp(\lambda t)$ (a good place to start for most linear differential equations with constant coefficients[1]), and, substituting into equation (5.21), we have,

$$\left[\lambda^2[m + \overline{a}(\infty)] + \lambda c + k\right] C = 0; \quad \exp(\lambda t) \neq 0. \tag{5.22}$$

In addition, since $C \neq 0$ to keep our solution from being a trivial solution such as $x = 0$,

$$\lambda^2[m + \overline{a}(\infty)] + \lambda c + k = 0. \tag{5.23}$$

We note then that if we can find solutions for λ to satisfy equation (5.23), we must conclude our trail solution will solve equation (5.21) and thus give us a result that would lead to the response we seek. The two solutions are

$$\lambda_1, \lambda_2 = -\frac{c}{2[m + \overline{a}(\infty)]} \pm \sqrt{\left(\frac{c}{2[m + \overline{a}(\infty)]}\right)^2 - \left(\frac{k}{m + \overline{a}(\infty)}\right)}. \tag{5.24}$$

We can see that there are three possibilities.

$$\left(\frac{c}{2[m + \overline{a}(\infty)]}\right)^2 > \frac{k}{m + \overline{a}(\infty)},$$

$$\frac{c}{2[m + \overline{a}(\infty)]} = \frac{k}{m + \overline{a}(\infty)},$$

$$\left(\frac{c}{2[m + \overline{a}(\infty)]}\right)^2 < \frac{k}{m + \overline{a}(\infty)}. \tag{5.25}$$

In the first instance, λ_1 and λ_2 are unequal, and are both real and negative. In the second, λ_1 and λ_2 are equal and real. In the third, λ_1 and λ_2 are complex, in fact, they are complex conjugates of each other. The steps from equation (5.21) to (5.25) can be cast in more general terms, to demonstrate their wider applicability. Dividing all terms in equation (5.21) by $(m + \overline{a}(\infty))$, we have

$$\ddot{x} + \frac{c}{(m + \overline{a}(\infty))}\dot{x} + \frac{k}{(m + \overline{a}(\infty))}x = 0. \tag{5.26}$$

Here we realize that, the coefficient on the x term is just the natural-frequency squared, equal to, ω_n^2. In addition, a "damping ratio" $\zeta > 1$ can be introduced such that we now have

$$\frac{k}{(m + \overline{a}(\infty))} = \omega_n^2,$$

$$\frac{c}{(m + \overline{a}(\infty))} = 2\zeta\omega_n. \tag{5.27}$$

---

[1] Note, however, that equation (5.19) does not really have constant coefficients if the actual form of F_R, the radiation force is used. Equation (5.21) and the trial solution mentioned earlier are only valid if the response to initial conditions is being observed in a dry test with the body outside of water.

Then, equation (5.24) can be transformed as

$$\lambda_1, \lambda_2 = -\zeta\omega_n \pm \omega_n \sqrt{\left(\zeta^2 - 1\right)}. \tag{5.28}$$

Thus, when $\zeta > 1$, both roots are real, unequal, and negative. If $\zeta = 1$, then the two roots are equal and negative. When $\zeta < 1$, the two roots are complex conjugate pairs. Systems or situations for which $\zeta > 1$ are referred to as overdamped systems. Those with $\zeta = 1$ are critically damped, and when $\zeta < 1$, we have underdamped behavior. Which of the three possibilities lead to natural oscillations in response to an initial "perturbation"? We see that only in the solution for $\zeta < 1$ can we see terms that include an exponential with an imaginary argument (or effectively, sine and cosine functions of time). Hence, we can conclude that natural oscillations are only possible when $\zeta < 1$. Moreover, the presence of real part $-\zeta\omega_n$ implies that the oscillations diminish with time, asymptotically approaching 0. The solutions in the three cases arising from equation (5.28) can be written as

$$x(t) = C_1 e^{-\lambda_1 t} + C_2 e^{-\lambda_2 t}, \ \zeta > 1;$$
$$x(t) = (C_1 + C_2 t) e^{-\lambda t}, \ \zeta = 1;$$
$$x(t) = e^{-\zeta\omega_n t} (C_1 \cos \omega_d t + C_2 \sin \omega_d t), \ \zeta < 1. \tag{5.29}$$

Here, $\omega_d = \omega_n \sqrt{1 - \zeta^2}$ is the "damped natural frequency," where $\omega_n = \sqrt{k/M}$ is the true natural frequency, with $M = m + \bar{a}(\infty)$. In each case within equation (5.29), the constants C_1 and C_2 can be evaluated using the two initial conditions involving the values of $x(t)$ and $\dot{x}(t)$ at $t = 0$.

We recall that, for a body in the ocean, it is not often easy to define a single instant as $t = 0$. However, forced oscillation tests performed in the laboratory to determine important motion parameters do allow measurement of initial conditions on position and velocity. In particular, this treatment can be used in the study of underwater vehicles when these are deeply submerged. The use of equation (5.21) is questionable when the body is immersed at or somewhat below the water surface, as body oscillations produce waves on the surface that apply a force on the body (the radiation force F_R in equation (5.19)), which alters the inherent hydrodynamic response of the body even when no other forces are present. For preliminary tests of the overall hardware, one may perform dry tests outside of water, in which case an analysis based on equation (5.21) is appropriate. Equation (5.21) and the steps through equation (5.12) are included in this chapter as an illustration of the broader mathematical framework relating to oscillating systems.

5.2.2   Forced Response

Since we have already evaluated the system response to nonzero initial conditions, we can now focus on finding the system response to the external forcing alone. At least four approaches are possible: (1) the method of undetermined coefficients (Kreyszig, Kreyszig, and Norminton 2011), (2) variation of parameters, (3) Laplace transformation, and (4) the convolution integral approach. The approaches (3)–(4) are briefly discussed here since they can be applied more easily to a wider class of forcing functions and represent techniques that may reappear later in this text. As before, we refer the reader to texts such as Kreyszig, Kreyszig, and Norminton (2011) and Greenberg (1978) for more detailed treatments.

5.2.3   Use of Laplace Transforms

When the system parameters attached to the dependent variable and its derivatives are (or can be approximated as) constants, and when the differential equations are linear (i.e., no powers or transcendental functions of the dependent variable in the equation), Laplace transformation can be used to obtain solutions, and is particularly convenient for analyzing multiple, cascading linear systems, systems with feedback, and so on. Application of Laplace transforms also leads to "transfer functions," which find broad usage in linear-system analysis and control. We begin by saying that the Laplace transform provides a straightforward way to transform differential equations into algebraic equations, providing they are linear and have constant coefficients. It achieves this transformation by mapping all variables and external forces to the so-called complex frequency domain. Thus, recall that the Laplace transform of a function of time $\mathcal{L}(\cdot)$ of $x(t)$ is defined as

$$\mathcal{L}(x(t)) \equiv X(s) = \int_0^\infty x(t)e^{-st}\,dt. \tag{5.30}$$

The Laplace transform of the derivative $\dot{x}$ is

$$\mathcal{L}(\dot{x}(t)) = sX(s) - x(0^-). \tag{5.31}$$

Further, the Laplace transform of the second derivative $\ddot{x}(t)$ is

$$\mathcal{L}(\ddot{x}(t)) = s^2 X(s) - sx(0^-) - \dot{x}(0^{-1}). \tag{5.32}$$

Recall that here the 0^- simply means just before $t = 0$ when we start the clock for our system, i.e., it represents our initial conditions. Assuming zero initial conditions for now, an equation of single mode (e.g., heave) for a deeply submerged body subjected to an oscillatory force (e.g., due to a slight unbalance in a motor driving a position-keeping thruster) can be written as

$$(m + a)\ddot{x} + C_v \dot{x} + kx = f(t). \tag{5.33}$$

For heave oscillation, the added mass $a = 2\overline{a}(\infty)$ for a heaving semi-submerged body of the same shape with exactly half of it submerged. C is a linearized viscous friction coefficient, while k here is due to the action of passive control surfaces on the body. Assuming zero initial conditions, i.e., $x(0) = 0$, $\dot{x}(0) = 0$, Laplace transformation of both sides yields

$$\left[(m + a)s^2 + Cs + k\right] X(s) = F(s). \tag{5.34}$$

where $f(0^-) = 0$ as well. The algebraic equation in (5.34) can be used to write

$$H(s) = \frac{X(s)}{F(s)} = \frac{1}{(m+a)s^2 + Cs + k}. \tag{5.35}$$

$H(s)$ is the transfer function for the heaving deeply submerged body. The solution $X(s)$ in the Laplace-frequency domain is then

$$X(s) = H(s)F(s). \tag{5.36}$$

In cases where $H(s)F(s)$ is in a form for which partial fraction expansion is convenient, it is straightforward to invert $X(s)$ to obtain $x(t)$, which is the desired solution quantifying the system response $x(t)$ to an external force $f(t)$.

For our second-order system in equation (5.33), let us suppose that the function $f(t)$ is a Dirac delta function representing an impulsive excitation. The Laplace transform of a delta function $\delta(t)$ located at $t = 0$ is

$$\int_0^\infty \delta(t)e^{-st}\,dt = e^{-s0} = 1, \tag{5.37}$$

which follows from the sifting property of the δ function discussed earlier. Thus, $F(s) = 1$ in equation (5.36). Thus,

$$X(s) = H(s)(1) = \frac{1}{(m+a)s^2 + Cs + k} = \frac{1/(m+a)}{s^2 + 2\omega_n\zeta s + \omega_n^2}. \tag{5.38}$$

Here, $X(s)$ is the Laplace transform of $x(t)$. The right-hand side can be expressed as

$$X(s) = \frac{1}{(s + \lambda_1)(s + \lambda_2)}, \tag{5.39}$$

where, as we have seen before, λ_1, λ_2 are real and negative when $\zeta > 1$, real, negative, and equal when $\zeta = 1$, and complex conjugates when $\zeta < 1$. In the first case,

$$X(s) = \frac{1}{(s + \lambda_1)(s + \lambda_2)}. \tag{5.40}$$

Here we point out the following Laplace transform relations. $H(t)$ denotes the Heaviside step function, having a value 0 until $t = 0$, and a value 1 for $t \geq 0$.

$$\mathcal{L}(H(t)) = 1/s; \quad \mathcal{L}^{-1}(1/s) = H(t)],$$
$$\mathcal{L}(f(t-T)) = F(s)e^{-sT}; \quad \mathcal{L}^{-1}(e^{-sT}) = \delta(t-T),$$
$$\mathcal{L}(e^{-at}) = \frac{1}{s+a}, \quad \mathcal{L}^{-1}(\frac{1}{s+a}) = H(t)e^{-at},$$
$$\mathcal{L}(t) = \frac{1}{s^2}, \quad \mathcal{L}^{-1}(\frac{1}{(s+a)^2}) = te^{-at}, \text{ etc.} \tag{5.41}$$

We next resort to partial fraction expansion to reduce the right-hand side of equation (5.40) to one of the commonly encountered Laplace transform pairs. Thus, let

$$X(s) = \frac{1}{(s + \lambda_1)(s + \lambda_2)} = \frac{A}{s + \lambda_1} + \frac{B}{s + \lambda_2}. \tag{5.42}$$

Here, A and B are constants to be determined. We use the method of residues here to determine A and B. To find A,

$$\lim_{s \to -\lambda_1} \frac{s + \lambda_1}{(s + \lambda_1)(s + \lambda_2)} = A + \lim_{s \to -\lambda_1} \frac{(s + \lambda_1)B}{(s + \lambda_2)}. \tag{5.43}$$

Evaluating the limits,

$$A = \frac{1}{\lambda_2 - \lambda_1}. \tag{5.44}$$

To find B,

$$\lim_{s \to -\lambda_2} \frac{s + \lambda_2}{(s + \lambda_1)(s + \lambda_2)} = \lim_{s \to \lambda_1} \frac{(s + \lambda_2)A}{(s + \lambda_1)(s + \lambda_2)} + B. \tag{5.45}$$

Once again, a straightforward evaluation of the limits in equation (5.45),

$$B = -\frac{1}{\lambda_2 - \lambda_1}. \tag{5.46}$$

Hence,

$$X(s) = \frac{1}{\lambda_2 - \lambda_1} \left(\frac{1}{s + \lambda_1} - \frac{1}{s + \lambda_2} \right). \tag{5.47}$$

The functions on the right-hand side can easily be inverse Laplace transformed (equation (5.41)) to write

$$x(t) = \frac{1}{\lambda_2 - \lambda_1} \left[e^{-\lambda_1 t} - e^{-\lambda_2 t} \right]. \tag{5.48}$$

By following a similar procedure, when $\lambda_1 = \lambda_2 = \lambda$, it can be seen that, when $f(t) = \delta(t)$,

$$x(t) = t e^{-\lambda t}. \tag{5.49}$$

When λ_1 and λ_2 are a complex conjugate pair, $-\zeta \omega_n \pm i \omega_d$, where $\omega_d = \omega_n \sqrt{1 - \zeta^2}$, the response to $f(t) = \delta(t)$ is

$$x(t) = \frac{1}{\omega_d} e^{-\zeta \omega_n t} \sin \omega_d t. \tag{5.50}$$

We will return to equation (5.50) in a moment. For now, consider another example of partial fraction expansion (as a step in differential equation solution by Laplace transformation) using the method of residues, for systems such as the critically damped system in equation (5.28), for which $\zeta = 1$ and the two solutions are equal. This is a case of repeated roots, which can arise in differential equations of higher-order (i.e., 3rd, 4th, Nth, etc.) systems. Thus, consider a transfer function defined by

$$H(s) = \frac{1}{(s + 1)(s + 2)^2}. \tag{5.51}$$

To separate this $H(s)$ into partial fractions, we proceed as follows, realizing that the three roots, or the three "poles" are at -1, -2, and -2.

$$H(s) = \frac{1}{(s + 1)(s + 2)^2} = \frac{A_1}{s + 1} + \frac{A_2}{s + 2} + \frac{A_3}{(s + 2)^2}. \tag{5.52}$$

To find A_1, we multiply both sides by $(s + 1)$ and evaluate the limits on both sides as $s \to -1$.

$$\lim_{s \to -1} \frac{(s + 1)}{(s + 1)(s + 2)^2} = \lim_{s \to -1} \left[\frac{(s + 1)A_1}{(s + 1)} + \frac{(s + 1)A_2}{(s + 2)} + \frac{(s + 1)A_3}{(s + 2)^2} \right]. \quad (5.53)$$

Evaluation of limits leads to the result, $A_1 = 1$. Next, multiplying both sides by $(s+2)^2$ and evaluating the limits as $s \to -2$,

$$\lim_{s \to -2} \frac{(s + 2)^2}{(s + 1)(s + 2)^2} = \lim_{s \to -2} \left[\frac{(s + 2)^2 A_1}{s + 1} + \frac{(s + 2)^2 A_2}{s + 2} + \frac{(s + 2)^2 A_3}{(s + 2)^2} \right]. \quad (5.54)$$

Again, evaluating the limits on both sides gives, $A_3 = -1$. Evaluation of A_2 is a little trickier, but use of the method of residues allows the following steps. Having multiplied both sides by $(s + 2)^2$, we differentiate both sides with respect to s and then evaluate the limit $s \to -2$ on the derivatives,

$$\lim_{s \to -2} \frac{d}{ds} \frac{1}{s + 1} = 0 + \lim_{s \to -2} \frac{d}{ds}(s + 2)A_2 + 0,$$

$$\lim_{s \to -2} \frac{-1}{(s + 1)^2} = -1 = A_2. \quad (5.55)$$

The partial fraction expansion thus becomes

$$H(s) = \frac{1}{(s + 1)(s + 2)^2} = \frac{1}{s + 1} - \frac{1}{s + 2} - \frac{1}{(s + 2)^2}. \quad (5.56)$$

$x(t)$ in equations (5.48)–(5.50) is the impulse response of our second-order system that represents a deeply submerged body. Physically, the impulse response function serves an important role in helping us understand the inherent dynamics of a body. Mathematically, it enables us to evaluate the response of a linear system such as equation (5.33) to an "arbitrary" excitation, providing it is Laplace transformable (i.e., does not violate conditions existence of a Laplace transform such as piecewise continuity). Examples of arbitrary excitations include irregular waves, wind loads, discontinuous loads and so forth, when a clearly defined initial time instant $t = 0$ can be identified.

5.3   Convolution Integral and Response to Arbitrary Excitation

The solution functions of equations (5.48)–(5.50) are all impulse response functions, and can be represented using a more general symbol $h(t)$. It is easy to see that, with $F(s) = 1$,

$$h(t) = \mathcal{L}^{-1}(H(s)F(s)) = \mathcal{L}^{-1}(H(s)). \quad (5.57)$$

Alternatively, we can write, using the definition of Laplace transforms,

$$\int_0^\infty h(t)e^{-st} dt = H(s). \quad (5.58)$$

Of course, $h(t)$ and $H(s)$ can represent any linear system, not just a second-order system we used as a starting point for this discussion. Now suppose that $F(s)$ represents the Laplace transform of an arbitrary Laplace transformable function $f(t)$. Then the Laplace transform $X(s)$ of the response of the system represented by $H(s)$ can be expressed, in a sort of generalization of equation (5.36), as

$$X(s) = H(s)F(s). \tag{5.59}$$

The solution $x(t)$ in the time domain represents the response of a linear system to an arbitrary excitation $f(t)$. The inverse Laplace transform of $X(s)$ leads to the convolution form

$$x(t) = \mathcal{L}^{-1}(X(s)) = \int_0^\infty h(\tau)f(t-\tau)d\tau. \tag{5.60}$$

As an example, the impulse response function corresponding to the Laplace domain or the s-domain $H(s)$ in equation (5.51) can be found via straightforward inverse Laplace transformation of the three terms on the right-hand side of equation (5.56). Thus,

$$h(t) = e^{-t} - e^{-2t} - te^{-2t}. \tag{5.61}$$

The convolution integral recognizes the following behavior of dynamic systems. Most systems with inertia have a memory, and most practical systems have a finite memory. $f(t)$ at $t = 0$ produces a response $x(t)$ of a system described by $h(t)$. $f(t)$ at a subsequent time instant also produces a response, which adds together with the response still playing out from the excitation at the previous time instant. The total response $x(t)$ is thus the sum total of all such responses, current and recent. The length of the impulse response function $h(t)$ determines the length of the system's memory into the past.

As an example, let us consider a function $g(t)$ defined as

$$g(t) = A\cos\omega_1 t + B\sin\omega_2 t. \tag{5.62}$$

Nothing about $g(t)$ as defined in equation (5.62) suggests that $f(t)$ has a definite start point at some value of t. In fact, it is nonzero as $t \to \pm\infty$. To make it consistent with the present initial-value framework, we multiply it by $H(t)$, the Heaviside function, which is zero until $t = 0$ and 1 thereafter. Thus, we define

$$f(t) = [A\cos\omega_1 t + B\sin\omega_2 t]\, H(t). \tag{5.63}$$

Now consider just a single-term impulse response function,

$$h(t) = e^{-at}. \tag{5.64}$$

The response $x(t)$ of the system represented by $h(t)$ in equation (5.64) can be found using

$$x(t) = \int_0^t h(\tau)f(t-\tau)d\tau = \int_0^t e^{-a\tau}\,(A\cos\omega_1(t-\tau) + B\sin\omega_2(t-\tau))\,H(t-\tau)d\tau. \tag{5.65}$$

Taking the Laplace transform of both sides,

$$X(s) = H(s)F(s) = \left(\frac{1}{s(s+a)}\right)\left[\left(\frac{B\omega_1}{s^2+\omega_1^2}\right) + \left(\frac{As}{s^2+\omega_2^2}\right)\right]. \tag{5.66}$$

$X(s)$ is amenable to partial fraction on the way to evaluation of $x(t)$. $x(t)$ is

$$x(t) = \mathcal{L}^{-1}(X(s)). \tag{5.67}$$

It is interesting to note that, as a general rule, upon Laplace transformation, a multiplication in frequency domain becomes a convolution in the time domain.

The convolution in equation (5.60) is known as a Laplace convolution, since it is defined using Laplace transformation, which is the transform technique commonly applied to initial value problems with an identifiable initial time $t = 0$. Though it is possible to see certain situations involving submerged and floating bodies as initial value problems (e.g., a self-excited ship or offshore-structure model in a wave tank; response of an underwater vehicle to a sequence of movements of an on-board manipulator arm, etc.), one deals with continuously acting loads such as surface waves, currents, and winds for bodies on or just under the ocean surface. For which the response to an arbitrary excitation may be found using an impulse response function defined such that a Fourier convolution now replaces the Laplace convolution of initial value problems. This is discussed briefly in Section 5.5. But first, we consider how the technique of Laplace transformation can be used to understand the response of systems described by N differential equations, or a system of N variables related together by N first-order differential equations. We note that these must be linear differential equations with constant coefficients for the following techniques to be valid.

5.4   Multi-Variable Systems

Consider a system of, say, six variables where each variable represents either the position or the velocity of three masses coupled together by springs and dampers. We know that such a system could be described by three second-order coupled differential equations of constant coefficients, or we could convert each second-order differential equation into two first-order equations, via the following technique. Starting with a single second-order differential equation,

$$m\ddot{x} + c\dot{x} + kx = f(t), \ x(0) = x_0; \ \dot{x}(0) = v_0. \tag{5.68}$$

We define $x_1(t)$ and $x_2(t)$ such that, $x_1(t) = x(t)$, and $x_2(t) = v(t) = \dot{x}(t)$. Thus,

$$\dot{x}_1 = x_2,$$
$$\dot{x}_2 = -\frac{k}{m}x_1 - \frac{c}{m}x_2 + \frac{1}{m}f(t). \tag{5.69}$$

Note that the two equations in equation (5.69) are both first order. In fact, equation (5.69) can be replaced by a single matrix equation,

$$\dot{\mathbf{x}} = \mathbf{A}\mathbf{x} + \mathbf{b}, \tag{5.70}$$

where

$$\mathbf{x} = \begin{Bmatrix} x_1 \\ x_2 \end{Bmatrix}; \ \mathbf{A} = \begin{bmatrix} 0 & 1 \\ -\frac{k}{m} & -\frac{c}{m} \end{bmatrix}; \ \mathbf{b} = \begin{Bmatrix} 0 \\ \frac{f(t)}{m} \end{Bmatrix}. \tag{5.71}$$

Equation (5.71) just describes the forced oscillation of the mass. Response to initial conditions is described separately. The first-order vector differential equation is referred to as a state-space system, with x_1 and x_2 describing the two states of the system. The technique can easily be generalized to consider oscillations of 5, 15 masses coupled together by springs and dampers. Since one oscillating mass needs a second-order differential equation, which leads to two first-order differential equations, M masses will give rise to 2M first-order differential equations. On the other hand, consider an Nth-order differential equation, which can be reduced to a system of N first-order equations. Such systems may arise when one writes out a differential equation model for a cascade of connected smaller systems. Given

$$\frac{d^n x}{dt^n} + a_1 \frac{d^{(n-1)} x}{dt^{(n-1)}} + \cdots + a_{n-1} \frac{dx}{dt} + a_n x = f(t). \tag{5.72}$$

This system is fully defined when N initial conditions are supplied. Equation (5.72) can be reexpressed as N first-order equations with the substitutions,

$$x_1 = x, \; x_2 = \dot{x}, x_3 = \dot{x}_2 = \ddot{x}, \ldots, \; \dot{x}_{(n-1)} = x_n. \tag{5.73}$$

It can be verified that these relations can be combined to write an N-variable state-space equation of the form

$$\dot{\mathbf{x}} = \mathbf{A}\mathbf{x} + \mathbf{b}, \; \mathbf{x}(0) = \mathbf{x}_0. \tag{5.74}$$

As before, the matrix $\mathbf{A}$ defines the essentials of a system's dynamics and captures the inner workings of the system. It is often referred to as the "system matrix." It is not difficult to write a general solution to the system by extending the basic ideas used in writing the solution to a single-variable system. Thus, consider a scalar system of the form

$$\dot{x} = ax + b; \; x(0) = x_0. \tag{5.75}$$

Here, the parameter a may be positive or negative, and b is an external force that is a function of time. First, the response of the system to its initial condition x_0 can be found via separation of variables in $\dot{x} = ax$ as

$$x_H(t) = x_0 e^{at}. \tag{5.76}$$

It can be seen that the impulse response function for this system is simply

$$h(t) = e^{at}. \tag{5.77}$$

The response to the input b, which may be a function of time, can be found using the convolution

$$x_P(t) = \int_0^t h(\tau) b(t - \tau) d\tau = \int_0^t e^{a\tau} b(t - \tau) d\tau. \tag{5.78}$$

Thanks to linearity, the complete response is the sum of the responses in equation (5.76) so that

$$x(t) = x_H(t) + x_P(t) = e^{at} x_0 + \int_0^t e^{a\tau} b(t - \tau) d\tau. \tag{5.79}$$

An alternative way to proceed from equation (5.75) would be to Laplace-transform both sides, find the solution in the Laplace domain, and then inverse-transform the solution to the time domain. Thus, consider first just the response to the initial condition x_0.

$$\dot{x} = ax; \ x(0) = x_0,$$
$$\mathcal{L} \Rightarrow sX(s) - x(0) = ax,$$
$$(s1 - a)X(s) = x_0,$$
$$X(s) = (s1 - a)^{-1}x_0,$$
$$\mathcal{L}^{-1} \Rightarrow x(t) = \mathcal{L}^{-1}(s1 - a)x_0,$$
$$x(t) = x_H(t) = e^{at}x_0. \tag{5.80}$$

The second step in equation (5.80) is a result of Laplace transformation. To find the response to the forcing $b(t)$,

$$\dot{x} = ax + b,$$
$$\mathcal{L} \Rightarrow sX(s) = aX(s) + b(s),$$
$$(s1 - a)X(s) = b(s),$$
$$X(s) = (s1 - a)^{-1}b(s). \tag{5.81}$$

The inverse transform of the last step in equation (5.81) is, based on the convolution theorem,

$$x(t) = x_P(t) = \int_0^t h(\tau)b(t - \tau)d\tau. \tag{5.82}$$

The complete solution can then be written as

$$x(t) = e^{at}x_0 + \int_0^t h(\tau)b(t - \tau)d\tau. \tag{5.83}$$

It is interesting that the solution in equation (5.79) generalizes easily to a multi-variable state-space system solution. This is made possible by the definition of a matrix exponential where the scalar parameter a in equation (5.79) is replaced by the matrix $\mathbf{A}$ (Friedland 1986). Recognizing that the series form expansion of the exponential function $\exp(at)$ is

$$e^{at} = 1 + at + \frac{1}{2!}a^2t^2 + \frac{1}{3!}a^3t^3 + \dots, \tag{5.84}$$

the matrix exponential can be expressed as

$$e^{\mathbf{A}t} = \mathbf{I} + \mathbf{A}t + \frac{1}{2!}\mathbf{A}^2t^2 + \frac{1}{3!}\mathbf{A}^3t^3 + \dots \tag{5.85}$$

with $\mathbf{I}$ denoting the identity matrix. The steps from equation (5.80) generalize to multi-variable systems to yield

$$\mathbf{x}(t) = \mathcal{L}^{-1}(s\mathbf{I} - \mathbf{A})x_0 + \mathcal{L}^{-1}\left[(s\mathbf{I} - \mathbf{A})^{-1}b(s)\right]. \tag{5.86}$$

Equation (5.86) results in the solution expression,

$$\mathbf{x}(t) = e^{\mathbf{A}t}\mathbf{x}_0 + \int_0^t e^{\mathbf{A}\tau}\mathbf{b}(t - \tau)d\tau. \tag{5.87}$$

It can be seen that both terms in the solution in equation (5.79) will "blow up" when $a > 0$. Equivalently, both terms in the solution in equation (5.85) will also blow up when any of the eigenvalues of the system matrix $\mathbf{A}$ are positive, or have positive real parts. When the solutions increase indefinitely with time, the system as a whole is unstable. When the solutions converge to a nonzero value or oscillate with a constant amplitude, the system is marginally stable. When all solutions asymptotically approach zero, the system is asymptotically stable. Practical applicability demands that a system be either asymptotically or marginally stable. A number of systems texts discuss stability in detail (e.g., Friedland 1986). Most dynamic systems (marine systems included) contain some energy dissipation through friction, viscosity, and so on, so it is reasonable to expect that their natural responses (i.e., response to initial conditions) are asymptotically stable. It is when the systems need to be controlled using feedback and when a number of subsystems with their own dynamics (i.e., hydraulic actuators, valves, motors, etc.) are included in the control action that a system could become unstable and drive itself (along with the control machinery) to destruction.

Dissipation or damping plays a direct role in the solution $\mathbf{x}(t)$. The first term in the solution (equations (5.87) and (5.83)) represents the system's response to initial conditions, i.e., its actions as determined by its initial displacement and velocity. When the system has no dissipation, as in the largely theoretical $\zeta = 0$, the system response to initial conditions will continue for ever. When $\zeta > 0$, i.e., when there is nonzero positive damping (i.e., energy loss at a rate determined by ζ), the response to initial conditions, the first term in equations (5.87) and (5.83), will eventually tend to zero. Only the action of the external force will determine the system's response thereafter. This is when the system is said to have reached "steady state." The transfer functions commonly used in systems and controls thus only represent the steady-state actions of a system. This point needs to be kept in mind when working on initial value problems in applications.

5.5   Fourier Convolution

For most ocean-going systems, external excitation is continuous, and it is hard to define a distinct temporal reference $t = 0$ for specifying the initial conditions on the system variables. For this reason, Fourier transformation and Fourier convolution are often more readily used in marine applications. For example, consider a deeply submerged moored buoyant sphere housing a current measurement system that is subjected to variable current forces. The equation for sway oscillation of such a sphere may be described by a second-order ordinary differential equation. Here, the sphere mass provides inertia, while the mooring line provides stiffness. Viscous friction damping occurs due to the small frictional drag forces acting on the sphere.

The damping force is here assumed to be linear, as is the stiffness provided by the moorings. With $f(t)$ representing the forcing and $x(t)$ the surge displacement of the sphere,

$$(m + a_s)\ddot{x} + C\dot{x} + Kx = f(t), \quad -\infty < t < \infty. \tag{5.88}$$

Taking the Fourier transform of both sides,

$$-\omega^2(m + a_s)x(i\omega) + i\omega Cx(i\omega) + Kx(i\omega) = f(i\omega). \tag{5.89}$$

Here,

$$x(i\omega) = \int_{-\infty}^{\infty} x(t)e^{-i\omega t}\,dt; \quad f(i\omega) = \int_{-\infty}^{\infty} f(t)e^{-i\omega t}\,dt. \tag{5.90}$$

Equation (5.89) is algebraic, and can be expressed using a frequency-response function $H(i\omega)$ as

$$x(i\omega) = H(i\omega)f(i\omega). \tag{5.91}$$

The frequency response function $H(i\omega)$ is given by

$$H(i\omega) = \frac{f(i\omega)}{x(i\omega)} = \frac{1}{-\omega^2(m + a_s) + i\omega C + K}. \tag{5.92}$$

$x(i\omega)$, $f(i\omega)$, and $H(i\omega)$ are all complex-valued, containing both amplitude and phase information. Now taking the inverse Fourier transform of $H(i\omega)$,

$$h(t) = \frac{1}{2\pi} \int_{-\infty}^{\infty} H(i\omega)e^{i\omega t}\,d\omega. \tag{5.93}$$

If we now take the inverse Fourier transform of both sides of equation (5.91), the following convolution form results.

$$x(t) = \int_{-\infty}^{\infty} h(\tau)f(t - \tau)\,d\tau. \tag{5.94}$$

Once again, product in the frequency domain transforms to convolutions in the time domain. Note that the limits on the convolution extend from $t \to -\infty$ to $t \to \infty$. Equation (5.94) represents the Fourier convolution form. The basic understanding that a product in the frequency domain inverse-transforms to a convolution in the time domain still holds. However, the presence of the negative t range in the convolution makes us think of causality when real time applications are to be derived based on the Fourier convolution-based approach necessary when the goal at hand cannot be phrased in terms of an initial value problem. In particular, if $h(t)$ is nonzero for $t <$ 0, $f(t)$ values from the future are needed in order to evaluate the response at the present time t. This question becomes relevant when real-time control of such systems is required. We return to this discussion later in Chapter 15, in the context of floating body oscillations, but a more detailed discussion can be found in Naito and Nakamura (1985), Falnes (1995), and Korde and Ringwood (2016).

5.6   Concluding Remarks

This was our first chapter devoted to reviewing ordinary differential equations. We started with first-order equations derived for simplified dynamic systems within the marine world for which variable separation was practical, and were able to use the method of separation of variables to obtain solutions, notwithstanding that the equations were nonlinear. Most of the chapter focused on linear differential equations, however, with constant coefficients to boot. Such equations assume small amplitudes of forcing and response, and are typically only strictly applicable over timescales over which system parameters can be assumed to be constant (even though they may be slowly varying). However, it is frequently observed (as in the case of floating bodies) that such equations lead to highly reliable results in practice. In the case of equations for which the independent variable was time, our attention was focused on representing time-invariant systems. Time-invariant systems have parameters or coefficients that are constant and not functions of time (as opposed to time-variant systems, for which the coefficients may change with time). A variety of general mathematical techniques are available for linear time-invariant systems analysis so that good insights can be developed into their response to external forcing. We reviewed traditional solution techniques, particularly highlighting the use of impulse response functions for arbitrary forcing. We also reviewed the technique of Laplace transformation, which is restricted to linear systems with constant coefficients (i.e., to linear time-invariant systems). We also discussed Laplace and Fourier convolutions and pointed out their different application domains. We reviewed systems of first-order equations derived from second- and higher-order equations, along with solutions based on the matrix exponential.

5.7   Self-Assessment

5.7.1

Derive the equation of motion for a jet-propelled underwater vehicle traveling in the positive x direction.

5.7.2

Assuming small oscillations, derive the equation of motion for a rectangular barge undergoing just heave and surge oscillations. What are some differences between the two oscillations?

5.7.3

Consider three masses oscillating in the vertical direction, about a static equilibrium configuration. The first mass is suspended from a vertical support by means of a spring.

The first and second masses are spring coupled, while a spring and a damper connect the second and third masses. Derive the equations of motion, and solve assuming that each mass is 1 kg, each stiffness is 10 N/m, and the damping constant for the damper is 2 Ns/m.

5.7.4

Solve the following equations using the method of your choice.

(a) For $x(0) = y(0) = 1$,

$$\dot{x} = 2x - 3y + e^{2t},$$
$$\dot{y} = x - y + e^{2t}. \tag{5.95}$$

(b) For $x(0) = 1, y(0) = 1, \dot{x} = 0, \dot{y} = 0$,

$$\ddot{x} + 2\dot{x} + 3\dot{y} + 4x = e^{t},$$
$$\ddot{y} + 3\dot{x} + 2\dot{y} + 3y = 0. \tag{5.96}$$

5.7.5

Solve the following equation with $x(0) = 1$.

$$\dot{x} + 6x = f(t), \quad f(t) = \sin 2t[H(t) - H(t-4)]. \tag{5.97}$$

6 Ordinary Differential Equations-II

In Chapter 5, we were mostly concerned with differential equations that are easily solvable. In particular, all of the equations we discussed could be solved by analytical means. Often, in marine systems as in any other dynamic systems, differential equations arise that cannot easily be solved analytically, in the sense that it is difficult for existing solution techniques to yield solutions or functional forms that satisfy them. Many fluid-flow and floating or submerged body problems result in models that require differential equations in which the coefficients attached to the dependent variable are not constant and depend on the independent variable. These coefficients do not contain the dependent variable (which would make the equations nonlinear), but only the independent variable. An immediate implication of the coefficients not being "constant" (even when the differential equation is linear) is that methods such as Laplace and Fourier transforms are no longer applicable in most cases. The standard initial guess solution $e^{\lambda t}$ (t being the independent variable) is also no longer a valid basic solution. However, because these equations are linear, linear superposition still works, so the solutions without external forcing can be added to the solutions with external forcing to derive the complete response. Examples of such equations include the Bessel equation which we encounter frequently in wave propagation and other problems where circular boundaries are involved, and the Legendre equation which frequently occurs when spherical domains are present. Often, the solution steps applicable to Legendre equations include solution of a Cauchy–Euler equation. In this chapter, we discuss differential equations with non-constant coefficients, focusing attention for the most part on the commonly occurring equations. We formulate much of our discussion based on the excellent treatments in Greenberg (1978) and Brown and Churchill (2008).

6.1 Cauchy–Euler Equations

These can occur in the solution of Laplace equations (which we will discuss in Chapter 7), under time-invariant conditions, for instance, when flow or temperature or density conditions are in steady state, so the only independent variables are the spatial variables and not time. Also, as mentioned earlier, they can arise in wave-motion problems in spherical domains, again, when we are studying just the spatial behavior. In light of these remarks, we adopt a notation where x now represents the independent variable while y represents the dependent variable. The general form of a Cauchy–Euler equation is

$$a_0 x^n \frac{d^n y}{dx^n} + a_1 x^{n-1} \frac{d^{n-1} y}{dx^{n-1}} + \cdots + a_n y = 0. \tag{6.1}$$

The Cauchy–Euler equations are interesting in that the power of x in each coefficient equals the order of differentiation of y. This fact is helpful in simplifying the solution process. We consider two instances of equation (6.1) (Greenberg 1978). First, the following:

$$x^2 \frac{d^2 y}{dx^2} + x \frac{dy}{dx} - 4y = 0. \tag{6.2}$$

The coefficient on the second derivative is x^2, while the coefficient on the first derivative is x. The equation, of course, is still linear as x is the independent variable, but the solution techniques of Chapter 5 based around a trial solution $e^{\lambda x}$ are here unlikely to produce good results. However, we can begin with a trial solution of the form $y = x^a$ and substitute this into equation (6.2), and simplify to find

$$\left(a^2 - 4\right) x^a = 0. \tag{6.3}$$

Equation (6.3) leads to the finding that $a = \pm 2$, which brings us close to the solution,

$$y = Ax^2 + Bx^{-2}, \tag{6.4}$$

with A and B to be determined from initial conditions. As a second example, we consider a slightly different form where

$$x^2 \frac{d^2 y}{dx^2} - x \frac{dy}{dx} + y = 0. \tag{6.5}$$

An alternative approach is used here, which involves a variable substitution $x = e^p$, or $p = \ln x$. This substitution leads to

$$\frac{dy}{dx} = \frac{dy}{dp} \frac{dp}{dx}, \text{ and}$$

$$\frac{d^2 y}{dx^2} = \frac{d^2 y}{dx^2} \left(\frac{dp}{dx}\right)^2 + \frac{dy}{dp} \frac{d^2 p}{dx^2},$$

$$\Rightarrow \frac{dy}{dx} = \frac{dy}{dp} \frac{1}{x}, \text{ and}$$

$$\frac{d^2 y}{dx^2} = \frac{d^2 y}{dp^2} \frac{1}{x^2} - \frac{dy}{dp} \frac{1}{x^2}. \tag{6.6}$$

Making substitutions based on equation (6.6) into equation (6.5), we find it reducing to

$$\frac{d^2 y}{dp^2} - 2 \frac{dy}{dp} + y = 0. \tag{6.7}$$

Equation (6.7) has constant coefficients attached to each term and so admits solutions of the type $y = e^{\lambda p}$. Substitution leads to $\lambda_1 = -1$ and $\lambda_2 = -1$. In other words, we have repeated roots. Thus,

$$y = (A + p)\, e^P \;\Rightarrow\; y = (A + \ln x)\, x. \tag{6.8}$$

The variable substitution $x = e^P$ will also work for an equation such as

$$x^2 \frac{d^2 y}{dx^2} - 5x \frac{dy}{dx} - 6y = 0, \tag{6.9}$$

for which it leads to the solution

$$y = A\frac{1}{x} + Bx^6. \tag{6.10}$$

This substitution works well for the Cauchy–Euler equations and is a convenient way to arrive at the solutions analytically. We have left A and B undetermined, knowing that their values will be fixed by the initial or boundary conditions on y.

6.2   Bessel Equation

Here, we consider an important equation that arises frequently when circular or cylindrical geometries are used in applications. Examples include cylindrical structures in waves, waves in approximately cylindrical basins, and so on. Indeed, one of the most archetypal examples of wave motion, that of waves emanating outward when a pebble falls into a calm pool of water, can be described effectively using Bessel functions (see Figure 6.1). There are Bessel functions of the first kind, and there are Bessel functions of the second kind, sometimes known as Neumann functions, and sometimes as Weber functions. Each kind comes in many orders, so descriptions such as Bessel function of the first or second kind and mth order encountered frequently. Bessel functions of the

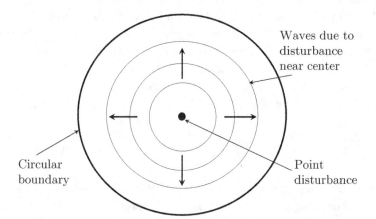

Figure 6.1 Illustration showing oscillations produced by a point disturbance within a circular boundary (e.g., by a pebble thrown into an approximately circular pond). The resulting oscillations within the boundary can be described using Bessel functions.

first kind are finite-valued throughout. Bessel functions of the first kind and order zero have a positive value at the spatial origin (e.g., $r = 0$) and approach a zero value far from the origin (i.e., $r \rightarrow \infty$), but higher orders are all zero at $r = 0$. Neumann functions start out being singular at the origin (like the logarithm function) but approach zero far from the origin. We can combine Bessel functions of first kind and Neumann functions into complex-valued functions to model outgoing waves or incoming waves depending on the sign of the Neumann function. Bessel functions of the first kind provide the real part, while the Neumann functions form the imaginary part. The combined complex-valued function is known as a Hankel function, of first kind and second kind, depending on the sign of the imaginary part. Hankel functions of the second kind model outward propagating waves such as those produced by the stone falling into a calm pool of water. If the origin is placed directly where the stone falls, the singularity of the Neumann function at the origin (i.e., becoming infinitely large) makes perfect sense in that a large amount of energy suddenly appears in a very small area to start the process of energy propagation over the water surface. We will use Bessel functions in Chapter 13.

While the Bessel functions mentioned earlier all contain oscillatory behavior and hence are well suited to describing spatial undulations, if the arguments used to express them are purely imaginary, we obtain modified Bessel functions. Modified Bessel functions are non-oscillatory, and in a sense work like the hyperbolic sine and cosine functions in representing evanescent behavior that is confined to the "near-field" of a system with an oscillatory response. Evanescent modes are local and confined to small regions of space near the origin of the wave motion. They do not convey energy into the far field, but keep it localized in a small region around the source of oscillation. When formulating a propagating wave solution in circular domains, both ordinary and modified Bessel functions are needed. Modified Bessel function solutions are chosen such that they will satisfy the evanescent wave field conditions. These are required to capture the full physical behavior (i.e., near-field, intermediate-field, and far-field) associated with waves and oscillations in cases in which both oscillatory and non-oscillatory solutions are admissible. Examples of such cases include floating bodies, beams, plates, and so on. In some cases, such as membranes and strings, only the propagating modes are sufficient. There are two kinds of modified Bessel functions: one starts at a value 1 or 0 at the spatial origin $r = 0$ and goes to infinity as $r \rightarrow \infty$, while the other starts at infinity and approaches zero as $r \rightarrow \infty$.

Bessel functions are solutions to Bessel equations. A Bessel equation can be written as

$$x^2 \frac{d^2 y}{dx^2} + x \frac{dy}{dx} + \left(x^2 - v^2\right) y = 0. \tag{6.11}$$

Equation (6.11) represents a Bessel equation of order v, where v may often be an integer, but does not have to be. Starting with the case of $v = 0$, we find that equation (6.11) somewhat simplifies to

$$x \frac{d^2 y}{dx^2} + \frac{dy}{dx} + xy = 0. \tag{6.12}$$

Recognize that this equation admits a series solution (with terms in the series asymptotically approaching zero) and that as a second-order (in the sense of the highest-order derivative being second order) must admit two independent solutions. The Bessel function of first kind and order zero forms the first solution, while the Bessel function of second kind and order zero (i.e., Neumann function of order zero) serves as the second. Thus,

$$y(x) = AJ_0(x) + BY_0 x. \tag{6.13}$$

For small arguments, i.e., when $x \to 0$ (Greenberg 1978),

$$J_0(x) \to 1,$$

$$Y_0(x) \to \frac{2}{\pi} \ln x. \tag{6.14}$$

As suggested earlier, Y_0 behaves like a logarithmic function as $x \to 0$, while J_0 approaches a finite value, namely 1.0. What type of behavior should we expect at large arguments, i.e., when $x \to \infty$? In fact both functions resemble damped sinusoids, slowly decreasing to zero as $x \to \infty$. More specifically,

$$J_0(x) \to \frac{\cos(x - \pi/4)}{\sqrt{\pi x/2}},$$

$$Y_0(x) \to \frac{\sin(x - \pi/4)}{\sqrt{\pi x/2}}. \tag{6.15}$$

Bessel equations and Bessel functions of integer order, i.e., when $\nu = n$ arise more frequently in commonly encountered situations. For a general n,

$$x^2 \frac{d^2 y}{dx^2} + x \frac{dy}{dx} + (x^2 - n^2) y = 0. \tag{6.16}$$

Equation (6.16) has general solutions of the type,

$$y(x) = A_n J_n(x) + B_n Y_n(x). \tag{6.17}$$

Also commonly encountered are Bessel equations of the form

$$x^2 \frac{d^2 y}{dx^2} + x \frac{dy}{dx} + (\alpha^2 x^2 - n^2) y = 0. \tag{6.18}$$

In this case, the general solution can be written as

$$y(x) = A_n J_n(\alpha x) + B_n Y_n(\alpha x). \tag{6.19}$$

Interestingly, $J_{-n}(x) = J_n(x)$, as equation (6.18) is seen to be insensitive to the sign of n. Another relevant property of J_n is

$$J_n(-x) = (-1)^2 J_n(x). \tag{6.20}$$

Thus, $J_1(-x) = -J_1(x)$, $J_2(-x) = J_2(x)$, and so on. Thus, odd and even values of n lead alternately to odd and even functions of x, respectively.

Figures 6.2 and 6.3 show plots of Bessel and Neumann functions for several n. Notice how the function values decrease with x, indicating a spread and corresponding decline in amplitude as we observe in waves emanating from the point on the water

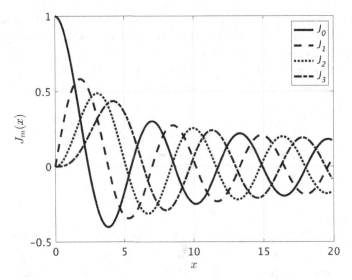

Figure 6.2 Figure showing the Bessel function of the first kind for orders 0 through 3.

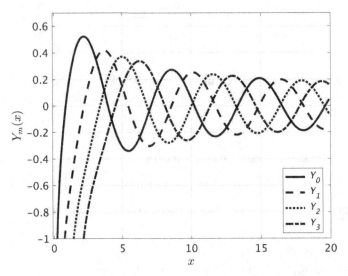

Figure 6.3 Figure showing the Neumann function (i.e., Bessel function of the second kind) for orders 0 through 3.

surface where a stone was dropped. This is typical of energy spreading over two-dimensional domains such as the water surface. In this respect, Bessel functions differ from sines and cosines that are frequently used to represent wave phenomena and wave propagation in one dimension, i.e., along a line, where energy does not suffer a radial spread. Notice also that the peaks and valleys as well as the zeros (points where the function value becomes zero) are all interlaced, and the peaks, valleys, and zeros for each n are distinct from those for any other $m \neq n$. In this respect, the Bessel functions resemble sines and cosines. Indeed, just as sines and cosines can be used to represent other periodic and nonperiodic functions, so too can Bessel functions be used in a similar manner in two-dimensional problems. Analogously, just as sines and cosines are linearly independent, so are Bessel functions. Because Bessel functions for different n are also mutually orthogonal just like sines and cosines, they can be used as "basis functions" and can form eigenfunctions for two-dimensional domains just like sines and cosines in one-dimensional domains. To be sure, sines and cosines can also be used in two-dimensional domains, but they are more convenient in rectangular domains (or domains with rectangular boundaries) while Bessel functions are much easier to use for circular domains. Last but not least, the points where an oscillatory Bessel function crosses the x axis (where x represents an arbitrary argument of the Bessel function) are referred to as "zeros" of that Bessel function. These zeros are very important in applications much the same way as the zeros of sines and cosines are. They can be used to determine the spatial eigenvalues for particular problems. Important to this determination is that the Bessel functions be linearly independent and mutually orthogonal. The orthogonality of Bessel functions can be expressed using

$$\int_0^L x J_n(\alpha x) J_m(\alpha x) dx = 0, \; n \neq m,$$

$$\int_0^L x J_n(\alpha x) J_m(\alpha x) dx = \frac{L^2}{2} [J_{n+1}(\alpha L)]^2, \; n = m. \tag{6.21}$$

The Bessel functions J_n and the Neumann function Y_n can be combined to form the Hankel function H_n in two ways. The first way is frequently used for an inward propagating wave, i.e., a wave field approaching inward from, for instance, wavemakers around the periphery of a circular wave tank. The second way represents an outward propagating wave, with a source at the center as indicated by the logarithmic singularity of the Neumann function. Thus,

$$H_n^{(1)}(\alpha x) = J_n(\alpha x) + i Y_n(\alpha x), \;\; \text{inward},$$

$$H_n^{(2)}(\alpha x) = J_n(\alpha x) - i Y_n(\alpha x), \;\; \text{outward}. \tag{6.22}$$

The superscript (1) within parentheses represents Hankel function of the first kind, while (2) denotes Hankel function of the second kind. A more general solution can be written by adding the two,

$$y(x) = A_n H_n^{(1)}(\alpha x) + B_n H^{(2)}(\alpha x). \tag{6.23}$$

A far-field approximation that is frequently used in marine applications involves an approximation for the outward propagating wave represented by $H^{(2)}$ (as would be the case for the radiated and scattered wave fields created by body oscillation near the water surface, and diffraction of incident waves). The disturbance component itself is outward propagating from a stationary body, while the incident wave field is inwardly propagating from afar. The far-field approximation for $H^{(2)}$ can be expressed as

$$H_n^{(2)}(\alpha x) \sim \frac{e^{-i(-\alpha x - \pi/4)}}{\sqrt{\pi x/2}}, \quad x \to \infty. \tag{6.24}$$

6.3   Modified Bessel Equation

One variant of equation (6.16) is

$$x^2 \frac{d^2 y}{dx^2} + x \frac{dy}{dx} - \left(\alpha^2 x^2 + n^2\right) y = 0. \tag{6.25}$$

A change of variables, $x = -i\xi$ leads to (Greenberg 1978)

$$\xi^2 \frac{d^2 y}{d\xi^2} + \xi \frac{dy}{d\xi} + \left(\alpha^2 \xi^2 - n^2\right) y = 0. \tag{6.26}$$

The general solution in x can be written as

$$y(x) = A_n I_n(\alpha x) + B_n K_n(\alpha x). \tag{6.27}$$

I_n is the modified Bessel function of the first kind, while K_n is the modified Bessel function of the second kind. In a very rough way, these functions perform the same role in circular domains as the hyperbolic sine and hyperbolic cosine functions. Thus, I_n is ever increasing and approaches ∞ in a manner reminiscent of the way a hyperbolic sine/cosine function would increase. In contrast, the function K_n starts near infinity and is ever decreasing, asymptotically approaching zero. As indicated earlier, I_n and K_n can be frequently used together to represent evanescent fields near boundaries (e.g., between two bodies near each other, etc.). K_n alone is useful for representing evanescent fields localized near a body, such as an isolated body being excited by waves. Plots of the first few modified Bessel functions are shown in Figures 6.4 and 6.5.

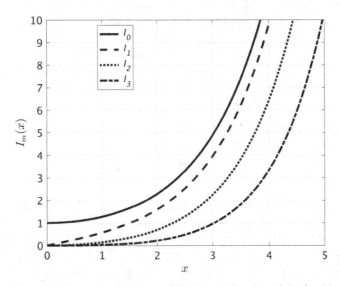

Figure 6.4 Figure showing the modified Bessel function of the first kind for orders 0 through 3.

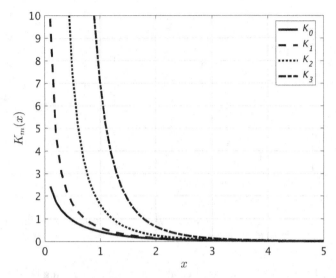

Figure 6.5 Figure showing the modified Bessel function of the second kind for orders 0 through 3.

6.4 Bessel Function Relations

A property that can be useful in working with Bessel functions is (Brown and Churchill 2008)

$$J_{n+1}(x) = -x^n \frac{d}{dx} \left[x^{-m} J_n(x) \right]. \tag{6.28}$$

Some other properties involving derivatives are also convenient. These are among the recurrence relations (Abramowitz and Stegun 1972) involving Bessel functions. For any general Bessel function (first kind is used here),

$$2J_n'(x) = J_{n-1}(x) - J_{n+1}(x),$$

$$\frac{2n}{x} J_n(x) = J_{n-1}(x) + J_{n+1}(x),$$

$$J'(x) = J_{n-1}(x) - \frac{n}{x} J_n(x),$$

$$J'(x) = -J_{n+1}(x) + \frac{n}{x} J_n(x). \tag{6.29}$$

These relations hold for fractional orders v, as well as for all other Bessel functions and Hankel functions. One example of the first relation is $J_0'(x) = -J_1(x)$.

6.5 Partial Wave Expansions

As we will see in the later chapters, floating bodies with circular symmetry make waves with their oscillations that propagate outward and can best be modeled using Hankel functions of the second kind. Similarly, fixed bodies with circular symmetry

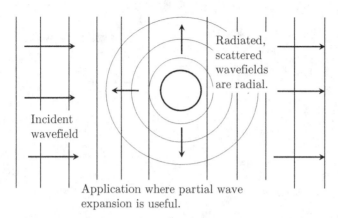

Application where partial wave
expansion is useful.

Figure 6.6 Illustration showing a situation where partial wave expansion is useful and would
consist of representing the plane wave with an infinite sum of Bessel functions.

create scattered waves around them which also can be modeled using Hankel functions
of the second kind, as they are also outgoing. When floating bodies with circular symmetry find themselves in wave fields that can best be represented using plane waves,
we have essentially two different wave topologies interacting with each other. The
interesting problems contained in these interactions would become much easier to
handle if we could find a way to represent plane waves using summations of radial
waves with circular wave fronts such as are represented using the Bessel function
combinations just mentioned (see Figure 6.6). The formal technique that allows us to
use a superposition of a number of (ideally infinite) Bessel functions to replace a plane
wave is "partial wave expansion," which offers a way to expand a plane wave into an
infinity of waves with circular wave fronts (see, e.g., Mei 1992). We will use partial
wave expansions again in Chapter 13. Consider a regular wave of amplitude A, angular
frequency ω, and wave number $k = 2\pi/\lambda$ (λ being the wave length) propagating from
left to right along the x axis, with z pointing vertically upward and y into the plane of
the paper.

$$\eta(x;t) = A\Re\left[e^{-i(kx-\omega t)}\right]. \tag{6.30}$$

With r denoting the radial distance coordinate and θ denoting the azimuth angle (i.e.,
angle measured counterclockwise from the positive x direction), $x = r\cos\theta$, and

$$\eta(r,\theta;t) = A\Re\left[e^{-i(kr\cos\theta-\omega t)}\right]. \tag{6.31}$$

The wave in equation (6.31) can now be approximated using

$$\eta(r,\theta) = A\sum_{0}^{\infty} \epsilon_n(-i)^n J_n(kr)\cos n\theta. \tag{6.32}$$

Equation (6.32) is the partial wave expansion of a plane wave propagating from left
to right along the x axis. It can be verified that, for an oppositely propagating wave
such as

$$\eta(x;t) = A\mathfrak{R}\left[e^{i(kx-\omega t)}\right],$$

$$(6.33)$$

the expansion would be

$$\eta(r,\theta) = A\sum_{0}^{\infty}\epsilon_n(i)^n J_n(kr)\cos n\theta.$$

$$(6.34)$$

Note the sign change attached to the imaginary variable i in the parentheses.

These expansions now allow us easily to combine incident, radiated, and scattered wave fields nth component by nth component, find forces and moments, evaluate oscillation amplitudes and phases, and so on.

6.6   Bessel-Reducible Equations

Consider an equation such as

$$x\frac{d^2y}{dx^2} + \frac{dy}{dx} + x^2y = 0.$$

$$(6.35)$$

Equation (6.35) can be written as

$$\frac{d}{dx}\left(x\frac{dy}{dx}\right) + x^2y = 0.$$

$$(6.36)$$

More generally, we can write equation (6.36) as

$$\frac{d}{dx}\left(x^a\frac{dy}{dx}\right) + bx^cy = 0.$$

$$(6.37)$$

Differential equations in this form can be reduced to an equivalent Bessel equation with a variable change (Greenberg 1978). Then, the solution looks like

$$y(x) = x^{\nu/\alpha}Z_\nu\left(x^{1/\alpha}\alpha\sqrt{b}\right).$$

$$(6.38)$$

In particular, we need to find parameters ν and α such that

$$\nu = \frac{1-a}{c-a+2}; \quad \alpha = \frac{2}{c-a+2}.$$

$$(6.39)$$

For equation (6.36), we see that $a = 1$, $b = 1$, and $c = 2$. Thus,

$$\nu = 0; \quad \alpha = 2/3.$$

$$(6.40)$$

Then, the solution for y can be written as

$$y(x) = x^0 Z_0\left(\frac{2}{3}x^{2/3}\right).$$

$$(6.41)$$

The general solution can now be written as

$$y(x) = AJ_0\left(\frac{2}{3}x^{2/3}\right) + BY_0\left(\frac{2/3^{2/3}}{x}x^{2/3}\right).$$

$$(6.42)$$

Notice that, even though the solution contains a J_0 and a Y_0, the argument is a little more complicated than a straightforward x as for a Bessel equation with order 0.

6.7   Legendre Functions

When applications involve spherical objects or spherical boundaries, we often need to use the Legendre equation to describe the angular dependence of the solution, especially, for the Laplace type partial differential equations we will cover in the next few chapters. Descriptions are thus seen to get increasingly less straightforward in going from sines and cosines for rectangular objects and domains, Bessel functions for circular or cylindrical objects/domains, and now Legendre functions for spherical domains. Legendre functions arise as solutions to the Legendre equation. Solutions containing Legendre functions can be considered a special case of the more general multipole expansions sometimes used in marine applications. There are Legendre functions of the first kind that are polynomials, and there are Legendre functions of the second kind that are infinite series. Both solutions are needed in applications. In describing the angular variation of the solution, a substitution $x = \cos\theta$ leads to a Legendre equation of the form

$$(1 - x^2)\frac{d^2y}{dx^2} - 2x\frac{dy}{dx} + n(n + 1)y = 0. \tag{6.43}$$

Here, n is a positive integer, though it can in general be a fraction and have negative values. In applications, n can be large, in fact, $n \to \infty$. For a discussion of Legendre functions, it is helpful to start with a consideration of the series solution. Note that equation (6.43) becomes a first-order equation when $x \to \pm 1$. With $x = \cos\theta$, we certainly expect that $|x| \le 1$. When writing a series solution about $x = 0$, we find it breaks up into two solutions y_1 and y_2 (Greenberg 1978), where

$$y_1 = 1 - \frac{n(n + 1)}{2!}x^2 + \frac{(n - 2)n(n + 1)(n + 3)}{4|}x^4 - \dots,$$

$$y_2 = x - \frac{(n - 1)(n + 2)}{3|}x^3 + \frac{((n - 3)(n - 1)(n + 2)(n + 4)}{5|} - \dots. \tag{6.44}$$

For even values of n, we see that y_1 will terminate at some point depending on n. Thus, for $n = 2$, the second term goes to zero. Since $(n - 2)$ appears in all subsequent terms in y_1, y_1 will terminate at $n = 2$. y_1 is thus a polynomial. For even values of n, the other solution y_2 has no numerator term in it that will become zero. Hence, y_2 will remain an infinite series. On the other hand, when n is odd valued, y_2 will terminate at some point. For instance, for $n = 3$, the third term in y_2 goes to zero, and since $n - 3$ appears in every term that follows, all subsequent terms in y_2 are zero, and y_2 now becomes a polynomial, while y_1 remains an infinite series. The polynomial solutions are represented as $P_n(x)$, and the series solutions as $Q_n(x)$. The general solution for any n can be expressed as

$$y(x) = A_n P_n(x) + B_n Q_n(x). \tag{6.45}$$

$P_n(x)$ represents the polynomial solution (i.e., the Legendre polynomials), while $Q_n(x)$ represents the Legendre series. The domain is typically bounded by $|x| \le 1$. At $|x| = 1$, Q_n diverges, and hence is generally unrealistic when finite boundary conditions are to be satisfied. The terms in both Legendre polynomials and Legendre series

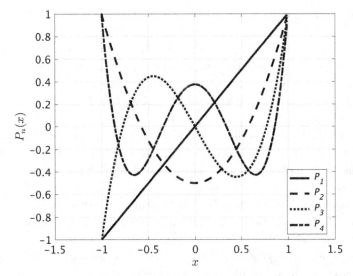

Figure 6.7 Figure showing the first four Legendre polynomials, P_1 through P_4.

are mutually orthogonal in $0 \le |x| \le 1$. The first few Legendre polynomials are listed below (Brown and Churchill 2008).

$$P_0 = 1,$$
$$P_1 = x,$$
$$P_2 = \frac{1}{2}\left(3x^2 - 1\right),$$
$$P_3 = \frac{1}{2}\left(5x^3 - 3x\right),$$
$$P_4 = \frac{1}{8}\left(35x^4 - 30x^2 + 3\right),$$
$$P_5 = \frac{1}{8}\left(63x^5 - 70x^3 + 15x\right). \tag{6.46}$$

Figure 6.7 plots the first four Legendre polynomials $P_n(x)$.

It is interesting to point out that all polynomials evaluate to a value of ± 1 at $x = \pm 1$. Specifically, all odd polynomials evaluate to -1 at $x = -1$ and to 1 at $x = -1$ and $x = 1$, respectively, while all even polynomials evaluate to 1 for both $x = 1$ and $x = -1$. Finally, the general expression for a Legendre polynomial is

$$P_n(x) = \frac{1}{2^n} \sum_{k=0}^{m} \frac{(-1)^k}{k!} \frac{(2n - 2k)!}{(n-2)!(n-k)!} x^{n-2k} \quad n = 0, 1, 2, 3, \ldots. \tag{6.47}$$

When n is even, $m = n/2$, while when n is odd, $m = (n-1)/2$. Finally,

$$P_n(-x) = (-1)^n P_n(x), \quad n = 0, 1, 2, \ldots. \tag{6.48}$$

6.8   Concluding Remarks

We have reviewed frequently occurring linear differential equations for which the coefficients are functions of the independent variable, and not constant. Allowing for that functional dependence increases the amount of work needed in solving the equations, but we are fortunate in being able to draw on established techniques and solutions that are general enough to apply in a range of marine applications. Specifically, our review considered Cauchy–Euler type equations and their solutions. For applications in cylindrical domains, we considered the frequently occurring Bessel equation, and its various forms, each leading to its own kinds of Bessel functions as solutions. Just as we are able to construct approximations for piecewise continuous functions in rectangular coordinates using sines and cosines, we can construct approximations of other more general functions that form the solution to the particular differential equation. In this instance, Bessel functions act as "basis functions" for the solutions representing the phenomena in cylindrical domains. Finally, for spherical domains, where the solution process is more complicated, we reviewed Legendre equations and Legendre functions as their solutions. Interestingly, for some problems defined over spherical domains, we use the Cauchy–Euler type equations for describing solution behavior in the radial domain, and the Legendre equations and Legendre functions in the angular domain. Properties of Legendre solutions were also reviewed.

The best way to understand these functions and their applications is to review the application problems at a deeper level, where many phenomena need to be described using partial differential equations. Chapter 7 reviews partial differential equations, with a focus on linear partial differential equations.

6.9   Self-Assessment

6.9.1

Find the radial eigenvalues and eigenfunctions for a circular membrane of radius R whose oscillations in the radial direction are being described using an nth-order Bessel equation.

6.9.2

Verify that the functions J_0 and J_1 are solutions of the Bessel equation.

6.9.3

Verify that the polynomials P_1 and P_2 solve the Legendre equation.

6.9.4

Using a series solution of the type

$$x(t) = \sum_{n=0}^{\infty} a_n t^n, \tag{6.49}$$

solve the equation

$$\ddot{x} + t^2 \dot{x} + 2tx = 0. \tag{6.50}$$

(a) with $x(0) = 1, \dot{x} = 0$,
(b) with $x(0) = 0, \dot{x} = 1$.

7   Partial Differential Equations-I

Water in the oceans forms a continuous medium. Forces, moments, displacements, velocities, and so on may all be functions of time, but in addition, they also depend on the spatial coordinates. Moreover, variations in space may sometimes be dependent on variations in time, and vice versa. Dynamic relations used in deriving ordinary differential equations now must be extended to account for spatial variations. Typically, this implies the need to use three additional coordinates. In the Cartesian coordinate system, we would want to use the four coordinates x, y, z, and t; in the cylindrical polar system, the coordinates r, θ, z, and t; while in the spherical coordinate system, we would use r, θ, and ϕ. All relations arising from Newton's second law, for instance, must in general be rewritten so that all four coordinates play a role in the dynamics. In this situation, ordinary differential equations using full derivatives in the time domain rarely provide a complete picture. We must use partial differential equations that relate multiple partial derivatives to express dynamic or static equilibrium. Being able to use partial differential equations enriches one's problem-solving ability. Specifically, using partial differential equations, we are able to understand and quantify wave propagation on the water surface, deep beneath the surface, or on the seafloor. Using partial differential equations, we are able to understand how a pollutant introduced by a ship or a factory discharge in coastal waters diffuses through the water, and compute the level of possible toxicity at any point in the fluid, so dangers to marine life and marine components can be assessed. Using partial differential equations, we may be able to predict the arrival times of tsunamis caused by near-by or distant earthquakes. Further, using partial differential equations, we may be able to predict the full response of a large floating structure or a marine riser or drill pipe or mooring lines to forcing applied by surface waves and currents. Using partial differential equations, we can predict the response of bays and harbors to waves approaching from an entrance/mouth. Indeed, partial differential equations arise almost routinely in fluid mechanics, as we will see in Chapters 10–14. They are also encountered in the development of analytical dynamics, as we will see in Chapter 16. Our attention in this chapter will be focused on linear partial differential equations for the most part. Nonlinear partial differential equations arise in the study of large-amplitude waves, and are sometimes addressed by means of perturbation expansions, as described in Chapter 12.

7.1   Commonly Encountered Types of Partial Differential Equations

The physical properties and behaviors of the variables being modeled are determined by the type of partial differential equation relationships they exhibit with or without external forcing. For instance, the way variables such as temperatures in a solid conductor, pollutant concentration, and movement in a fluid evolve in time, and how their spatial distribution must change in a manner consistent with their temporal evolution, and so on. Indeed, as a pollutant spreads in spatial extent, its concentration at each point goes on decreasing with time. This type of behavior is described by partial differential equations that are first order in time and second order in space, as we will see when we review the derivation of such equations. We will find that they belong to the category of *parabolic equations*. Parabolic equations in different contexts can be seen to admit solutions that are characteristic of parabolic systems.

Eventually, a system such as one just described will attain steady state. Once this happens, no further temporal changes are evidenced, and the temporal derivative goes to zero. The resulting second-order equation in space satisfies the conditions for an elliptic equation or elliptic system. The equation is still a partial differential equation, because changes are typically in the three-dimensional spatial domain, sometimes in two-dimensional spatial domain, if some symmetry conditions are satisfied (e.g., axial symmetry). Laplace equations fall into this category, and steady-state density or temperature distribution in a part of the ocean may be approximated by an elliptic equation.

For many systems with spatially distributed inertia and stiffness, any displacement from equilibrium produces an opposing force due to the stiffness or restoring forces that the system responds with. In addition, any motion imparted to any of the particles comprising the system will sometimes result not only in local velocity of the particles but also in local acceleration, due to inertia. There may also be some dissipation forces that take energy away from the particles, or the particles would go on oscillating indefinitely. Such systems are generally described as hyperbolic systems. Acoustic waves traveling in water and structural oscillations of a ship hull or large plate floating or slightly submerged below the water surface are typically described by hyperbolic equations. Ocean surface waves are more complicated in that they are not described by hyperbolic equations. We will return to this point in Chapter 12. However, the important point to remember is that standard solutions exist for each of the three types: parabolic, elliptic, and hyperbolic.

A simple test on the coefficients informs us whether a particular partial differential equation is parabolic, elliptic, or hyperbolic (Greenberg 1978). Consider a general form of second-order equations such as

$$A\frac{\partial^2 u}{\partial x^2} + 2B\frac{\partial^2 u}{\partial x \partial y} + C\frac{\partial^2 u}{\partial y^2} = 0. \tag{7.1}$$

We may compute the discriminant $B^2 - AC$ as a step toward determining the type to second-order partial differential equations. Further, we can say that

$$B^2 - AC = 0 \Rightarrow \text{parabolic system,}$$
$$B^2 - AC > 0 \Rightarrow \text{hyperbolic system,}$$
$$B^2 - AC < 0 \Rightarrow \text{elliptic system.} \qquad (7.2)$$

In physical problems, partial differential equations describe how a quantity gets distributed over a spatial domain. In the case of elliptic equations, for instance, they help us quantify the spatial distribution over a slab of metal of, say, temperature, if we are only given the boundary conditions for the slab, i.e., the temperature at the boundary, rate of heat flow across the boundary, or a combination of both. For parabolic equations, we additionally can evaluate the rate at which such a distribution will evolve and spread until it reaches steady state. A hyperbolic equation describes how a single packet of disturbance will travel through a medium as time evolves, at what speed, in which direction, and so on. In fact, for a parabolic system with $B^2 - AC = 0$, a disturbance, such as a quantity of pollutant, only evolves in one direction, i.e., it spreads outward with time. For a hyperbolic system with $B^2 - AC > 0$, two possibilities exist: a disturbance generally travels in two directions. For example, if we quickly pluck a string near its midpoint, that disturbance will travel in both directions along the length of the string. In the case of elliptic systems with $B^2 - AC < 0$, no real propagation possibility exists. This is not surprising as there is no time dependence implied in an elliptic equation.

We consider these three types of systems in Sections 7.2–7.4. We will first review the three principal categories of partial differential equations, and then discuss the domains over which these equations are defined (i.e., whether finite, infinite, or semi-infinite), boundary and initial conditions for finite and semi-infinite equations, and so forth. Next, we will discuss solution techniques such as separation of variables and transform techniques. Often, the solution procedures attempt to convert partial differential equations into ordinary differential equations, which in turn can thereafter be converted into algebraic equations. These steps lead us closer to the solutions and/or provide insights into the physical behavior described by the full partial differential equations.

7.2   Parabolic Equations

Such equations can be used to model the dynamic behavior of systems when, for instance, a pollutant finds its way into a body of water such as a bay, perhaps via a discharge line from a nearby factory. It will gradually spread through the region surrounding it, as material, in this case, the pollutant, flows from a point of high concentration to points of lower or zero concentration. Heat flow in solids typically behaves in a similar way, as heat flows from a warmer region to colder regions. The process by which a substance spreads from a region of excess concentration to other regions is termed *diffusion*, while the process by which heat spreads in a solid is referred to as *conduction*. The two processes are analogous. We consider the diffusion process in the following text.

The mathematical description (i.e., the partial differential equation) is simply a statement of mass conservation in this case. The total amount of substance per unit time crossing a surface defining a region must equal the total change in concentration of that substance within the volume bounded by the surface area. Thus, letting u denote the mass concentration (i.e., mass per unit volume), we can write an expression for the increase in total mass per unit time as (Greenberg 1978)

$$\int_V \frac{\partial u}{\partial t} dV. \tag{7.3}$$

We assume that the volume V is itself not changing with time. The net concentration flow across the surface S binding the volume V is

$$\int_S k \frac{\partial u}{\partial n} dS. \tag{7.4}$$

Here, k is the *diffusivity* or *diffusion constant* and has units of area divided by time. This step is a statement of Fourier's law of conduction or Fick's law of diffusion. We must have

$$\int_S k \frac{\partial u}{\partial n} dS = \int_V \frac{\partial u}{\partial t} dV. \tag{7.5}$$

We realize that

$$\frac{\partial u}{\partial n} = \nabla u \cdot \mathbf{n}. \tag{7.6}$$

In addition, by divergence theorem, we can recast the surface integral into a volume integral as

$$\int_S k \nabla u \cdot \mathbf{n} dS = \int_V k \nabla \cdot \nabla u dV. \tag{7.7}$$

We can now see that equation (7.3) and the right-hand side of equation (7.7) are both volume integrals. Therefore,

$$\int_V \frac{\partial u}{\partial t} dV = \int_V k \nabla \cdot \nabla u dV. \tag{7.8}$$

Equation (7.8) can be rearranged into a single-volume integral as

$$\int_V \left(\frac{\partial u}{\partial t} - k \nabla^2 u \right) dV = 0. \tag{7.9}$$

Because V is just an arbitrary volume, the integrand itself must be zero. Thus emerges the diffusion equation

$$\frac{\partial u}{\partial t} = k \nabla^2 u. \tag{7.10}$$

For a one-dimensional domain,

$$\frac{\partial u}{\partial t} = k \frac{\partial^2 u}{\partial x^2}. \tag{7.11}$$

How do we know that the diffusion equation is parabolic? For easy visualization, if we rewrite the one-dimensional diffusion equation as

$$\frac{\partial u}{\partial t} - k\frac{\partial^2 u}{\partial x^2} = 0, \tag{7.12}$$

letting the time variable t stand in for the variable y in equation (7.1) and x for the variable x,

$$A = -k, \quad \text{and } B = 0, \, C = 0. \tag{7.13}$$

Therefore, $B^2 - AC = 0$, implying that equation (7.11) and its more general version (7.10) are parabolic. The solution $u(x,t)$ to a system such as equation (7.11) depends both on the initial values $u(x,t = 0)$ and on the values of $u(x,t)$ or its derivative $\partial u/\partial x$ over the two boundaries, $x = 0$ and $x = L$. The diffusion equation is thus an initial/boundary value problem, and leads to a unique solution only when the initial condition and the boundary condition are known.

7.3   Elliptic Equations

In the diffusion or heat conduction process described by equation (7.11), when steady state is reached, $\frac{\partial u}{\partial t} = 0$ so that the temperature distribution of the pollutant density distribution in Section 7.2 can at steady state be described by

$$\frac{\partial^2 u}{\partial x^2} = 0. \tag{7.14}$$

Potentially interesting distributions could result when the temperature and concentrate distribution phenomena are two to three dimensional. In such cases,

$$\nabla^2 u = 0. \tag{7.15}$$

In a rectangular domain described using a Cartesian coordinate system (x, y, z),

$$\nabla^2 u = \frac{\partial^2 u}{\partial x^2} + \frac{\partial^2 u}{\partial y^2} + \frac{\partial u^2}{dz^2} = 0. \tag{7.16}$$

The Laplacian looks considerably different in cylindrical polar and spherical polar coordinate systems. For instance, in the cylindrical polar coordinate system (r, θ, z),

$$\nabla^2 u = \frac{\partial^2 u}{\partial r^2} + \frac{1}{r}\frac{\partial u}{\partial r} + \frac{1}{r^2}\frac{\partial^2 u}{\partial \theta^2} + \frac{\partial^2 u}{\partial z^2} = 0. \tag{7.17}$$

In a spherical polar coordinate system, on the other hand,

$$\nabla^2 u = \frac{\partial^2 u}{\partial r^2} + \frac{2}{r}\frac{\partial u}{\partial r} + \frac{1}{r^2 \sin^2 \phi}\frac{\partial^2 u}{\partial \theta^2} + \frac{1}{r^2}\frac{\partial^2 u}{\partial \phi^2} + \frac{1}{r^2 \tan \phi}\frac{\partial u}{\partial \theta} = 0. \tag{7.18}$$

Equations (7.15)–(7.18) are elliptical. To see this, we can write a two-dimensional Laplace equation as

$$\frac{\partial^2 u}{dx^2} + \frac{\partial^2 u}{dy^2} = 0. \tag{7.19}$$

Here, $A = 1$, $B = 0$, and $C = 1$ so that $B^2 - 4AC = -4 < 0$, which shows that the equation is elliptic. The solution to a Laplace equation such as equation (7.16) is unique when either u or its spatial derivative at the domain boundaries are known.

7.4 Hyperbolic Equations

We return to the example of a string stretched between two end points. If the string is plucked briefly somewhere along its length, the string responds by opposing the instantaneous displacement with a restoring force, which is downward directed for an upward initial deflection. It then experiences a deflection under the action of the restoring force, and by the time the string returns to its equilibrium position, it has already gathered momentum due to its inertia, and so continues its displacement in a downward direction, which in turn leads to an upward restoring force. This pattern continues and travels in both directions along the string, as no point on the string can be displaced without disturbing every other point, given that the string is a continuum of connected points. The dynamic interaction described here is expressed by the following relation,

$$\frac{\partial^2 u}{\partial t^2} = c^2 \frac{\partial^2 u}{\partial x^2}. \tag{7.20}$$

Equation (7.20) is the so-called wave equation, and c is analogous to a wave speed. For a taut string, $c = \sqrt{T/\rho}$, where T is the tension in the string in force units (N), and ρ is the mass per unit length (kg/m) for the string. For a uniform membrane with no structural rigidity (unlike a plate), equation (7.20) becomes

$$\frac{\partial^2 u}{\partial t^2} = c^2 \nabla^2 u. \tag{7.21}$$

The membrane may be rectangular or circular. Here again, $c = \sqrt{T/\rho}$, but T now is tension applied at the membrane boundary and is therefore expressed as force per unit length (N/m). ρ is the membrane mass per unit area (kg/m^2). For a rectangular membrane,

$$\nabla^2 = \frac{\partial^2}{\partial x^2} + \frac{\partial^2}{\partial y^2}, \tag{7.22}$$

and for a circular membrane,

$$\nabla^2 = \frac{\partial^2}{\partial r^2} + \frac{1}{r}\frac{\partial}{\partial r} + \frac{1}{r^2}\frac{\partial^2}{\partial \theta^2}. \tag{7.23}$$

Sometimes, thin floating structures held together under tension forces produced by buoyant spheres can be modeled as membranes or thin plates. Action of external oscillatory forces such as those applied by waves can then be understood using the unforced and forced versions of equation (7.21).

It is interesting to see the two wave propagation directions embedded in equation (7.20), by rewriting it as follows:

$$\frac{\partial^2 u}{\partial t^2} - c^2 \frac{\partial^2 u}{\partial x^2} = \left(\frac{\partial u}{\partial t} + c \frac{\partial u}{\partial x} \right) \left(\frac{\partial u}{\partial t} - c \frac{\partial u}{\partial x} \right) = 0. \tag{7.24}$$

Equation (7.24) immediately leads to two situations:

$$\frac{\partial u}{\partial t} + c \frac{\partial u}{\partial x} = 0,$$

$$\frac{\partial u}{\partial t} - c \frac{\partial u}{\partial x} = 0. \tag{7.25}$$

Equation (7.25) contains two descriptions. The top equation for a right-propagating disturbance u and the bottom equation for a left-propagating disturbance. The disturbance travels in two opposite directions at the speed c.

To verify that the full-wave equation is hyperbolic, consider equation (7.20). Here, we can use $A = 1$ and $C = -c^2$, with $B = 0$. $B^2 - AC$ then equals c^2, which is positive, implying that equation (7.20) is hyperbolic. The solution to equations such as equation (7.20) is unique if two initial conditions are known, i.e., $u(x,t = 0)$ and $\partial u / \partial t(x,t = 0)$ (displacement and velocity at each point), and two boundary conditions are also known, for instance, $u(x = 0,t)$ and $u(x = L,t)$.

7.5   Solution Techniques

The partial differential equations we have considered so far are all linear, which fact alone makes a number of potential solution techniques available. Particularly important to note is that the commonly encountered solution techniques tend to drive toward a linear superposition of a large number of component solutions, which of course would not be an option with nonlinear partial differential equations, unless some type of approximation or variable transformation is used. We consider in the following text the commonly used technique of separation of variables, which begins by only looking for solutions of a certain form. We will first demonstrate this technique using parabolic equations (e.g., the diffusion problem), before moving on to elliptic equations (Laplace equation), and then to hyperbolic equations (such as the wave equation). Much of the discussion pertaining to the diffusion equation and the Laplace equation is based on Churchill (Brown and Churchill 2008).

7.5.1   Separation of Variables: Rectangular Domains

We will first consider the three types of partial differential equations as they arise in rectangular regions, before turning our attention to other realist domains such as cylindrical and spherical. A qualifier, "largely" is implicit in our thought process, as a largely rectangular region will be approximated as a rectangular region in our treatment. Similarly, largely cylindrical and spherical regions will also be approximated as cylindrical and spherical in our treatment. We begin with the diffusion equation, which we recall is a parabolic equation.

Diffusion Equation

Recall that the diffusion equation is parabolic and describes how a quantity of substance added into a fluid at a certain point and a certain time will distribute itself through the fluid and how quickly or slowly.

$$\frac{\partial u}{\partial t} = k\nabla^2 u. \tag{7.26}$$

The equation is parabolic because the disturbance only spreads outward, i.e., from points of higher concentration to lower concentration, just as heat in a conducting medium only flows from a warmer region to a colder region, unless some action is performed by an external agency to reverse the heat flow "direction." The dispersal rate and final distribution of the newly added material when steady state or equilibrium is reached (i.e., when the time rate of change of concentration approaches zero) are determined by conditions at the fluid boundary. We, therefore, expect our solution to be a function of time and space. Further, the solution function must also satisfy the fact that the temporal rate of change approaches zero with time. Furthermore, the solution function must also satisfy the boundary conditions for the fluid domain, and the initial condition at $t = 0$. Given these two observations, we can look for a solution that simply contains a product of two functions: one a function of time only and another a function of space only. In other words, let us look for solutions of the type

$$u(\mathbf{r};t) = \varrho(\mathbf{r})T(t), \tag{7.27}$$

where ϱ is a function of $\mathbf{r}$ only, $\mathbf{r}$ being the position vector for a point in the domain being studied. $T(t)$ is a function of time only and has no spatial dependence. In fact, for clarity of discussion, let us begin with a one-dimensional domain (e.g., a pipe), where the only influential spatial variable is x, along the domain length. In this case, $\nabla^2 \equiv \partial^2/\partial x^2$, and we may use

$$u(x,t) = X(x)T(t). \tag{7.28}$$

We expect this solution to be valid over the entire domain from $x = 0$ to $x = L$, where L defines the length of the pipe or tube we are observing. To define the problem further, let us suppose that no material can flow through the boundaries $x = 0$ and $x = L$. This means that the rate of change of concentration across the two boundaries must be zero. Let us further specify an initial condition, where we define the initial concentration, i.e., its functional form over x at $t = 0$.

$$\frac{\partial u}{\partial x} = 0,\ x = 0;\quad \frac{\partial u}{\partial x} = 0,\ x = L,$$
$$u(x,t) = f(x),\ t = 0. \tag{7.29}$$

Figure 7.1 depicts this situation.

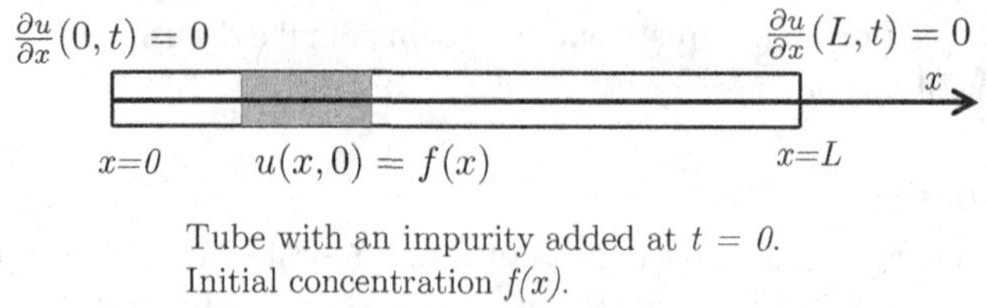

$\frac{\partial u}{\partial x}(0, t) = 0$ $\qquad\qquad\qquad\qquad$ $\frac{\partial u}{\partial x}(L, t) = 0$

x

$x{=}0$ $\qquad$ $u(x, 0) = f(x)$ $\qquad\qquad\qquad$ $x{=}L$

Tube with an impurity added at $t = 0$.
Initial concentration $f(x)$.

Figure 7.1 A fluid-filled tube closed at two ends, so no material flows in or out, but with an initially introduced concentration of an impurity, specified by $f(x)$.

Now using the expansion $u(x, t) = X(x)T(t)$ in equation (7.26) with the knowledge that the Laplacian is one dimensional, the temporal and spatial derivatives result in

$$\frac{\partial u}{\partial t} = X(x)\frac{dT}{dt}; \quad \frac{\partial^2 u}{\partial x^2} = T(t)\frac{d^2 X}{dx^2}. \tag{7.30}$$

This implies that

$$X(x)\frac{dT}{dt} - kT(t)\frac{d^2}{dx^2} = 0. \tag{7.31}$$

For a nontrivial solution (i.e., $u(x, t) \neq 0$), we may divide the two sides by $ku(x, t)$ to obtain

$$\frac{X(x)}{kT}\frac{dT}{dt} - \frac{d^2 X}{dx^2} = 0; \Rightarrow \frac{1}{kT}\frac{dT}{dt} = \frac{1}{X}\frac{d^2 x}{dx^2}. \tag{7.32}$$

Because k is a constant by definition and T is defined as a function of t only, the left-hand side of the second part of equation (7.32) can only be a function of t or constant, but must have no dependence on x. Similarly, because X is defined to be a function of x only, the right-hand side of the second equation can be a function x at best or a constant. But the two sides are equal to each other. This is only possible when the two are equal to a constant, and the same constant at that.

Thus, let

$$\frac{1}{kT}\frac{dT}{dt} = -\lambda, \quad \frac{1}{X}\frac{d^2}{dx^2} = -\lambda, \tag{7.33}$$

where λ is our separation constant. We now have two ordinary differential equations in place of one partial differential equation. Both are linear, and both have constant parameters.

$$\frac{dT}{dt} + k\lambda T = 0,$$

$$\frac{d^2 X}{dx^2} + \lambda X = 0. \tag{7.34}$$

What do the initial and boundary conditions look like in terms of our separated functions T and X?

$$T(t)\frac{dX}{dx} = 0, \ x = 0; \quad T(t)\frac{dX}{dx} = 0, \ x = L,$$

$$T(0)X(x) = f(x). \tag{7.35}$$

There are three possibilities for the types of solutions we can expect. If $\lambda = 0$, then $T(t) = C$ and $X(x) = Ex + F$. However, application of the first boundary condition leads to $E = 0$, while the second boundary condition leaves F undetermined. A solution where both X and T are constants then results. If $\lambda < 0$,

$$T(t) \sim e^{k\lambda t},$$
$$X(x) \sim D_1 e^{\sqrt{\lambda}x} + D_2 e^{-\sqrt{\lambda}x}. \tag{7.36}$$

We note then that $\lambda < 0$ leads to a $T(t)$ that will increase exponentially with time, which is physically not possible for diffusion and conduction problems. The spatial behavior can be expressed as

$$X(x) \sim C_1 \cosh \sqrt{\lambda}x + C_2 \sinh \sqrt{\lambda}x. \tag{7.37}$$

If we apply the boundary conditions, $\frac{dX}{dx}(0) = 0$ and $\frac{dX}{dx}(L) = 0$, we find that $C_1 = 0$ because of the first boundary condition (since $\cosh 0 \neq 0$), and $C_2 = 0$ because, again, $\cosh \sqrt{\lambda}L \neq 0$. The two boundary conditions and the temporal behavior thus lead to a trivial solution for $\lambda < 0$, and eliminate $\lambda < 0$ as a possibility. The most interesting possibility is $\lambda > 0$, for which

$$T(t) \sim e^{-k\lambda t} \tag{7.38}$$

is the temporal behavior, which shows an exponential decrease with time, which is more plausible, for instance, with the concentration of a newly added substance decreasing with time as it diffuses through the fluid. The spatial behavior can be described as

$$X(x) = A\cos \alpha x + B \sin \alpha x, \text{ with } \alpha = \sqrt{\lambda}, \tag{7.39}$$

where α is introduced for convenience, and $\alpha^2 = \lambda$. Recall from Chapter 5 that the expression (7.39) follows from the expression $X(x) = D_1 e^{i\alpha x} + D_2 e^{-i\alpha x}$. As a footnote, the X equation subject to the boundary conditions at the two ends represents a so-called Sturm–Liouville problem, for which a nontrivial solution only exists at specific values of λ.

   If we now apply the two boundary conditions, we find, from the first boundary condition at $x = 0$,

$$B = 0. \tag{7.40}$$

The second boundary condition at $x = L$ requires more discussion, and gives

$$A\alpha \sin \alpha L = 0. \tag{7.41}$$

Since λ, and hence $\alpha \neq 0$, the only way a nontrivial solution can exist (i.e., $A \neq 0$) is if

$$\sin \alpha L = 0. \tag{7.42}$$

Thus, the possibilities span a countable infinity, in that any α_n that satisfies equation (7.42) represents a solution. In other words,

$$\alpha_n = \frac{n\pi}{L}, \; n = 0, 1, 2, 3, \ldots,$$

$$\text{or } \lambda_n = \left(\frac{n\pi}{L}\right)^2, \; n = 0, 2, 3, \ldots. \tag{7.43}$$

λ_n here are referred to as eigenvalues. Each eigenvalue λ_n is associated with an eigenfunction X_n, where

$$\lambda_0 = 0, \; X_0 = 1; \; \lambda_n = \left(\frac{n\pi}{L}\right), \; X_n(x) = \cos\frac{n\pi x}{L}. \tag{7.44}$$

The eigenfunction X_0 is consistent with what we found in our earlier discussion related to the $\lambda = 0$ possibility. With these λ_n in hand, we can now write the temporal evolution for the nth solution as

$$T_n(t) \sim \exp\left(-\frac{n^2\pi^2 k}{L^2}t\right), \tag{7.45}$$

which leads to $T_0(t) = 1$, a constant, as expected. Now combining the temporal and spatial solutions,

$$u_0(x,t) = X_0(x)T_0(t) = 1,$$

$$u_n(x,t) = X_n(x)T_n(t) = \exp\left(-\frac{n^2\pi^2 k}{L^2}t\right)\cos\frac{n\pi x}{L}, \; n = 1, 2, 3, \ldots. \tag{7.46}$$

The complete solution can now be constructed via a linear superposition of the solutions for all n. Attaching a constant A_n to each component,

$$u(x,t) = A_0 + \sum_{n=1}^{\infty} \infty A_n \exp\left(-\frac{n^2\pi^2 k}{L^2}t\right)\cos\frac{n\pi x}{L}. \tag{7.47}$$

How do we make our constants A_n specific to our particular problem? This is where the initial condition, for $t = 0$, comes in. Since $T_n(0) = 1$ for all n,

$$u(x,0) = f(x) = A_0 + \sum_{n=1}^{\infty} A_n \cos\frac{n\pi x}{L}. \tag{7.48}$$

It may be recalled that the right-hand side of $u(x,0)$ is just the Fourier cosine series expansion for a given function, in this case, $f(x)$. Thus, we can find A_0 as

$$A_0 = \frac{1}{L}\int_0^L f(x)dx \tag{7.49}$$

and

$$A_n = \frac{1}{L}\int_0^L f(x)\cos\frac{n\pi x}{L}dx, \; n = 1, 2, 3, \ldots. \tag{7.50}$$

The complete solution for the diffusion equation for the given boundary and initial conditions is now fully known by inserting A_0 and A_n from equations (7.49) and (7.50) into $u(,t)$ in equation (7.47).

Laplace Equation

The Laplace equation is elliptic, and hence describes the internal distribution of a quantity when that quantity or its spatial derivative are known or held fixed at the domain boundaries. For instance, we may use the Laplace equation to quantify the temperature distributions over two rectangular plates being welded together, once steady state is reached. These temperature distributions will enable us to compute the thermal stresses and strains left in place by a welding process. In such a problem, we can quantify the boundary conditions knowing whether the plate edges are insulted to cut off heat transfer, or whether they may transfer heat across an edge in order to maintain a constant temperature at that edge. When heat flux is zero across a boundary (i.e., edge), the derivative of temperature across that boundary is zero. When the temperature is constant at a boundary, heat may flow across it, but we have a way to specify the temperature at that boundary. In the context of Laplace equations, the first type of boundary conditions (i.e., with a known or specified derivative of temperature across it) are *Neumann conditions*, while the second type (i.e., with a known value or function representing the temperature at the boundary) are *Dirichlet conditions*.

Let us consider the Laplace equation over a rectangular domain $L \times B$,

$$\frac{\partial^2 u}{\partial x^2} + \frac{\partial^2 u}{\partial y^2} = 0, \ 0 \le x \le L; \ 0 \le y \le B. \tag{7.51}$$

The boundary conditions may be specified as

$$\frac{\partial u}{\partial x}(0, \ y) = 0, \ u(L, \ y) = 0.$$

$$-K\frac{\partial u}{\partial y} = f(x), \ y = 0 \quad \frac{\partial u}{\partial y} = 0, \ y = B. \tag{7.52}$$

If we now look for a solution of the form, $u(x, y) = X(x)Y(y)$, a substitution of this expression in equation (7.51) leads to

$$Y(y)\frac{d^2 X}{dx^2} + X(x)\frac{d^2 Y}{dy^2} = 0. \tag{7.53}$$

For a nontrivial solution $u(x, y) \ne 0$, a division of both sides of equation (7.53) gives

$$\frac{1}{X}\frac{d^2 X}{dt^2} + \frac{1}{Y}\frac{d^2 Y}{dy^2} = 0. \tag{7.54}$$

The first term in equation (7.54) is a function of x only, while the second is a function of y only. For their sum to be zero for arbitrary x and y, they must both equal a constant, such that

$$\frac{1}{X}\frac{d^2 X}{dx^2} = -\lambda^2,$$

$$\frac{1}{Y}\frac{d^2 Y}{dy^2} = \lambda^2. \tag{7.55}$$

This results in two second-order ordinary differential equations,

$$\frac{d^2X}{dx^2} + \lambda^2 X = 0,$$
$$\frac{d^2Y}{dy^2} - \lambda^2 Y = 0. \tag{7.56}$$

The first of equations (7.56) has the solution

$$X(x) = A\cos\lambda x + B\sin\lambda x. \tag{7.57}$$

The boundary conditions, $u(0,y) = X(0)Y(y) = 0$ and $u(L,y) = X(L)Y(y) = 0$, lead to

$$A(1) + B(0) = 0 \Rightarrow A = 0,$$
$$B\sin\lambda L = 0 \Rightarrow \lambda_n = \frac{n\pi}{L}. \tag{7.58}$$

Here again, λ_n are the eigenvalues in the x direction. For the solution $Y(y)$, we use

$$Y(y) = C\left[\exp\left(\frac{n\pi y}{L}\right) + \exp\left(\frac{n\pi(2B - y)}{L}\right)\right]. \tag{7.59}$$

$Y(y)$ in equation (7.59) satisfies the boundary condition $\frac{\partial}{\partial y}u(x,B) = \frac{dY}{dy}(B) = 0$. Letting C be

$$C = \frac{1}{2}\exp\left(-\frac{n\pi B}{L}\right), \tag{7.60}$$
$$Y(y) = \sinh\frac{n\pi(B - y)}{L}. \tag{7.61}$$

The overall solution can now be expressed as

$$u(x,y) = \sum_{n=1}^{\infty} B_n\sin\frac{n\pi(B - y)}{L}\sin\frac{n\pi x}{L}. \tag{7.62}$$

Now we can apply the boundary condition at $y = 0$ to find

$$\frac{\partial}{\partial y}u(x,0) = f(x) = \sum_{n=1}^{\infty} B_n\cosh\frac{n\pi(B)}{L}\sin\frac{n\pi x}{L}. \tag{7.63}$$

The negative sign and the factor $n\pi/L$ have been absorbed into B_n. Assuming that $f(x)$ is piecewise continuous,

$$B_n = \frac{2}{L\cosh\frac{n\pi}{L}} \int_0^L f(x)\sin\frac{n\pi x}{L}dx. \tag{7.64}$$

The overall solution is now specific to the given boundary condition at $y = 0$.

Wave Equation

Having applied the method of variable separations to the parabolic and elliptic partial differential equations earlier, let us now turn our attention to the wave equation. Note that this equation does not, in general, capture the behavior of ocean surface waves, for which we require the treatment of Chapter 12. To begin our discussion here, we treat one-dimensional wave propagation in the following text, for which the describing equation is

$$\frac{\partial^2 w}{\partial^2 t} - c^2\frac{\partial^2 w}{\partial x^2} = 0. \tag{7.65}$$

As discussed in Section 7.4, this equation describes transverse waves on a string, with $w(x;t)$ denoting the vertical displacement of the string at a point x and at time t. The waves will travel in both directions at speed c all the way out to $\pm\infty$ if the string were infinite in length. Here, we only consider a finite string (i.e., with finite length) of length L, held between two rigid supports. A constant tension is applied between the two supports. What we must expect then is that the string boundaries will either reflect or transmit, or both, the wave field that exists along the string length. Since the sting itself only exists between the two supports at the boundary, only reflections back from the boundaries are realistic. At steady state, we must therefore expect a standing wave pattern, which we will investigate in the following text. This is equivalent to considering the vibrations occurring on the string, since vibrations of a continuous structure held between boundary supports are equivalent to standing waves that get set up between the boundary supports. $c = \sqrt{T/\rho}$ is still the wave speed at which the two components of the standing wave travel in either direction.

Let us consider a string with the following boundary and initial conditions:

$$w(0,t) = 0; \quad w(L,t) = 0.$$

$$w(x,0) = f(x); \quad \frac{\partial w}{\partial t}(x,0) = 0. \tag{7.66}$$

The function $f(x)$ is piecewise continuous, and $f(0) = f(L) = 0$. Note that since equation (7.65) has a second-order time derivative, two initial condition are needed. We start by deciding to look for solutions of the form

$$w(x,t) = X(x)T(t), \tag{7.67}$$

where $X(x)$ is a function of position x alone and $T(t)$ is a function of time t only. Substituting the form in equation (7.67) into a slightly rearranged wave equation of (7.65),

$$\frac{1}{c^2} X \frac{d^2 T}{dt^2} = T \frac{d^2 X}{dx^2}. \tag{7.68}$$

Since we are looking for a nontrivial solution $w(x,t) \neq 0$, division of both sides of equation (7.68) by $w(x,t)$ results in

$$\frac{1}{c^2 T} \frac{d^2 T}{dt^2} = \frac{1}{X} \frac{d^2 X}{dx^2} = -\lambda^2. \tag{7.69}$$

We thus have two ordinary differential equations: one for t and the other for X.

$$\frac{d^2 X}{dx^2} + \lambda^2 X = 0,$$

$$\frac{d^2 T}{dt^2} + \lambda^2 c^2 T = 0. \tag{7.70}$$

Equations (7.70) have solutions of the form,

$$X(x) = A \cos \lambda x + B \sin \lambda x,$$

$$T(t) = C \cos c\lambda t + D \sin c\lambda t. \tag{7.71}$$

Application of the boundary conditions at $x = 0$ and $x = L$ leads to

$$A = 0,$$

$$\sin \lambda L = 0 \tag{7.72}$$

so that

$$\lambda_n = \frac{n\pi}{L}, \quad n = 1, 2, 3, \ldots, \infty. \tag{7.73}$$

Here, λ_n are the eigenvalues for the string, with $\sin(n\pi x/L)$ being the eigenfunctions. $n = 0$ again leads to a trivial solution. There are thus countably infinite eigenvalues and eigenfunctions. Since we have the initial condition, $\partial y(x,0)/\partial t = 0$ or

$$\frac{dT}{dt}(t = 0) = \lambda_n c \left(-C \cos \lambda_n t + \lambda_n c D \sin \lambda_n t \right) \Big|_{t=0} = 0, \Rightarrow C = 0. \tag{7.74}$$

The solution for $T(t)$, within a constant, is thus

$$T(t) = \cos \left(\frac{n\pi c}{L} t \right). \tag{7.75}$$

The overall solution can now be written as

$$w(x,t) = \sum_{n=1}^{\infty} B_n \sin \left(\frac{n\pi x}{L} \right) \cos \left(\frac{n\pi cx}{L} \right). \tag{7.76}$$

The constants B_n can be found using the other initial condition, $y(x,0) = f(x)$. Since

$$w(x,0) = \sum_{n=1}^{\infty} B_n \sin \left(\frac{n\pi x}{L} \right) = f(x), \tag{7.77}$$

$$B_n = \frac{2}{L} \int_0^L f(x) \sin \left(\frac{n\pi x}{L} \right) dx, \quad n = 1, 2, 3, \ldots, \infty. \tag{7.78}$$

The complete solution can now be written for the given boundary and initial conditions. It is instructive to see how this technique extends to a two-dimensional problem. Let us consider the free-vibration solution for a rectangular membrane with pinned boundaries. The equation of motion is

$$\frac{\partial^2 w}{\partial t^2} - c^2 \left(\frac{\partial^2 w}{\partial x^2} + \frac{\partial w}{\partial y^2} \right) = 0. \tag{7.79}$$

The domain is defined by the boundaries, $[0, L]$ and $[0, B]$. The boundary is pinned at all four edges, which means that the membrane displacement at the edges is zero. Thus,

$$w(x,0;t) = w(x,B;t) = w(0,y;t) = w(L,y;t) = 0. \tag{7.80}$$

In addition, we have two initial conditions at $t = 0$ (one for deflection and one for velocity) where

$$w(x,y;0) = f(x,y), \quad \frac{\partial}{\partial t} w(x,y;0) = 0. \tag{7.81}$$

The variable separation approach here becomes interesting, as we now have three functions: one is a function of t, one is a function of x, and one is a function of y. Thus, we look for a solution that is of the form

$$w(x,y;t) = X(x)Y(y)T(t). \tag{7.82}$$

Substitution of the separation form into equation (7.79) and division by $w(x,y;t)$ leads to

$$\frac{1}{c^2 T} \frac{d^2 T}{dt^2} = \left(\frac{d^2 X}{dx^2} + \frac{d^2 Y}{dy^2} \right) = \lambda^2. \tag{7.83}$$

Equation (7.83) immediately leads to

$$\frac{d^2 T}{dt^2} + \lambda^2 c^2 T = 0. \tag{7.84}$$

Meanwhile, equation (7.83) can be further separated using

$$\frac{d^2 Y}{dy^2} = -\lambda^2 - \frac{d^2 X}{dx^2} \frac{d^2 Y}{dy^2} = -\mu^2. \tag{7.85}$$

With some algebra, equation (7.85) now leads to

$$\frac{d^2 X}{dx^2} + (\lambda^2 - \mu^2)X = 0 \tag{7.86}$$

and

$$\frac{d^2 Y}{dy^2} + \mu^2 Y = 0. \tag{7.87}$$

We thus have three ordinary second-order differential equations, whose solutions are well known to us. In particular, for the given boundary conditions (equations (7.80) and (7.81)), we have

$$X(x) \sim \sin \beta x, \quad Y(y) \sim \sin \mu y, \tag{7.88}$$

where we have used $\beta^2 = \lambda^2 - \mu^2$. Only $\beta > 0$ values are considered. For the given boundary conditions,

$$\beta_n = \frac{n\pi}{L}, \ n = 1,2,3,\dots,$$

$$\mu_n = \frac{m\pi}{B}, \ m = 1,2,3,\dots,. \tag{7.89}$$

Realizing that

$$\lambda_{nm}^2 = \beta_n^2 + \mu_m^2, \tag{7.90}$$

equation (7.84) becomes, with the second initial condition (i.e., the velocity condition) invoked,

$$\frac{d^2T}{dt^2} + (\beta_n^2 + \mu_n^2)c^2T = 0, \quad (d/dt)T(0) = 0. \tag{7.91}$$

Equation (7.91) leads to a solution for $T(t)$ of the form (within a constant),

$$T_{nm}(t) \sim \cos\left(\sqrt{(\beta_n^2 + \mu_m^2)}\right) ct. \tag{7.92}$$

The complete solution for vertical deformation of the membrane can now be written as

$$w(x,y;t) = \sum_{n=1}^{\infty}\sum_{m=1}^{\infty} C_{nm} \sin\beta_n x \sin\mu_m y \cos\left(\sqrt{(\beta_n^2 + \mu_m^2)}\right) ct. \tag{7.93}$$

We can use the first initial condition (i.e., the displacement condition), which informs us that $w(x,y;0) = f(x,y)$. Equation (7.93) at $t = 0$ now provides

$$w(x,y;0) = f(x,y) = \sum_{n=1}^{\infty}\sum_{m=1}^{\infty} C_{nm} \sin\beta_n x \sin\mu_m y. \tag{7.94}$$

Our task now is to relate $f(x,y)$ to successive Fourier series expansions in the x and y directions. We can proceed in the following steps. First, over the interval $[0, L]$, let

$$\frac{2}{\pi}\int_0^L f(x,y) \sin\beta_n x dx = \sum_{m=1}^{\infty} C_{nm} \sin\mu_m y. \tag{7.95}$$

It is easy to see that the left side must, in general, be a function of y so that for any chosen n,

$$\frac{2}{\pi}\int_0^L f(x,y) \sin\beta_n x dx = F_n(y) = \sum_{m=1}^{\infty} C_{nm} \sin\mu_m y. \tag{7.96}$$

Now we can expand $F_n(y)$ the interval $[0, B]$ to find

$$C_{nm} = \frac{2}{\pi}\int_0^B F_n(y) \sin\mu_m y dy. \tag{7.97}$$

With the arbitrary constant C_{nm} thus fully determined, the entire solution can be written as

$$w(x, y; t) = \sum_{n=1}^{\infty} \sum_{m=1}^{\infty} C_{nm} \sin \beta_n x \sin \mu_m y \cos \left(\sqrt{(\beta_n^2 + \mu_m^2)} \right) ct, \qquad (7.98)$$

with C_{nm} given as

$$C_{nm} = \frac{4}{\pi^2} \int_0^B \left(\int_0^L f(x, y) \sin \beta_n x dx \right) \sin \mu_m y dy. \qquad (7.99)$$

7.5.2   Separation of Variables: Non-Rectangular Domains

The need to use non-rectangular coordinate systems may arise sometimes in practical problems. Specifically, in some cases, it may be more convenient to use cylindrical or spherical coordinate systems. For example, when considering wave generation over a circular bay or a circular tank, it may be more convenient to model those regions as cylindrical domains and to use cylindrical coordinate systems instead. In the study of large-scale flow patterns over the earth as a whole, a spherical coordinate system may be more convenient. Here, we discuss a few examples. To begin with, consider a situation where a very shallow circular tank is partitioned into two equal halves by means of a wall made of an insulating material. The semicircular region on top is slowly brought to a temperature distribution. Our goal is to determine the temperature distribution we can expect in the top half of the tank, assuming that the mid-plane insulator wall is maintained at a constant temperature that serves as a datum 0 for the rest of the tank. Thus, we have a situation that is best described by the Laplace equation,

$$\nabla^2 u = 0. \qquad (7.100)$$

We can use a cylindrical coordinate system (r, θ, z) here, for which the Laplacian is given by

$$\nabla^2 u = \frac{\partial^2 u}{\partial r^2} + \frac{1}{r} \frac{\partial u}{\partial r} + \frac{1}{r^2} \frac{\partial^2 u}{\partial \theta^2} + \frac{\partial^2 u}{\partial z^2}. \qquad (7.101)$$

Since our tank is very shallow, we may assume that changes in the z direction are much smaller in relation to other quantities in equation (7.101). Thus, setting the last term in equation (7.101) to zero, we can express the Laplace equation of equation (7.100) as

$$\frac{\partial^2 u}{\partial r^2} + \frac{1}{r} \frac{\partial u}{\partial r} + \frac{1}{r^2} \frac{\partial^2 u}{\partial \theta^2} = 0. \qquad (7.102)$$

Multiplying both sides by r^2, we find

$$r^2 \frac{\partial^2 u}{\partial r^2} + r \frac{\partial u}{\partial r} + \frac{\partial^2 u}{\partial \theta^2} = 0. \qquad (7.103)$$

Equation (7.103) is the Laplace equation in a cylindrical coordinate system. Our interest is in knowing what sort of a temperature distribution we should expect to see in our partitioned tank, when the straight boundary is maintained at a constant temperature that acts as the reference, or the zero for the rest of the half tank, when the circular boundary is slowly brought to a particular, nonzero temperature distribution. It is important to note the qualifier "slowly" in the previous sentence, as if that process were performed rapidly, correct description of it would require that we account for the interaction between the temporal changes and the spatial changes, as set by the conductivity of the material, in this case, water. That would considerably alter the resulting spatial distribution of temperature (before and) at equilibrium, which we could model from start to finish using the diffusion equation.

For the present problem, however, all changes occur slowly, so the temporal derivative terms are negligible, and the spatial distribution can be evaluated using the Laplace equation model. We look for a solution $u(r,\theta)$ that is bounded in the domain $0 \le r \le R, 0 \le \theta \le \pi$, meaning that there exists a finite number M such that $u(r,\theta)$ is less than M over the entire region $0 \le \theta \le \pi$. The two spatial variables of relevance are r, the radial-position coordinate, and θ, the azimuthal or angular-position coordinate. Let us suppose that the tank radius is A, and that our tank half is the top half of the cylinder defined by the straight side and the circular side. The two intersect at $\theta = 0$ and at $\theta = \pi$. For the half-tank situation described earlier, we can specify the boundary conditions as

$$u(r,0) = 0, \quad u(r,\pi) = 0, \quad \text{for } 0 \le r \le A,$$
$$u(a,\theta) = f(\theta), \tag{7.104}$$

where $f(\theta)$ is some known function that matches the temperatures along the circular boundary. Moving forward with variable separation in the (r,θ) system, let us look for a solution of the form $u(r,\theta) = R(r)\Theta(\theta)$. A substitution of this expression into equation (7.103) provides

$$r^2\Theta(\theta)\frac{d^2R}{dr^2} + r\Theta(\theta)\frac{dR}{dr} + R(r)\frac{d\Theta}{d\theta^2} = 0. \tag{7.105}$$

Dividing through by $u(r,\theta)$ (knowing that we are only interested in nontrivial solutions), we see that

$$\frac{1}{R}\left(r^2\frac{d^2R}{dr^2} + r\frac{dR}{dr}\right) = -\frac{1}{\Theta}\frac{d^2\Theta}{d\theta^2} = \lambda^2. \tag{7.106}$$

We can see that λ^2 must be a constant independent of r and θ, since the two left sides must equal a constant if they must equal each other. Therefore, two ordinary differential equations result, one in r and the other in θ.

$$r^2\frac{d^2R}{dr^2} + r\frac{dR}{dr} - \lambda^2R = 0,$$
$$\frac{d^2\Theta}{d\theta^2} + \lambda^2\Theta = 0. \tag{7.107}$$

The second of equation (7.107) is straightforward to solve, given the boundary conditions (from the first of equation (7.104)),

$$\Theta(0) = 0, \quad \Theta(\pi) = 0. \tag{7.108}$$

The solution to the θ equation is

$$\Theta(\theta) = C_1 \cos \lambda\theta + C_2 \sin \lambda\theta. \tag{7.109}$$

The condition $\Theta(0) = 0$ implies that $C_1 = 0$, and the condition $\Theta(\pi) = 0$ implies that $\sin \lambda\theta = 0$. Therefore, we have

$$\lambda_n\theta = n\pi, \; n = 0, \; 1, \; 2,\ldots. \tag{7.110}$$

The eigenvalue $n = 0$ can be eliminated as it trivially leads to $\Theta(\theta) = 0$. The first of equation (7.107) is a Cauchy–Euler equation, which we have seen in Chapter 6. Here, we again use the helpful substitution

$$r = e^s. \tag{7.111}$$

It can be verified that a substitution of the relation (7.111) into

$$r^2\frac{d^2 R}{dr^2} + r\frac{dR}{dr} - n^2 R = 0, \tag{7.112}$$

with the finding from equation (7.110) that $\lambda = n$, leads to

$$\frac{d^2 R}{ds^2} - n^2 R = 0. \tag{7.113}$$

Equation (7.113) leads to the solution

$$R(s) = Ce^{ns} + De^{-ns}. \tag{7.114}$$

Given that, per our substitution, $r = e^s$, the solution is

$$R(r) = Cr^n + Dr^{-n} = Cr^n + \frac{D}{r^n}. \tag{7.115}$$

Since our solution must be bounded, the second term needs to be set to zero by choosing $D = 0$. This leaves

$$R(r) = r^n \tag{7.116}$$

within a constant. The overall solution can now be written as

$$u(r,\theta) = \sum_{1}^{\infty} A_n r^n \sin n\theta. \tag{7.117}$$

The constants A_n are determined by the remaining boundary condition (we have already used the first of equation (7.104) in deriving the θ solution)

$$f(\theta) = \sum_{1}^{\infty} A_n a^n \sin n\theta \tag{7.118}$$

so that, using the orthogonality of the sine function,

$$A_n = \frac{2}{a^n} \int_{0}^{\pi} f(\theta) \sin n\theta d\theta. \tag{7.119}$$

The overall solution for the temperature distribution we sought can now be written by substituting A_n from equation (7.119) into $u(r,\theta)$ of equation (7.117).

Let us consider another example, that of a thin circular membrane of radius a and thickness h that is stretched at a constant tension applied at the boundary. The density of membrane material is ρ. We would like to determine the natural response of this membrane to any initial conditions, knowing the boundary conditions, i.e., how the membrane is supported at its boundary. We begin with the equation of motion

$$\rho h \frac{\partial^2 w}{\partial t^2} - T\nabla^2 w = 0, \tag{7.120}$$

where we recall that the Laplacian $\nabla^2 w$ for our circular geometry is defined, in cylindrical polar coordinates, as

$$\nabla^2 w = \frac{\partial^2 w}{\partial r^2} + \frac{1}{r}\frac{\partial w}{\partial r} + \frac{1}{r^2}\frac{\partial^2 w}{\partial \theta^2} + \frac{\partial^2 w}{\partial z^2}. \tag{7.121}$$

Since the membrane is assumed to be thin, changes in w in the z direction are practically very small, and hence can be neglected. The last term in the expression in equation (7.121) can therefore be dropped, leaving just r and θ as the pertinent coordinates over which to quantify membrane oscillations here and in general. The membrane has an initial displacement and velocity, and it is clamped at the boundary. Thus, we have

$$w(r,\theta;0) = f(r,\theta), \quad \frac{\partial}{\partial t}(r,\theta;0) = 0,$$

$$w(a,\theta;t) = 0. \tag{7.122}$$

Next, dividing both sides of equation (7.120) by ρh, with the understanding that the tension T is the tension magnitude per unit length along the circumference, we are able to define a parameter c which has the significance and units of wave speed, even though as we have seen earlier, for a finite domain such as the present one, only standing waves (which we see as vibrations) are expected.

Next, we adopt a variable separation approach by first letting

$$w(r,\theta;t) = W(r,\theta)T(t). \tag{7.123}$$

A substitution into equation (7.121) leads to

$$W\frac{d^2 T}{dt^2} - c^2 T\nabla^2 W = 0. \tag{7.124}$$

Dividing through by a nontrivial $w(r,\theta;t)$ (i.e., nonzero) and rearranging,

$$\frac{1}{c^2 T}\frac{d^2 T}{dt^2} = \frac{1}{W}\nabla^2 W = -\lambda^2, \tag{7.125}$$

where λ is the separation constant. We are thus led to two ordinary differential equations,

$$\frac{d^2 T}{dt^2} + c^2\lambda^2 T = 0,$$

$$\nabla^2 W + \lambda^2 W = 0. \tag{7.126}$$

The first of equation (7.126) has a solution of the form

$$T(t) = A \cos \omega t + B \sin \omega t, \tag{7.127}$$

where we have defined $\omega = c\lambda$. A further separation of variables can be applied to $W(r,\theta)$, as

$$W(r,\theta) = R(r)\Theta(\theta). \tag{7.128}$$

This leads to the result,

$$\Theta \left(\frac{d^2 R}{dr^2} + \theta \frac{1}{r} \frac{dR}{dr} \right) + R \frac{1}{r^2} \frac{d^2 \Theta}{d\theta^2} + \lambda^2 R \Theta = 0. \tag{7.129}$$

Dividing through by $W(r,\theta) \neq 0$,

$$r^2 \frac{1}{R} \left(\frac{d^2 R}{dr^2} + \frac{1}{r} \frac{dR}{dr} + \lambda^2 R \right) = -\frac{1}{\Theta} \frac{d^2 \Theta}{d\theta^2} = \alpha^2. \tag{7.130}$$

We are thus led to two ordinary differential equations,

$$r^2 \frac{d^2 R}{dr^2} + r \frac{dR}{dr} + r^2 \lambda^2 R - \alpha^2 R = 0,$$

$$\frac{d^2 \Theta}{d\theta^2} + \alpha^2 \Theta = 0. \tag{7.131}$$

The second of equation (7.131) has a solution of the form

$$\Theta(\theta) = C_1 \cos \alpha \theta + C_2 \sin \alpha \theta. \tag{7.132}$$

Since $w(r,\theta;t)$ needs to be continuous throughout the membrane, $w(r,\theta;t) = w(r,\theta + 2n\pi;t)$. Therefore, $\Theta(\theta)$ needs to be a periodic function, meaning that α needs to be an integer, or $\alpha = n$, with $n = 0, 1 2, 3, \ldots, \infty$. The second of equation (7.131) then becomes

$$r^2 \frac{d^2 R}{dr^2} + r \frac{dR}{dr} + \left(\lambda^2 r^2 - n^2 \right) R = 0. \tag{7.133}$$

Equation (7.133) is an nth-order Bessel equation with a solution for each n, with $\lambda > 0$,

$$R_n(r) = D_n J_n(\lambda r) + E_n Y_n(\lambda r). \tag{7.134}$$

We recall that J_n represents the Bessel function of first kind and order n, while Y_n denotes the Neumann function of order n. Further, whereas J_n is continuous and differentiable throughout the interval $0 \leq r \leq R$, Y_n goes to infinity as $r \to 0$. Inclusion of Y_n in the solution would make $w(r,\theta;t)$ singular, nonintegrable, and nondifferentiable at $r \to 0$. For this reason, we force $E_n = 0$ and only limit our procedure to solutions with a nonzero D_n. This gives us, for a chosen n,

$$W_n(r,\theta) = D_n J_n(\lambda r) (C_{n1} \cos n\theta + C_{2n} \sin n\theta). \tag{7.135}$$

It follows then that

$$w_n(r,\theta;t) = D_n J_n(\lambda r) (C_{1n} \cos n\theta + C_{2n} \sin n\theta) (A \cos \omega t + B \sin \omega t). \tag{7.136}$$

If we now invoke the initial condition on the membrane velocity, i.e., $\partial w/\partial t = 0$ for each n, at $t = 0$, we must have $B = 0$, being the coefficient of the $\cos \omega t$ term in $\partial w/\partial t$. Next we apply the boundary condition at $r = s$, which requires that $w(a, \theta; t) = 0$. The only way this condition can be satisfied is if

$$J_n(\lambda a) = 0. \tag{7.137}$$

Now there is a countable infinity of points where J_n is zero, which we referred to as the zeros of J_n. Using the symbol j_{nm} to denote the mth zero of the nth-order Bessel function J_n,

$$\lambda a = j_{nm}, \Rightarrow \lambda_{nm} = \frac{j_{nm}}{a}; \; m = 1, 2, 3, \ldots, \infty. \tag{7.138}$$

λ_{nm} are the eigenvalues of w_n, with

$$\omega_{nm} = c\lambda_{nm}, \Rightarrow \omega_{nm} = \sqrt{T/\rho h} \lambda_{nm}; \; n = 0, 1, 2, 3, \ldots, \infty, m = 1, 2, 3, \ldots, \infty. \tag{7.139}$$

The overall solution can then be written as a double summation,

$$w(r, \theta; t) = \sum_{n=0}^{\infty} \sum_{m=1}^{\infty} J_n(\lambda_{nm} r) \cos \omega_{nm} t \left(C_{1nm} \cos n\theta + C_{2nm} \sin n\theta \right). \tag{7.140}$$

Recalling now the initial condition in equation (7.122) that $w(r, \theta; 0) = f(r, \theta)$,

$$C_{1nm} = \int_0^{2\pi} f(r, \theta) J_n(\lambda_{nm} r) \cos n\theta d\theta, \tag{7.141}$$

and

$$C_{2nm} = \int_0^{2\pi} f(r, \theta) J_n(\lambda_{nm} r) \sin n\theta d\theta. \tag{7.142}$$

Here, we have used the orthogonality of J_n and of $\cos n\theta$. The overall solution specific to the prescribed initial and boundary conditions can now be written, using C_{1nm} and C_{2nm} just determined. This will be the natural response for the deflection of the circular membrane at any point and at any time $t \geq 0$. For each eigenvalue λ_{nm}, we have a natural frequency ω_{nm} for membrane oscillations. Each eigenvalue λ_{nm} corresponds to eigenfunctions $J_n(\lambda_{nm} r) \cos n\theta$ and $J_n(\lambda_{nm} r) \sin n\theta$. $n = 0$, $m = 1, 2, 3, \ldots, \infty$ represent the symmetric oscillation modes with no θ dependence.

7.5.3   Response to Forcing

By way of illustration, let us consider a narrow rectangular membrane strip stretched between two supports that are held fixed. We have seen how we would evaluate the oscillations of the string that result from a given initial displacement. In the process, we were able to define the eigenfunctions and eigenvalues of the string for the specified boundary conditions. The response of the string was then written as a linear superposition of its eigenfunctions, and the constants attached to the eigenfunctions were determined by expanding the initial displacement function along the eigenfunctions, much as we would expand an arbitrary vector in space into its three Cartesian components.

The eigenfunction expansion approach can be used to find the response of a flexible strip of thickness h and material density ρ (kg/m^3) to an arbitrary forcing function distributed along the length of the string. Suppose that the distributed force $q(x,t)$ (in N/m^2) is a piecewise continuous function (i.e., integrable). A tension force T (in N/m) acts between the two ends. Then, the equation of motion can be written as

$$\rho h \frac{\partial^2 w}{\partial t^2} - T \frac{\partial^2 w}{\partial x^2} = q(x,t). \tag{7.143}$$

We have seen that the strip has eigenfunctions of the form, $\phi_n(x) = \sin n\pi x/L$. Let us expand the forced response $w(x,t)$ of the strip along the eigenfunctions ϕ_n as

$$w(x,t) = \sum_{m=1}^{\infty} C_m(t)\phi_m(x). \tag{7.144}$$

Similarly, we may expand $q(x,t)$ using the same eigenfunctions as

$$q(x,t) = \sum_{m=1}^{\infty} f_m(t)\phi_m(x), \tag{7.145}$$

where

$$f_m(t) = \frac{2}{L} \int_0^L q(x,t)\phi_m(x)dx. \tag{7.146}$$

Further, as we have seen, we can define a wave speed c in this case using

$$c^2 = \frac{T}{\rho h}. \tag{7.147}$$

Substituting the relations (7.144)–(7.147) into equation (7.143),

$$\sum_{m=1}^{\infty} \ddot{C}_m(t)\phi_m(x) + \sum_{m=1}^{\infty} c^2 \beta_m^2 \phi_m(x) = \sum_{m=1}^{\infty} f_m(t)\phi_m(x). \tag{7.148}$$

Here we have defined $\beta_m = m\pi/L$. Next, multiplying both sides by $\phi_n(x)$, integrating over $x = 0$ to $x = L$, and realizing that the eigenfunctions ϕ_n are orthogonal,

$$\ddot{C}_n(t) + \omega_n^2 C_n(t) = f_n(t). \tag{7.149}$$

If the applied load is harmonic, i.e., if $f_n(t) = F_n e^{i\omega t}$, then the amplitude C_{an} can be evaluated as

$$C_{na} = \frac{F_n}{-\omega^2 + \omega_n^2}, \tag{7.150}$$

and the complete response can be expressed as

$$w(x,t) = \sum_{n=1}^{\infty} \Re \left[\frac{F_n}{-\omega^2 + \omega_n^2} \sin \beta_n x e^{i\omega t} \right]. \tag{7.151}$$

It is easy to observe that resonance will occur for any forcing function whose frequency ω closely approaches one of the modal natural frequencies ω_n. Typically, the summation is truncated at a large-enough N that ensures that all physically meaningful modes have been captured in the summation.

7.6   Green's Functions for Finite Domains

One of the frequently used solution techniques used in the study of waves and bodies excited by ocean waves is the method of Green's functions. The method works if linear superposition applies; in other words, when wave amplitudes and oscillation amplitudes are both small, and can be described by linear differential equations. Here, we illustrate the development of Green's functions for a finite domain, keeping in mind that the same principles apply when domains are infinite, as we will see in Chapters 11 and 13.

We have already seen what an impulse-response function of a linear system is. It is basically the output or response of a linear system such as a spring-mass-damper system to an impulsive load applied somewhere on the system. The independent variable is time, and the dependent variable is the system response. Using the impulse-response function, we can synthesize the system's response to an arbitrary excitation (as long as it is physically realizable and expressible using functions that are integrable) via convolution, where the impulse response function forms one of the two functions appearing in the convolution, the excitation function being the other.

A Green's function in a sense generalizes the idea of an impulse-response function to a system with two or more independent variables, such as a string, rod, membrane, or a plate, or the water surface in the ocean. Thus, in addition to the load being concentrated into a very small instant of time, it is also concentrated to a very small point within the domain. The Green's function for a system is just the response at any point and at any time to an impulsive point load applied somewhere on the system. Here we note that a force that acts only for an instant, as represented by a Dirac delta function, is a singularity in the time domain. Similarly, a force that acts only at a point (as represented by a Dirac delta function in space) is also a singularity. For body motion problems, the singularity may take the form of a point force or a point moment. For fluid flow problems, the singularity may take the form of a point source.

As an illustration of the Green's function approach, let us again consider the narrow thin rectangular flexible strip in equation (7.144). Letting $G(x, \xi; t, \tau)$ be the response of the vertical deflection of the strip at point x and time t in response to an impulsive point load in the form of a Dirac delta function located at $x = \xi$ and applied at $t = \tau$, we have

$$\frac{\partial^2 G}{\partial t^2} - c^2 \frac{\partial^2 G}{\partial x^2} = \delta(x - \xi)\delta(t - \tau). \tag{7.152}$$

Let us now express the Green's function to be determined here for a pair (ξ, τ) as a superposition of the natural modes $\phi_n(x)$. In other words, let

$$G(x,\xi;t,\tau) \sum_{m=1}^{\infty} C_m(t)\phi_m(x). \tag{7.153}$$

Then,

$$\sum_{m=1}^{\infty} \ddot{C}_m(t)\phi_m(x) + \sum_{m=1}^{\infty} c\beta_m^2 \phi_m(x) = \delta(x-\xi)\delta(t-\tau). \tag{7.154}$$

Now multiplying both sides by $\phi_n(x)$, integrating over x, applying the orthogonality of ϕ_n, and utilizing the sifting property of the Dirac delta function,

$$\ddot{C}_n + \omega_n^2 C_n = \frac{2}{L}\phi_n(\xi)\delta(t-\tau). \tag{7.155}$$

If we include a small dissipation in equation (7.155) and introduce a modal damping ratio ζ_n,

$$\ddot{C}_n + 2\zeta_n\omega_n C_n + \omega_n^2 C_n = \frac{2}{L}\phi_n(\xi)\delta(t-\tau). \tag{7.156}$$

The second-order differential equation for C_n can be solved using methods we have already discussed. We find then that

$$C_n(t) = Ce^{-\zeta_n\omega_n t}\cos(\omega_{dn}t - \theta_n), \tag{7.157}$$

where

$$\omega_{dn} = \omega_n\sqrt{1-\zeta_n^2}; \quad \tan\theta_n = \frac{2\zeta_n\omega_n}{-\omega^2 + \omega_n^2}. \tag{7.158}$$

Setting $C = 1$ without loss of generality, we now have the Green's function for the finite strip as

$$G(x,\xi;t,\tau) = e^{-\zeta_n\omega_n t}\cos(\omega_{dn}t - \theta_n)\phi_n(\xi). \tag{7.159}$$

The Green's function can now be used to evaluate the response $w(x,t)$ to an arbitrary load $q(x,t)$ as

$$w(x,t) = \int_0^L \int_0^t q(x-\xi,t-\tau)G(x,\xi;t,\tau)d\tau d\xi. \tag{7.160}$$

We see the integral in equation (7.160) as a spatiotemporal convolution integral that utilizes the response of our system to a spatiotemporal impulse function to construct the response to an arbitrary excitation. This overall idea is later generalized to infinite domains, where we lose the benefit of having discrete eigenfunctions and eigenvalues to work with and use Fourier transforms instead.

7.7   Concluding Remarks

In this chapter, we considered some of the commonly occurring partial differential equations important to almost all disciplines where multiple independent variables are involved. We focused on linear partial differential equations, stressing features particular to parabolic, elliptic, and hyperbolic equations. The first-order temporal derivative for parabolic equations represents a diffusion or conduction type behavior in which no temporal oscillations are produced. The spatial behavior is captured using sin and cos functions for rectangular domains, while other spatial variations are represented by functions such as ρ^n and $J_n(\alpha r)$. Hyperbolic equations on the other hand have a second-order temporal derivative, in addition to a second-order spatial derivative (in some cases, for instance, for thin plates, fourth-order spatial derivatives are involved), and in this case, oscillatory behavior is seen in both the temporal and the spatial domains. We also studied elliptic equations, which we here used to represent steady-state behavior as t tends to ∞ when all temporal changes had gone to zero. Boundary conditions then determined the distribution of the variables being modeled (e.g., temperature). In all three cases, we used the variable separation technique. The solutions we obtained contained infinite series, which could be considered as Fourier series expansions with the different eigenfunctions forming the basis functions, much the same way as the sin and cos functions do in "standard" trigonometric Fourier series expansions.

It should be emphasized that this chapter on partial differential equations was devoted to finite domains, for which boundary conditions played a prominent role. Indeed, we found that they define spatial eigenfunctions with corresponding eigenvalues. We found that there is a countable infinity of eigenvalues and eigenfunctions. In the case of hyperbolic equations, we found that the spatial eigenvalues were closely related to temporal natural frequencies. Even though we could define "wave speeds" for these problems, the boundary conditions ensured that we would only observe standing waves, which for structures like beams, rods, membranes, and plates can be interpreted as vibrations. The eigenfunctions then define mode shapes. Each mode shape then corresponds to particular natural frequency, and vice versa. Finally, we used eigenfunction (modal) expansions to find the forced response of an example system, and extended that approach to determine the Green's function as a step toward evaluating the forced response to arbitrary excitation.

7.8   Self-Assessment

7.8.1

Write down the analytical models for the following situations:

(a) Pollutant transfer over a long, narrow tube of length L, given some initial distribution.
(b) A pressure wave traveling through the tube above.
(c) Steady-state temperatures if one of the tube ends were maintained at a known temperature.
(d) In each case, develop the solutions.

7.8.2

Find the eigenvalues and eigenfunctions for the tube in situation (b) above.

7.8.3

During a ship manufacturing operation, heat is applied at the center of a cold rectangular plate at a constant rate for a given length of time so that the temperature there rises steadily while the heat is being applied. No heat is allowed to flow through any of the boundaries. How would we describe the temperature distribution over the plate while heat is being applied, shortly after heat application stops, and a long time afterward.

7.8.4

Find and describe the response of a circular membrane of radius R and stretched over a fixed boundary. The membrane is being excited by a force of unit magnitude distributed over a circular area of radius a around a location (r_0, θ_0) on the membrane. Assume that the excitation is oscillatory at frequency ω.

8 Partial Differential Equations-II

Our discussion so far has centered around three primary classes of partial differential equations: (i) parabolic, (ii) elliptic, and (iii) hyperbolic. Thus, we have had a chance to study the diffusion equation, Laplace equation, and the wave equation. In all of our work, so far, our interest was in learning how the physical processes unfolded within a finite spatial domain, and we knew the values of our quantities of interest or their rates of change at the domain boundaries. This knowledge in turn enabled us to look for, and expect to find, particular types of behavior over the rest of the domain. In other words, our knowledge of the boundary conditions allowed us to evaluate eigenfunctions and eigenvalues specific to the known boundary conditions and the process being modeled. The eigenfunctions could be used as "basis functions" in function space, and so a linear superposition of these functions (as many as needed for the required precision) could be used to express more general functions that satisfied the partial differential equation and the boundary conditions. The eigenvalues expressed the spatial frequencies (i.e., the number of repetitions) underlying the spatial patterns expressed by the eigenfunctions.

The phenomena surrounding marine applications occur over spatial scales large enough to be treated as infinite. Waves traveling through the open ocean are one example, where the domain is infinite, both on the surface and often below the surface. Pollutants dumped into the ocean thus almost have an infinity of space to spread over. How does our phenomenon behave at the boundaries? For that matter, where are the boundaries? In many cases, the best we can do is to say that the boundaries are at infinity, and expect to find solutions that will have a certain asymptotic behavior at infinity. For instance, when a body oscillates on the water surface and radiates waves, we know the waves will travel all the way to infinity, if only asymptotically. What we know for certain is that waves radiating from the body travel outward and away from the body. Hence, infinitely far along the water surface, those waves must still be outward propagating. The seafloor is often nearly an infinite depth below the water surface, but still, no water particle can move through it. In other words, the water particle velocity component through the seafloor is zero. Unless the seafloor is at a depth small enough to be considered finite, it does not make sense to look for eigenvalues and eigenfunctions in the z direction. Yet we know that we can model and quantify surface wave motion in the deep ocean. This chapter considers some of the solution techniques that can be used in infinite domains for the partial differential equations we have considered in Chapter 7, as a preparation for the type of equations we come across in marine applications in Chapters 12 and 13. We begin with the parabolic (diffusion) equation, moving on then to the elliptic (Laplace) equation, before taking up the hyperbolic (wave) equation in infinite domains. Most of our treatment in the following text is based on Greenberg (1978) and Graff (1991).

8.1   Diffusion Equation in an Infinite Domain

Here, we consider a diffusion equation that models the spread of a pollutant in a thin, infinitely long tube. The problem is one-dimensional in space, and hence the solution steps are easier to visualize. More general, three-dimensional diffusion will require the use of all three spatial coordinates, and is not treated here. With $u(x,t)$ defining the concentration/density of the pollutant, we write the one-dimensional diffusion equation as

$$\frac{\partial u}{\partial t} = k\frac{\partial^2 u}{\partial x^2}, \quad -\infty < x < \infty. \tag{8.1}$$

We let both u and $\partial u/\partial x$ approach zero as x approaches either infinity. Further, we know the initial functional variation of u at the start of the process. Thus,

$$u \to 0, \frac{\partial u}{\partial x} \to 0, \quad x \to \pm\infty,$$

$$u(x,0) = f(x). \tag{8.2}$$

Because the spatial domain is infinite, we can start by converting the second-order spatial derivative into algebraic expressions, so as to turn the partial differential equation into an ordinary differential equation. Thus,

$$\int_{-\infty}^{\infty} \frac{\partial u}{\partial t}e^{-i\beta x}dx = k\int_{-\infty}^{\infty} \frac{\partial^2 u}{\partial x^2}e^{-i\beta x}dx. \tag{8.3}$$

We use the symbol β here for spatial frequency to distinguish it from the oft-used ω for temporal frequency. Using $u\hat{u}(i\beta,t)$ to denote the spatial Fourier transform of $u(x,t)$ and using the properties of Fourier transforms of derivatives, we find that Fourier transformation reduces equation (5.89) to an ordinary differential equation written as

$$\frac{d\hat{u}}{dt} = -\kappa^2\beta^2\hat{u}, \tag{8.4}$$

where we have introduced $\kappa^2 = k$ for convenience. Equation (8.4) is a first-order ordinary differential equation with a constant coefficient, and has a solution,

$$\hat{u}(\beta,t) = Ae^{-\kappa^2\beta^2 t}. \tag{8.5}$$

We can evaluate the constant A using the initial condition for this problem. Since $f(x)$ is specified in the x-domain while $\hat{u}$ in equation (8.5) is written in the spatial-frequency or β-domain, we must either inverse Fourier transform the solution $\hat{u}(t)$ back into the x-domain, or transform the initial condition $f(x)$ into the β-domain. Assuming that the latter option is more straightforward for the specified $f(x)$, we use

$$\hat{f}(\beta) = \int_{-\infty}^{\infty} f(x)e^{-i\beta x}dx \tag{8.6}$$

to evaluate the constant A, which is simply

$$A = \hat{f}(\beta). \tag{8.7}$$

Inserting $\hat{f}(\beta)$ for A and applying the inverse Fourier transform on $\hat{u}$,

$$u(x,t) = \frac{1}{2\pi} \int_{-\infty}^{\infty} \hat{u}(\beta,t)e^{i\beta x}d\beta. \tag{8.8}$$

Using equations (8.6) and (8.7), we can rewrite equation (8.8) as

$$u(x,t) = \frac{1}{2\pi} \int_{-\infty}^{\infty} \hat{f}(\beta)e^{-\kappa^2\beta^2 t}e^{i\beta x}d\beta. \tag{8.9}$$

Multiplying the two exponential terms together,

$$u(x,t) = \frac{1}{2\pi} \int_{-\infty}^{\infty} \hat{f}(\beta)e^{i\beta x - \kappa^2\beta^2 t}d\beta. \tag{8.10}$$

We recognize here that $\hat{f}(\beta)$ is as defined by equation (8.6). To avoid confusion, we need to introduce a dummy variable ξ that stands in for the x in equation (8.6). This is a reasonable step, as ξ defines a different, self-contained operation that is independent of the inverse Fourier transformation of equation (8.8). Thus, we now have

$$u(x,t) = \frac{1}{2\pi} \int_{-\infty}^{\infty}\int_{-\infty}^{\infty} f(\xi)e^{-i\beta\xi}d\xi e^{i\beta x - \kappa^2 t}d\beta. \tag{8.11}$$

Since ξ and x vary independently of each other, the two integrations are independent of each other and both have limits independent of one another, the two integrals can be interchanged. $u(x,t)$ can therefore be reexpressed as

$$u(x,t) = \frac{1}{2\pi} \int_{-\infty}^{\infty} f(\xi)d\xi \int_{-\infty}^{\infty} e^{-i\beta(x-\xi)-\kappa^2\beta t}d\beta. \tag{8.12}$$

From Fourier transform tables such as can be found in Gradshteyn and Ryzhik (1994), we can write

$$\int_{-\infty}^{\infty} e^{-i\beta(x-\xi)-\kappa^2\beta t}d\beta = \frac{e^{-(\xi-x)^2/4\kappa^2 t}}{2\kappa\sqrt{\pi t}}. \tag{8.13}$$

We can define the quantity on the right side as a function $U(\xi - x)$. $u(x,t)$ can now be written as

$$u(x,t) = \int_{-\infty}^{\infty} f(\xi)U(\xi - x,t)d\xi. \tag{8.14}$$

Equation (8.14) in fact represents a convolution integral. Hence, if the initial condition $f(x)$ is the Dirac delta function prescribed as

$$f(x,0) = \delta(x), \qquad (8.15)$$

$$u(x,t) = U(x,t), \qquad (8.16)$$

$U(x,t)$ represents the impulse-response function of our system.

8.2   Laplace Equation on a Semi-Infinite Slab

It is interesting to consider the steady-state distribution of the quantity of interest, say temperature over a slab, if one of its sides is maintained at a particular temperature distribution with the help of an insulation (i.e., isothermal boundary condition). It is assumed, however, that the slab is large, long enough in one direction to be assumed to be infinite in extent. It is also assumed to be infinite in the perpendicular direction, though because we are able to hold one of its edges at particular temperature distribution, we consider it to be semi-infinite in the direction perpendicular to that edge. These conditions could approximate selective heating of plates assembled for a ship hull in order to relieve residual stresses induced during automated welding operations.

Assuming that the slab is infinite in the x direction and semi-infinite in the y direction (Greenberg 1978), we have (see Figure 8.1)

$$\nabla^2 u = \frac{\partial^2 u}{\partial x^2} + \frac{\partial^2 u}{\partial y^2} = 0, \qquad (8.17)$$

subject to the boundary condition,

$$u(x,0) = f(x), -\infty < x < \infty. \qquad (8.18)$$

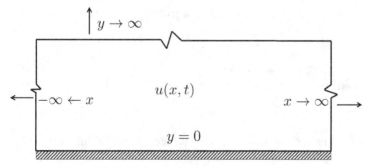

Semi-infinite slab

Figure 8.1 Schematic showing a semi-infinite slab; infinite in the x direction and infinite in the $y > 0$ direction only. $u(x,t)$ denotes the temperature distribution over the slab.

Since the domain is infinite in the x direction, we may take the Fourier transform of equation (8.17) (i.e., all terms therein) to find

$$\int\limits_{-\infty}^{\infty} \frac{\partial^2 u}{\partial x^2} e^{-ikx}\, dx + \int\limits_{-\infty}^{\infty} \frac{\partial^2 u}{\partial y^2} e^{-ikx}\, dx = 0. \tag{8.19}$$

One integration by parts of the first term leads to

$$\frac{\partial u}{\partial x} e^{-ikx} \Big|_{-\infty}^{\infty} - \int\limits_{-\infty}^{\infty} -ik \frac{\partial u}{\partial x} e^{-ikx}\, dx. \tag{8.20}$$

Here, we assume that $\partial u/\partial x(x, y) \to 0$, as $\to \pm\infty$, and hence the boundary term goes to zero. Another integration by parts of the second term produces

$$-\left[-ikue^{-ikx} \Big|_{-\infty}^{\infty} - \int\limits_{-\infty}^{\infty} -k^2 u e^{-ikx}\, dx \right]. \tag{8.21}$$

Assuming further that $u(x, y) \to 0$ as $x \to \pm\infty$, we find that the boundary term again vanishes. We are, therefore, left with

$$\int\limits_{-\infty}^{\infty} \frac{\partial^2 u}{\partial x^2} e^{-ikx}\, dx = -k^2 \hat{u}(ik, y). \tag{8.22}$$

The Fourier transformation in equation (8.19) thus results in

$$-k^2 \hat{u} + \frac{d^2 \hat{u}}{dy^2} = 0. \tag{8.23}$$

A straightforward solution to equation (8.23) is

$$\hat{u}(y) = A e^{|k|y} + B e^{-|k|y}. \tag{8.24}$$

We further assume that $u(x, y) \to 0$ as $y \to \infty$, its Fourier transform with respect to x also approaches zero as $y \to \infty$. We can, therefore, see that, since the exponential of large negative numbers asymptotically approaches zero, we must have $A = 0$, leaving

$$\hat{u}(y) = B e^{-|k|y}. \tag{8.25}$$

Fourier transforming equation (8.18), we have $\hat{u}(0) = \hat{f}(ik)$. This implies that $B = \hat{f}(ik)$. Hence,

$$\hat{u} = \hat{f}(ik) e^{-|k|y}. \tag{8.26}$$

Inverse Fourier transforming both sides, we see that the product becomes a convolution in the x-domain. The inverse Fourier transform of the exponential is (from tables such as in Gradshteyn and Ryzhik (1994))

$$\frac{1}{2\pi} \int\limits_{-\infty}^{\infty} e^{-|k|y} e^{ikx}\, dk = \frac{y}{\pi(x^2 + y^2)}. \tag{8.27}$$

Since the inverse Fourier transform of $\hat{f}(k)$ is just $f(x)$, we have

$$u(x,y) = \frac{y}{\pi} \int_{-\infty}^{\infty} U(\xi - x)f(\xi)d\xi, \tag{8.28}$$

where

$$U(\xi - x, y) = \frac{y}{\pi(\xi - x)^2 + y^2}. \tag{8.29}$$

$U(x,y)$ and equation (8.28) enable the temperature distribution over the semi-infinite slab to be evaluated given the boundary condition that the temperature is to be held such that it is an arbitrary function of x.

8.3   Wave Equation on an Infinite Domain (One-Dimensional)

In this section, we consider the hyperbolic wave equation as defined on an infinite domain. Once again, we point out that ocean surface waves are in general not amenable to this description. However, the insights we gain from the present study are valuable in understanding ocean surface-wave propagation. For clarity's sake, we will first consider an infinite string. Recall that the wave equation for a string is a hyperbolic equation. Written for an infinite domain, it looks like

$$\frac{\partial^2 w}{\partial t^2} - \frac{T}{\rho} \frac{\partial^2 w}{\partial x^2} = 0, \quad -\infty < x < \infty; \ -\infty < t < \infty. \tag{8.30}$$

Note that both x and t vary between $\pm\infty$, which indicates that the propagation occurs over an infinite length, and that it "has been going on forever." Equation (8.30) in fact contains two wave equations,

$$\frac{\partial w}{\partial t} \pm c \frac{\partial w}{\partial x} = 0, \ c = \sqrt{T/\rho}, \ -\infty < x < \infty, \ -\infty < t < \infty. \tag{8.31}$$

The equation with the positive sign on c is the right-propagating wave for both $x < 0$ and $x > 0$. The right-propagating wave can be represented using a differentiable function $f(x - ct)$. Direct substitution into the first of equations (8.31) (i.e., with the positive sign on c) shows that f satisfies the equation. With $g(x + ct)$ representing a left-traveling wave, a general solution to equation (8.30) can be expressed as

$$W(x,t) = \frac{1}{2} \left[f(x - ct) + g(x + ct) \right], \tag{8.32}$$

It is interesting to note that the right-propagating wave $f(x - ct)$ is an incoming wave to an observer at $x = 0$ facing into the plane of the paper and looking leftward. It is an outgoing wave to the same observer if looking rightward. Conversely, the left-propagating wave $g(x + ct)$ is an incoming wave to our observer when looking to the right and an outgoing wave when looking to the left.

Next, let us consider a waveform described by a function $w(x,t) = \cos(kx - \omega t)$. This is a right-traveling waveform, since as t increases, x must increase, for any particular point on the waveform. If we substitute this w in the wave equation for right-traveling waves, we find

$$\frac{\partial w}{\partial t} + c\frac{\partial w}{\partial x} = \omega\sin(kx - \omega t) - ck\sin(kx - \omega t) = 0. \tag{8.33}$$

For equation (8.33) to be satisfied for arbitrary x and t combinations, we must have

$$\omega - ck = 0. \tag{8.34}$$

ω is the temporal frequency and k is the spatial frequency (or wave number). The two are related together by the *dispersion relation* $\omega - ck = 0$. This relation tells us that the wave speed c, or *phase velocity* c_p, is given by

$$c_p = \omega/k = c. \tag{8.35}$$

If we write $\omega = ck$ and differentiate both sides with respect to k,

$$c_g = \frac{d\omega}{dk} = c. \tag{8.36}$$

c_g as defined in equation (8.36) is the *group velocity* of the traveling waves. A single wave such as described using a single combination of k and ω travels at a phase velocity c_p. A wave group comprised of multiple wave frequencies travels at the group velocity c_g. A rectangular pulse, for instance, represents a group of multiple frequencies and amplitudes (as we have seen in the context of Fourier transforms). If the group velocity equals the phase velocity, all wave components comprising the pulse will travel at the same speed as the group defining the pulse. The pulse will therefore retain its shape exactly as it was defined at $t = 0$. On the other hand, if the group velocity differed from the phase velocity, the individual wave components defining the pulse at $t = 0$ will all travel at different velocities and the pulse will lose its shape.

When $c_g = c_p$, thus, propagation is nondispersive, shapes as initially defined are retained through the propagation process. When $c_g \neq c_p$, propagation is dispersive, and shapes as initially defined break up over time. For waves propagating over an infinite string, we find that the phase velocity and group velocity are equal. We next consider wave propagation on infinite beams. We will find that traveling flexural waves are dispersive. We will also see in Chapter 12 that water waves are dispersive except in shallow waters where long waves become nondispersive.

A beam resting on ground is an interesting model for railroad tracks through level country (see Figure 8.2). The propagation dynamics are very different from those describing water waves. The remarkable point to note, however, is that the general solutions for waves on beams resemble the solution we tested in equation (8.33). They are also similar to the general solutions that describe surface water waves. The ground on which the beam is resting typically has some compliance and can be assumed to have finite flexibility. With κ denoting the stiffness of the ground and $w(x,t)$ denoting the vertical displacement of the beam at any point x and time t, the vertical deflection of the beam can be described using

$$\rho A\frac{\partial^2 w}{\partial t^2} + EI\frac{\partial^4 w}{\partial x^4} + \kappa w = 0; \quad -\infty < x < \infty, \ -\infty < t < \infty. \tag{8.37}$$

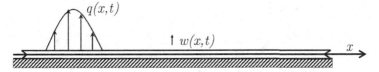

Infinite beam on ground (assumed elastic).

$q(x,t) = 0$, for dispersion relation.

Figure 8.2 Schematic showing a long beam idealized as an infinitely long beam. The beam rests on a foundation (e.g., level ground) that is assumed to be elastic, having a stiffness κ in N/m/m^2 (i.e., N/m per unit area).

Note that this equation is of second order in time and fourth order in space. The ground stiffness is assumed to be constant at κ. ρ and A are the volumetric density of the beam material and the cross-sectional area of the beam, respectively. E is the Young's modulus of elasticity, and I is the second moment of the beam cross section. A uniform, isotropic beam of constant cross-sectional area is thus assumed. We look here for the conditions under which the beam on elastic ground will support propagating wave motion. Let us examine the behavior of the system in equation (8.37) when excited by a wave of the form

$$w(x,t) \sim e^{-i(kx-\omega t)}. \tag{8.38}$$

A direct substitution into the three terms of equation (8.37) leads to the condition that, for a function like $w(x,t)$ in equation (8.38) to be a solution, the following condition needs to be satisfied:

$$- \rho A \omega^2 + EIk^4 + \kappa = 0. \tag{8.39}$$

This in fact is our dispersion relation for wave propagation along a beam on an elastic foundation. To clarify thoughts, if we consider the more straightforward situation without the elastic foundation, we can see easily that

$$\omega^2 = \frac{EI}{\rho A}k^4. \tag{8.40}$$

This is clearly a nonlinear dispersion relation, so we can expect that a wave group will propagate differently than a single wave. We can see that

$$\frac{\omega}{k} = c_p = \sqrt{\frac{EI}{\rho A}}k. \tag{8.41}$$

The phase velocity c_p thus depends on k, and in fact, for these waves, the longer waves travel slower than the shorter waves (c_p being proportional to $k = 2\pi/\lambda$). Taking the derivative of both sides of equation (8.40) with respect to k,

$$\omega \frac{d\omega}{dk} = 2\frac{EI}{\rho A}k^3. \tag{8.42}$$

From equation (8.40) we have

$$\frac{EI}{\rho A} = \frac{\omega^2}{k^2}. \tag{8.43}$$

Making this substitution in the right-hand side of equation (8.42) and canceling appropriate terms,

$$\frac{d\omega}{dk} = 2\frac{\omega}{k}. \tag{8.44}$$

Realizing that the derivative on the left side is just the group velocity c_g for the waves in equation (8.38), we find that, here,

$$c_g = 2c_p. \tag{8.45}$$

For the beam, then, the wave group travels twice as fast as the individual component waves. Relative to the wave group, the individual component waves appear to travel in the opposite direction. Waves traveling on a beam are thus definitely dispersive.

How does the elastic foundation complicate the propagation (Steele 1994)? Returning to equation (8.39),

$$\omega = \sqrt{\left[\frac{EI}{\rho A}k^4 + \frac{\kappa}{\rho A} \right]}. \tag{8.46}$$

The beam now has an interesting behavior, in that when no waves exist, i.e., when $k = 0$, there is still a nonzero frequency ω_0 (contrast this with what we would find from equation (8.40)), where

$$\omega_0 = \sqrt{\frac{\kappa}{\rho A}}. \tag{8.47}$$

This is the natural frequency of the beam behaving as a rigid body oscillating on a stiffness κ. For the type of waves we see in equation (8.38), propagation is only possible when $\omega > \omega_0$. If an oscillating source on such a beam oscillated at a frequency $\omega < \omega_0$, the beam oscillations in response will be confined to a small region around the location of this source, and these oscillations are trapped in that region.

The story of wave propagation is thus an exciting one, and as we see here, all that is needed in order to capture it is the partial differential equation that describes the dynamics of the motion.

8.4   Infinite Domains and the Method of Stationary Phase

We continue our discussion of wave propagation over infinite domains, but now turn our attention to an important approximation technique that is frequently used with ocean surface waves as they pertain to marine applications. This is the method of stationary phase, originally applied by Kelvin (Thomson 1887), based on an approximation used by Stokes (1883). Our discussion here is based on the treatment of Graff (1991). The method of stationary phase is particularly applicable to dispersive waves. Dispersive waves exhibit a group behavior because waves at different frequencies travel at different speeds. Thus, as we have seen before, an impulse at any point will break down into a wider packet, which will eventually separate further as different wave components travel away at different speeds. When a structure or a fluid is subjected to excitation that consists of a wide range of frequencies, it becomes interesting to see how the waves arising from that excitation will resolve themselves far away from the source of excitation. Situations such as these arise frequently in marine applications, particularly in the context of floating body oscillations. Often, when a body oscillates over a wide frequency band, such as may be excited by a wave group impinging on it, and with its oscillations radiates waves on the water surface, we are often interested in knowing what the radiated waves look like far away from the body, i.e., in the far field. This is where the method of stationary phase proves useful (e.g., Mei 1992). For illustration, we consider a general oscillation response in an infinite beam that is expressed as an inverse Fourier transform,

$$w(x,t) = \frac{1}{2\pi} \int\limits_{-\infty}^{\infty} W(\omega)e^{-i(k(\omega)x - \omega t)}d\omega. \tag{8.48}$$

We consider the exponent in the exponential term further, and rewrite it slightly,

$$-i(k(\omega)x - \omega t) = -ix\left(k(\omega) - \omega\frac{t}{x}\right) = -xh(\omega). \tag{8.49}$$

In the far field where $x \to \infty$, and a long time into the future, $\frac{t}{x} \to 1$, so the use of the notation $h(\omega)$ makes sense. If we now consider the far-field integral,

$$\lim_{x\infty} w(x,t) = \frac{1}{2} \int\limits_{-\infty}^{\infty} W(\omega)e^{-ixh(\omega)}d\omega, \tag{8.50}$$

and in particular, write out the real part of the term in the exponential, we have

$$\Re\left[e^{-ixh(\omega)}\right] = \cos(xh(\omega)). \tag{8.51}$$

We can see that as x becomes larger, the argument of the cos term becomes larger and larger so that $\cos(xh(\omega))$ exhibits more and more rapid swings from maximum to minimum, effectively implying that within an integral over ω, the swings will cancel out and contribute very little to the integral. Except when $h(\omega)$ happens to be such that $xh(\omega)$ varies very slowly, and is in fact close to a valley that makes $\cos xh(\omega) \to 1$. For a small range of $xh(\omega)$ on either side of the lowest point of this valley or a peak, the function $\cos xh(\omega)$ undergoes slow changes (see Figure 8.3). This is the region that will make the dominant contribution to the integral. In fact, for an arbitrarily large x, it is really $h(\omega)$ that determines how $\cos xh(\omega)$ will vary. Thus, for $h(\omega)$ to be at a valley or a peak where $\omega = \omega_0$, we expect that

$$\frac{dh(\omega)}{d\omega} = 0, \ \omega = \omega_0. \tag{8.52}$$

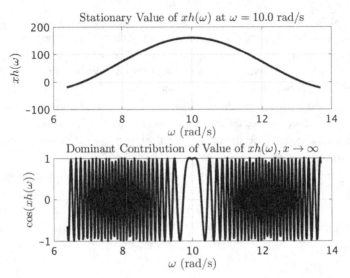

Figure 8.3 Illustration demonstrating the justification for the stationary phase approximation. Note how the contributions away from the stationary point will cancel each other out.

Because $\cos(xh(\omega))$ varies slowly with ω near $\omega = \omega_0$, it is reasonable to approximate $h(\omega)$ using a Taylor-series expansion about $\omega = \omega_0$. Thus,

$$h(\omega) = h(\omega_0) + \frac{dh}{d\omega}(\omega - \omega_0) + \frac{1}{2!}\frac{d^2h(\omega)}{d\omega^2}(\omega - \omega_0)^2 + \ldots, \qquad (8.53)$$

listing terms up to second order. In light of equation (8.52), $h(\omega)$ then approximates to

$$xh(\omega) \approx xh(\omega_0) + \frac{1}{2}x\frac{d^2h(\omega)}{d\omega^2}(\omega - \omega_0)^2. \qquad (8.54)$$

In light of equation (8.54), it is seen that in the far field, only the behavior around $\omega = \omega_0$ remains significant. Hence, the far-field approximation for the integral in equation (8.50) can be written as

$$w(x,t) \approx \int_{-\infty}^{\infty} W(\omega_0) \exp\left[ixh(\omega_0) + \frac{1}{2}ix\frac{d^2h(\omega)}{d\omega^2}(\omega - \omega_0)^2\right] d\omega. \qquad (8.55)$$

This reduces to

$$w(x,t) = W(\omega_0)\exp[ixh(\omega_0)] \int_{-\infty}^{\infty} \exp\left[\frac{1}{2}ix\frac{d^2h(\omega_0)}{d\omega^2}(\omega - \omega_0)^2 d\omega\right]. \qquad (8.56)$$

The integral on the right can be found using integration tables such as given in Gradshteyn and Ryzhik (1994),

$$\int_{-\infty}^{\infty} \exp\left[\frac{1}{2}ix\frac{d^2h(\omega_0)}{d\omega^2}(\omega - \omega_0)^2 d\omega\right] = \sqrt{\left[\frac{2\pi}{xh''}\right]}\exp\left(\frac{i\pi}{4}\right). \qquad (8.57)$$

The far-field solution for $w(x,t)$ can now be expressed as

$$w(x,t) = \sqrt{\left[\frac{2\pi}{xh''}\right]} W(\omega_0) \exp\left[ixh(\omega_0) + \frac{i\pi}{4}\right], \qquad (8.58)$$

where we have dropped the $\approx$ qualification with the understanding that it is the far-field evaluation that is of interest. We have also used the double prime to represent the second derivative. Substituting back for $h(\omega)$, we find first that

$$\frac{dh}{d\omega} = 0 = -\frac{1}{ix}\left(\frac{dk}{d\omega} - \frac{t}{x}\right) = 0, \ \omega = \omega_0. \qquad (8.59)$$

Equation (8.59) thus suggests that

$$\frac{d\omega}{dk} = \frac{x}{t}, \qquad (8.60)$$

thereby defining group velocity c_g. This also implies that the part of the disturbance that arrives at a point $x \to \infty$ does so at the group velocity c_g and consists of the dominant frequency ω_0 and nearby frequencies. The full expression for the far-field solution can be written by substituting back for $h(\omega)$ as

$$w(x,t) = -\frac{2\pi i \omega_0 e^{i\pi/4}}{xk''} W(\omega_0) e^{-i(k(\omega_0)x - \omega_0 t)}. \qquad (8.61)$$

A further substitution $-i = \exp(-i\pi/2)$ in equation (8.61) leads to the final result,

$$w(x,t) = \frac{2\pi e^{-i\pi/4}}{xk''} W(\omega_0) e^{-i(k(\omega_0)x - \omega_0 t)}. \qquad (8.62)$$

8.5   Concluding Remarks

We continued here our discussion of the three representative partial differential equations, representing different classes of physical processes, namely diffusion (and/or conduction), steady-state spatial distributions driven by boundary conditions, and wave propagation. Our interest was in phenomena and changes that occur over large domains, or domains that can be considered large relative to the spatial scales of interest for particular phenomena or their source of excitation. Indeed, our domains were large enough to be approximated as infinite in extent. Examples include diffusion of pollutants dumped mid-ocean, heat transfer in large water reservoirs or lakes wherein one portion of the shoreline is held at a constant temperature, flexural waves propagating over long railroad tracks, and so on. In problems where some parts of the boundary of an otherwise "infinitely large" region were accessible, the domains were thought to be semi-infinite. Although none of the geometries just mentioned are strictly speaking infinite, they can be assumed to be infinite relative to the spatial scales of interest.

It is important to realize that when the domains here are infinite, we no longer have boundary conditions, i.e., values or variations of the phenomena of interest at domain boundaries. Recall that we used the known boundary conditions to arrive at eigenvalues and eigenfunctions for the problem. We were then able to use these quantities to gain general insights into the spatial distributions and temporal response of our systems in face of arbitrary excitation. This was because we could use the eigenfunctions and eigenvalues to express our system's response in space and time (much like a Fourier series expansion). For infinite domains, the idea of Fourier series expansions generalizes to Fourier transforms. Finite number of discrete eigenvalues give way to an infinite number of eigenvalues that are continuous. Finite, discrete eigenfunctions generalize to Fourier transforms. In most of the problems we considered in this chapter, Fourier transforms enabled us to reduce the partial differential equations to ordinary differential equations describing the temporal behavior of our Fourier transforms in response to excitation expressed as Fourier transforms of the external phenomena (forcing). Solution in the correct physical domain was then obtained via an inverse Fourier transformation of both sides (response and excitation).

For wave propagation over infinite domains, these techniques enabled us to derive the dispersion relations for each process. These relations enable us to determine how groups of waves of different frequencies propagate, how individual wave frequencies propagate, and so forth. In particular, we could evaluate the dispersion relation for an infinite beam, and we were able to conclude that in this case, individual wave frequency components travel slower than a group of waves of different frequencies. This behavior is different from the behavior we encounter in the case of surface waves on water, wherein individual wave frequencies travel faster than the wave group. We will continue this discussion in Chapter 12.

8.6   Self-Assessment

8.6.1

Develop the model and the solution for pollutant transfer over a long, narrow tube of a very small radius, if there is a steady fluid flow (i.e., advection) in the tube at a speed u. Discuss how diffusion and advection work side by side to determine the result along with any potential implications for pollutant transfer through rivers.

8.6.2

Consider a very long fluid-filed tube with an inner diameter R_i and an outer diameter R_0. A length $0 \leq z \leq l$ of the tube is subjected to a radially uniform sound pressure at frequency ω and amplitude P_0. What is the response of the fluid in the tube assuming zero initial conditions? Justify any assumptions made.

8.6.3

Find the response of an infinitely large circular membrane subjected to a harmonic load of magnitude q_0 over the central region $r \leq a$? Compare this with the response of a finite membrane of radius R. For the infinite membrane, what is the approximate response in the far field?

Part III

Mathematics of Scaling

Here, we review the mathematics of dimensions, units, and sizes. Correct use of mathematical relations among diverse physical quantities requires that their true physical nature be captured accurately. Units in which the physical quantities are expressed are crucial in applications, and what underlies correct use of units and ensures their consistency is dimensional analysis. Additionally, numerous marine applications involving engineered structures include some sort of testing prior to deployment at sea, and often, such testing is performed using small-scale physical models of the structures and of their operating environment. environment. How do we ensure that our small-scale physical models really represent the full-size physical structure? How do we relate our results from such smaller scale to the full-scale structure being designed for the real-sea environment? The methods presented in this part will help us answer these and other these and other similar questions.

9 Dimensional Analysis

9.1 Introduction

All physical quantities observed or measured result in real numbers. These real numbers are assigned a dimensional signature called unit. If a measurement is independent of a particular system or scale of measurement, such as MKS, SI, or Imperial unit system, then a physical quantity is dimensionless. But if a measurement depends on a particular scale of measurement used, then the quantity measured is dimensional. For example, velocity equals 15 m/s is a dimensional quantity, whereas velocity equals $v/\sqrt{gh} = 15$ is a dimensionless quantity, where v is the dimensional velocity, g is the (dimensional) gravitational acceleration, and h is the (dimensional) length.

One of the advantages of the use of dimensionless variables in experiments is that the number of variables that a physical quantity depends on can be reduced. This reduction, of course, allows one to rapidly complete the measurement process. For example, if the dimensionless number, called the Reynolds number, $Re = vl/v$, is used in experiments to measure the force, F, on an object, where l is the length and v is the kinematic viscosity of the fluid, then we can measure $F = F(Re)$ by varying the Re number only, rather than varying v, l, and v separately and fixing the other two.

The study of the principles of such measurements is called dimensional analysis. Physical modeling is the most important area that dimensional analysis can be applied to. By physical modeling, we refer to the technique of reproducing a physical phenomenon on a *greater* or *smaller* scale. For example, if we are interested in measuring the resistance of a ship to determine the power requirements, we would conduct resistance experiments with a small-scale ship model in a towing tank to determine the forces experienced by the prototype ship. Another example is the motions of offshore platforms that can be measured by means of model tests in a test basin (see, e.g., Chakrabarti 1994). In coastal engineering, e.g., one often needs to conduct experiments before finalizing the design of an harbor (see, e.g., Hughes 1993).

In most fluid mechanics problems, the three dimensions, length, mass, and time are the fundamental dimensions. Other physical quantities can then be measured in terms of these three fundamental dimensions. However, some physical quantities such as temperature and mechanical energy may occasionally be used as the fundamental dimensions.

If a measurement results in a real number, which can directly be compared to one of these fundamental dimensions, then the measurement is called a direct measurement. For example, a distance of 5 m is a direct measurement since its unit is one of the fundamental units, namely the length. On the other hand, if a measurement is a result of various comparisons that give a real number, whose dimension is a combination of two or more fundamental units, then the measurement is called a derived measurement. A typical example is the velocity, since both length and time have to be measured to obtain the velocity. In derived measurements, there is always a function that expresses the relation between the direct measurements. In the case of velocity measurements, e.g., this function is $f(x,t) = dx/dt$.

9.2   Similarity and Types of Forces

Like in many other fields of engineering, we frequently need to resort to experimental techniques even if we can analytically and/or numerically solve fluid-mechanics problems in ocean engineering. Typical examples of such problems are motions and resistance of offshore platforms and ships, design of harbors and breakwaters, forces on cylindrical objects such as piles and pipelines, and scouring (sediment pile-up on the sea/river floor around coastal and offshore structures). The scale model of any prototype system has to satisfy certain conditions called the law of similarity or similitude so that the behavior of prototype phenomenon can accurately be reproduced. The similarity can refer to one or more of

1. geometric similarity (refers to length),
2. kinematic similarity (refers to velocity), and
3. dynamic similarity (refers to force).

The aforementioned problems have to obey one or more of these similarity laws in order that one can properly conduct experiments.

It is necessary at this point to introduce various force mechanisms that are present in the flow of real fluids. This necessity arises because, under certain conditions of the specific problem being investigated, some force mechanism may be dominant over the other. As a result, we may isolate, or even neglect, the smaller forces by choosing a proper similarity law.

The three principal types of internal forces mechanism in a real fluid are:

1. inertial force,
2. gravitational force, and
3. viscous force.

The inertial force is due to the particle acceleration. It is proportional to $\rho u_1 (\partial u_1/\partial x_1)$, where ρ is the fluid mass density, and u_1 is the particle velocity component in the x_1 direction (other terms are ignored without loss of generality). The gravitational force is due to the weight of the fluid itself and is proportional to ρg. The viscous force is due to the difference between the shear forces acting on a fluid element, and is proportional to $\mu \left(\partial^2 u_1 / \partial x_1^2 \right)$, where μ is called the dynamic viscosity coefficient (again other terms are left out). These forces are included in the Navier–Stokes Equations given by

$$\frac{\partial u_i}{\partial t} + u_j \frac{\partial u_i}{\partial x_j} = -\frac{1}{\rho} \frac{\partial p}{\partial x_i} - g\delta_{i2} + \frac{\mu}{\rho} \frac{\partial^2 u_i}{\partial x_j \partial x_j}, \quad i, j = 1, 2, 3. \tag{9.1}$$

In equation (9.1), u_i are the three components of the particle velocity vector, p is the pressure, and δ_{ij} is the Kronecker delta, and it is understood that Einstein's summation convention and a right-handed rectangular Cartesian coordinate system, $Ox_1x_2x_3$, are employed (see Section 1.5).

The ratios of the internal forces (per unit volume) with one another give us the relative importance or dominance of these forces. For example, we can write

$$\frac{\text{Inertial force}}{\text{Gravitational force}} \propto \frac{\rho u_1 \frac{\partial u_1}{\partial x_1}}{\rho g} = \frac{u_1}{g} \frac{\partial u_1}{\partial x_1} \propto \frac{V^2}{gl}. \tag{9.2}$$

In equation (9.2), V is a characteristic velocity, l is a characteristic length of the problem, and ρ and g are the fluid mass density and gravitational acceleration, respectively. It is seen that the ratio is proportional to V^2/gl.

Similarly, we can write

$$\frac{\text{Inertial force}}{\text{Viscous force}} \propto \frac{\rho u_1 \frac{\partial u_1}{\partial x_1}}{\mu \frac{\partial^2 u_1}{\partial x_1^2}} \propto \frac{Vl}{\nu}, \tag{9.3}$$

where $\nu = \mu/\rho$ is called the kinematic viscosity coefficient. The third ratio, i.e., the ratio of the gravitational force to viscous force, can be obtained by a combination of equations (9.2) and (9.3), and therefore it need not be considered. The square root of equation (9.2) is called the Froude number, Fr, and equation (9.3) itself is called the Reynolds number, Re. So, for small Fr, for instance, we can say that the flow is gravity dominated, and for large Re, for instance, we can say that the flow is inertia dominated or viscosity is negligible (inviscid fluid assumption is a good one).

If we require that the flow about a body be similar fully both in the scale model and prototype, then it is necessary that both the Fr and Re numbers be the same for the model and prototype. In other words, the Froude number of the model scale must be the same as the Froude number of the prototype scale, and similarly for the Reynolds number. Unless under very special circumstances, it is not practically possible to scale a prototype by holding both the Fr and Re numbers the same. To see this, consider the Froude number and the Reynolds number which must both be kept constant for the model and prototype scales, i.e.,

$$Fr = \frac{V_m}{\sqrt{g_m l_m}} = \frac{V_p}{\sqrt{g_p l_p}}, \tag{9.4}$$

and

$$Re = \frac{V_m l_m}{\nu_m} = \frac{V_p l_p}{\nu_p}, \tag{9.5}$$

where the subscripts m and p refer to the model and prototype, respectively. Equation (9.4) requires that if $l_m < l_p$, then $V_m < V_p$, assuming that $g_m = g_p$. Whereas equation (9.5) requires that if $l_m < l_p$, then $V_m > V_p$, assuming that $v_m = v_p$. Therefore, unless the gravitational acceleration is considerably increased during the experiments (this is not uncommon considering, e.g., some soil-mechanics experiments conducted in a centrifugal apparatus) or v_m is considerably decreased or some combination of the two, the ratio of Fr to Re cannot be held constant simultaneously.

One of the two similarity laws, i.e., Froude's law (or similitude), given by equation (9.4), or Reynolds' law (or similitude), given by equation (9.5), has to be used depending on the particular application in mind. Froude's law or scale is generally used in conjunction with the experiments on ship and platform motions and resistance, open channel flows, and surface waves. Reynolds' law or scale, on the other hand, is used in experiments related to deeply submerged bodies, pipe flow, lubrication, among others.

Another force mechanism is the surface tension. For example, if we consider the ratio of the inertial force to surface tension force (per unit volume), we obtain

$$\frac{\text{Inertial force}}{\text{Surface tension force}} \propto \frac{\rho u_1 \frac{\partial u_1}{\partial x_1}}{\sigma / L^2} \propto \frac{\rho V^2}{L \sigma / L^2} = \frac{\rho L V^2}{\sigma}, \qquad (9.6)$$

where σ is the surface tension (lb/ft or N/m). The last equality is called the Weber number, i.e.,

$$We = \frac{\rho L V^2}{\sigma}. \qquad (9.7)$$

There are a number of other important dimensionless numbers used in offshore engineering: Strouhal number, $St = fD/U$, where f is the vortex shedding frequency and D is the characteristic length, e.g., diameter, and Keulegan–Carpenter number, $KC = UT/D$, where T is the wave period, are two examples. Some of the dimensionless numbers used are shown in Table 9.1, where U is the velocity, g the gravitational acceleration, L or D the characteristic length, f the cyclic frequency, v the kinematic viscosity of water, T the wave period, H the wave height, h the water depth, E Young's modulus, p the pressure, and σ the surface tension.

Table 9.1 Some of the frequently used dimensionless numbers.

Dimensionless Number	Symbol	Definition
Froude number	Fr	$U/\sqrt{gL}$
Reynolds number	Re	UL/v
Strouhal number	St	fD/U
Keulegan–Carpenter number	KC	UT/D
Euler number	Eu	$p/\rho U^2$
Cauchy number	Cy	$\rho U^2/E$
Ursell number	Ur	HL^2/h^3
Weber number	We	$\rho U^2 L/\sigma$

9.3 Dimensional Homogeneity

An equation that represents a physical phenomenon is dimensionally homogeneous if the equation does not depend on the fundamental units of measurement. For example, Newton's second law, $F = mg$, for the force acting on a body of mass, m, under the action of gravity, g, is a dimensionally homogeneous equation since, regardless of what unit system we use, it will give us the correct value of the force in that unit system. On the other hand, it is possible that this equation can be put in a form which is not dimensionally homogeneous. To see this, employ, for instance, the metric system and write $F = 9.82$ m ($g = 9.82$ m/s^2). This equation is valid only in the metric system but not in another system such as the Imperial unit system. Therefore, it is very important in physical sciences not to associate dimensions to numbers in equations that represent fundamental physical laws. Nevertheless, it is not unusual to see such dimensionally inhomogeneous equations in some fields of engineering. These equations are, generally, empirical equations that depend on experimental data. They contain one or more constants (real numbers) that are dimensional. A typical example is the Chezy formula

$$S_e = \frac{U^2}{C_c^2 R},$$ (9.8)

where S_e is the slope of the energy gradient, R is the hydraulic radius, C_c is the Chezy coefficient, and U is the uniform velocity. C_c is a dimensional coefficient, and therefore equation (9.8) is not dimensionally homogeneous. This equation is frequently used in tidal-dynamics problems in estuaries. There are two basic problems with such nonhomogeneous equations. First, they are restricted to the range of variables covered by the experimental work upon which they are based. Second, they cannot be used in dimensional analysis.

The original view of dimensional homogeneity was first included in the 1822 book, *The Analytical Theory of Heat*, of Joseph Fourier, a French mathematician (see, e.g., Adiutori 2005). Many physical laws, such as Ohm's law or Hooke's law, were written correctly only after the work of Fourier had been understood and appreciated by the engineering community.

In direct measurements, where a physical quantity can be measured in terms of one of the fundamental units, one has to postulate the fundamental property that the ratio of the two numbers obtained by use of two different unit systems should be the same within the same fundamental unit category. For example, if we use the SI system (meter- kilogram-second) in our measurements and denote the fundamental units by L_1, M_1, and T_1, then a new set of measurements obtained by use of the British system (feet-slug-second) whose fundamental units are denoted by L_2, M_2, and T_2 will be related to the first measurements by

$$L_1 = \alpha L_2, \quad M_1 = \beta M_2, \quad T_1 = \gamma T_2.$$ (9.9)

Depending on which unit system L_1 and L_2 (and similarly for M and T) are defined in, the value of α (and similarly of β and γ) will change. For example, if L_1 is in meter and L_2 is in feet, then $\alpha = 0.3048$, but if L_1 is in feet and L_2 is in meter, then $\alpha = 3.2808$. The fundamental property postulated then assures that α, β, and γ are the same or constants in all comparisons between the fundamental units of two systems under consideration.

This property is easily extended to derived measurements as follows. Let us suppose that, for a particular derived measurement, we obtained p length measurements, $l_1, l_2, \ldots, l_p, q$ mass measurements, $m_1, m_2, \ldots, m_q$, and r time measurements, $t_1, t_2, \ldots, t_r$. All these measurements are assumed to be governed by a given function X as follows:

$$X = f(l_1, l_2, \ldots, l_p; m_1, m_2, \ldots, m_q; t_1, t_2, \ldots, t_r). \tag{9.10}$$

Now if we change the scale (or units) of measurements to l_i', m_j', and t_k', then we can write

$$l_i' = \alpha l_i, \quad i = 1, 2, \ldots, p,$$

$$m_j' = \beta m_j, \quad j = 1, 2, \ldots, q, \tag{9.11}$$

$$t_k' = \gamma t_k, \quad k = 1, 2, \ldots, r,$$

since the fundamental postulated property holds for direct measurements, i.e., α, β, and γ remain the same. And in the new scale (prime system)

$$X' = f(l_1', l_2', \ldots, l_p'; m_1', m_2', \ldots, m_q'; t_1', t_2', \ldots, t_r'). \tag{9.12}$$

Then the fundamental assumption is that, for any given dimensional category, the ratio of the measurements before and after a change of scale in the set of fundamental units is the same for all physical systems in the category. In other words,

$$\frac{X'}{X} = \text{constant}. \tag{9.13}$$

If this fundamental assumption is valid, then any function f must satisfy the condition:

$$f(l_1', .., l_p'; m_1', .., m_q'; t_1', .., t_r') = \alpha^a \beta^b \gamma^c f(l_1, .., l_p; m_1, .., m_q; t_1, .., t_r), \tag{9.14}$$

in a derived measurement. In other words, the constant in equation (9.13) is $\alpha^a \beta^b \gamma^c$. Here, the exponents a, b, and c represent the dimensions of the function f and will be discussed next. The equation (9.14) is known as the homogeneity theorem. It is important to note that the form of the function f remains the same in equations (9.10) and (9.12).

Let us next show by way of an example that equation (9.13) is true once the fundamental property holds. Take $X = 20$ ft/s, $x_1 = 40$ ft, $x_2 = 2$s in a derived measurement where $X = f(x_1, x_2) = 20$ft/s. Suppose we change the scale to metric and obtain $(\alpha = 0.3048, \gamma = 1.0)$ $X' = f(x_1', x_2') = f(\alpha x_1, \gamma x_2) = 40\alpha/2\gamma$ m/s $= 6.096$ m/s. Therefore, $X'/X = 0.3048$. To show that this ratio, i.e., 0.3048 is a constant in any set of measurements in the velocity category, consider the result of another independent measurement, say, X'',

$$X'' = f(x_1'', \ x_2'') = 30\text{ft/s}.$$

We now change the scale to metric. Then, as before, $X''' = f(x_1''', x_2''') = f(\alpha x_1'', \gamma x_2'') = 9.144\text{m/s}$. Therefore, $X'''/X'' = 0.3048$. Thus, it is clear that

$$\frac{X'}{X} = \frac{X'''}{X''} = \text{constant}, \qquad a = 1, \quad b = 0, \quad c = 1. \tag{9.15}$$

One can show that the constant of proportionality in equation (9.13) or (9.15) is equal to $\alpha^a \beta^b \gamma^c$, where a, b, and c are the dimensions of the set of physical quantities $\{X\}$ as we see in detail in Section 9.4.

9.4 The Algebra of Dimensions

The dimension of a physical category, X, measured in some system will be shown by $[X]$. Let x be a real number that corresponds to the physical measurement in the same system whose fundamental units are given by L_1, M_1, and T_1 (e.g., meter, kilogram, second). Suppose now that we make the same measurement by use of a different unit system whose fundamental units are given by L_1', M_1', and T_1' (e.g., feet, pounds, seconds). This new measurement produces a real number, say, x'. The two fundamental unit sets are related by (see equation (9.9))

$$L_1 = \alpha L_1', \quad M_1 = \beta M_1', \quad T_1 = \gamma T_1'. \tag{9.16}$$

In general, x and x' are derived measurements and the relation between them is implied by the homogeneity theorem, i.e.,

$$x' = \alpha^a \beta^b \gamma^c x, \tag{9.17}$$

where a, b, and c (the same constants in equation (9.14)) represent the dimension of X:

$$[X] = (a, b, c) = L^a M^b T^c. \tag{9.18}$$

X is said to be dimensionless if $a = b = c = 0$, i.e.,

$$[X] = (0, 0, 0) = L^0 M^0 T^0 = I. \tag{9.19}$$

For example, if X is the length and $x = 15$ in the ft-slug-sec system, then $X = 15\,\text{ft}, [X] = (1, 0, 0)$ or $[X] = L^1 M^0 T^0$. If we measure X in another system, say, in the metric system, then x' will be different from x but the dimension of X will obviously remain the same. Noting that, from equations (9.16) and (9.17), $L_1 = 1$ ft $= \alpha\, 1$ m $\Rightarrow \alpha = 0.3048$, $x' = (0.3048)^1 \beta^0 \gamma^0 x = 4.572$ since $x = 15$. Therefore, $X = 4.572$ m.

The aforementioned argument, which was carried out for direct measurements, can now be extended to derived measurements. Suppose we measure X in L_1, M_1, and T_1 units and obtain x, where x depends on n direct measurements, $x_1, x_2, \ldots, x_n$. Then the derived measurement x can be expressed by an equation like

$$x = f(x_1, x_2, \ldots, x_n). \tag{9.20}$$

Note that all x_i are measured in L_1, M_1, and T_1 units. Now if we change the units from L_1, M_1, and T_1 to L'_1, M'_1, and T'_1, we have

$$x' = f(x'_1, x'_2, \ldots, x'_n). \tag{9.21}$$

In other words, f remains the same, as it is the requirement for a function to be dimensionally homogeneous. And, therefore, f must satisfy the relation

$$f(\alpha^{a_1} \beta^{b_1} \gamma^{c_1} x_1, \ldots, \alpha^{a_n} \beta^{b_n} \gamma^{c_n} x_n) = \alpha^a \beta^b \gamma^c f(x_1, \ldots, x_n), \tag{9.22}$$

where a, b, and c are the dimensions of X or f in the L_1, M_1, and T_1 system, and a_i, b_i, and $c_i, i = 1, \ldots, n$, are the dimensions of x'_i in the L'_1, M'_1, and T'_1 system.

9.4.1   Example

Suppose that we measure velocity (a derived measurement) by measuring distance and time and obtain $X = 20\text{ft/s} = f(x_1, x_2)$ in the ft-slug-sec system or $[X] = (1, 0, -1)$, $a = 1, b = 0$, and $c = -1$. In equation (9.20), x_1 corresponds to length and x_2 corresponds to time measurements. Therefore, $b_1 = c_1 = a_2 = b_2 = 0$ in equation (9.22). Now, if we change the scale to the metric system, we have, from equations (9.16) and (9.17), $x_1' = \alpha^{a_1} x_1$, $x_2' = \gamma^{c_2} x_2$, where $\alpha = 0.3048, \gamma = 1.0$. Since $a_1 = c_2 = 1.0$, we have $x' = f(0.3048 x_1, x_2)$, which is equal to $0.3048 f(x_1, x_2) = 6.096$. Therefore, $X = 6.096$ m/s.

9.5   Dimensional Independence

Let us suppose now that we have three physical quantities (e.g., length, velocity, and acceleration), $x_i, i = 1, 2, 3$, whose dimensions are a_i, b_i, c_i. These three variables are dimensionally independent if there is no relation in the form

$$\sum_{i=1}^{3} A_i \left\{ \begin{array}{c} a_i \\ b_i \\ c_i \end{array} \right\} = \left\{ \begin{array}{c} 0 \\ 0 \\ 0 \end{array} \right\}, \tag{9.23}$$

when the arbitrary constants, A_i are not all zero. Conversely, x_i are dimensionally dependent when at least one A_i is not zero and

$$\sum_{i=1}^{3} A_i [X_i] = 0. \tag{9.24}$$

As an example, consider x_1, x_2, and x_3 as real numbers that correspond to length, velocity, and acceleration, respectively, i.e., $[X_1] = (1, 0, 0)$, $[X_2] = (1, 0, -1)$, and $[X_3] = (1, 0, -2)$. These dimensions can be thought of as vectors in the L, M, and T space as we will see shortly. Then from equation (9.23), we have

$$A_1 \left\{ \begin{array}{c} 1 \\ 0 \\ 0 \end{array} \right\} + A_2 \left\{ \begin{array}{c} 1 \\ 0 \\ -1 \end{array} \right\} + A_3 \left\{ \begin{array}{c} 1 \\ 0 \\ -2 \end{array} \right\} = \left\{ \begin{array}{c} 0 \\ 0 \\ 0 \end{array} \right\}, \tag{9.25}$$

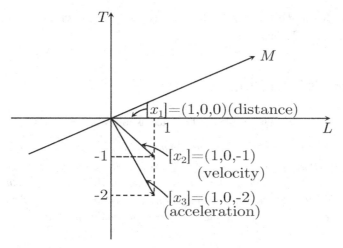

Figure 9.1 Some coplanar vectors in the dimension space.

or

$$A_1 + A_2 + A_3 = 0, \quad 0 = 0, \quad -A_2 - 2A_3 = 0. \tag{9.26}$$

From equation (9.26), $A_2 = -2A_3$ and $A_1 - 2A_3 + A_3 = A_1 - A_3 = 0$ or $A_1 = A_3$. Because the system of equations are underdetermined, A_3 is arbitrary. Therefore, for any nonzero value of A_3, equation (9.25) is automatically satisfied. Therefore, length, velocity, and acceleration are dimensionally dependent.

It is important to note that we did not have to carry out the above calculations to see that x_1, x_2, and x_3 are dimensionally dependent, since these three variables, whose dimensions can be thought of as vectors in the L, M, and T dimension space, are coplanar, i.e., they lie on the same rectangular Cartesian plane as shown in Figure 9.1. Recall from a theorem in linear algebra that any set of three coplanar vectors is linearly dependent. Another way of looking at the problem then is forming the matrix

$$\begin{bmatrix} a_1 & a_2 & a_3 \\ b_1 & b_2 & b_3 \\ c_1 & c_2 & c_3 \end{bmatrix} \begin{Bmatrix} A_1 \\ A_2 \\ A_3 \end{Bmatrix} = 0, \tag{9.27}$$

and requiring that the determinant of the coefficient matrix that contains a_i, b_i, and c_i be not zero if the set x_1, x_2, and x_3 is linearly independent. If the determinant is zero, then x_1, x_2, and x_3 are linearly dependent. Here we use the terms "linearly" and "dimensionally" interchangeably. In the aforementioned example, this determinant vanishes, i.e.,

$$det[X] = \begin{vmatrix} 1 & 1 & 1 \\ 0 & 0 & 0 \\ 0 & -1 & -2 \end{vmatrix} = +1(0) - 1(0) + 1(0) = 0.$$

The significance of dimensional independency is that we can select a set of dimensionally independent variables to form a basis to which all other variables are compared. Note that in a two-dimensional space, x_1 and x_2 are dimensionally dependent if the two vectors lie on the same line (but not necessarily in the same direction) in the $L-M, M-T$, or $L-T$ plane.

9.6   Buckingham's Pi Theorem

This theorem (Buckingham 1914) can be stated as follows:

Any physical law that is given as a functional relationship between n+1 quantities, $x, x_1, x_2, \ldots, x_n$, e.g., $x = f(x_1, x_2, x_3, \ldots, x_n)$, can be expressed as a relation between n+1−k dimensionless quantities, denoted by $\Pi, \Pi_1, \ldots, \Pi_{n-k}$, under a change of scale, where k is the maximum number of dimensionally independent quantities among $x, x_1, \ldots, x_n$ and $k \leq$ maximum number of fundamental dimensions.

With regard to the Pi theorem, several remarks are necessary:

1. If $k = n$, then there is only one Π which is constant. For example, for a function in the form of $R = f(l, \rho, g)$, $n = 3$, with the basis l, ρ, g ($k = 3$), which is a dimensionally independent set, we have $[R] = LMT^{-2}$ (force), $[l] = L$(length), $[\rho] = L^{-3}M$ (density), $[g] = LT^{-2}$ (gravitational acceleration), and $LMT^{-2} = [l]^p [\rho]^q [g]^r$. Therefore, equating the exponents of L, M, and T, we obtain $1 = p - 3q + r$, $1 = q$, $-2 = -2r$, and therefore $p = 3$, $q = 1$, $r = 1$. Hence, $\Pi = R/\{l^3 \rho g\} = $ constant $= G(1, 1, 1)$.
2. If a different dimensionally independent set is used, one will obtain different Πs.
3. There must be enough number of quantities contained in a properly formulated physical law to construct a dimensionless variable. If this cannot be done, then one should be alerted to the fact that original formulation was incomplete or wrong.
4. If the result of the use of the Pi theorem contradicts with the experimental data, then two things are possible:

   1. the experimental data are wrong or
   2. some relevant variables that should have been included in the dimensional analysis were left out by mistake.

There is no third possibility since the Pi theorem is a very well established theorem based on a firm mathematical ground. Unfortunately though, dimensional analysis cannot provide which physical quantities are missing in the formulation.

9.7   Dimensionless Equations

It is sometimes advantageous to work with dimensionless equations rather than dimensional equations whose solutions produce dimensional physical quantities. For instance, if we need to write a computer program to solve a set of algebraic, ordinary, or partial differential equations, it would be very efficient if the numerical solutions do not depend on a particular unit system. As an example, consider a problem where the solution of a physical quantity depends on the wave amplitude, wave length, and mass density. If we use dimensional variables in our program, the physical quantity that we solve for will be correct only for the wave amplitude, wave length, and mass density that we input. For other values of these parameters, we must run the program again, especially if we have not explored the dimensional relation between the particular physical quantity and the parameters that it depends on.

On the other hand, it is sometimes possible to make a mistake in working with dimensionless variables and equations. This is because one has generally no idea if there is any error made in deriving a dimensionless equation since the dimensions of different terms in an equation can no longer be compared. For example, if we use dimensional variables and derive an equation like

$$\frac{\partial u_1}{\partial t} + u_1 \frac{\partial u_1}{\partial x_1} + u_2 \frac{\partial^2 u_1}{\partial x_1^2} = 3, \tag{9.28}$$

where u_1 and u_2 are the velocity components, x_1 is the spatial coordinate, and t is the time, we immediately know that something is wrong with this equation because the dimensions of each term are not the same. The dimensions of the first and second terms on the left-hand side of equation (9.28) is L/T^2, while the third term on the left-hand side has a dimension of $1/T^2$, and the term on the right-hand side is a dimensionless number unless we are told what its dimension is. Assuming that the first and second terms are correct, we know that we must find our error in deriving the other terms.

On the other hand, if equation (9.28) contained dimensionless variables such as $\bar{u}_1 = u_1/\sqrt{gh}, \bar{u}_2 = u_2/\sqrt{gh}, \bar{x}_1 = x_1/h, \bar{t} = t\sqrt{g/h}$, then we cannot determine if there are any dimensional errors in an equation like

$$\frac{\partial \bar{u}_1}{\partial \bar{t}} + \bar{u}_1 \frac{\partial \bar{u}_1}{\partial \bar{x}_1} + \bar{u}_2 \frac{\partial^2 \bar{u}_1}{\partial \bar{x}_1^2} = 3. \tag{9.29}$$

Let us next look into nondimensionalizing some equations which, e.g., are to be used in the numerical analysis of a given problem. As an example, consider a linear and unsteady shallow-water wave equation for constant water depth given by

$$\frac{\partial^2 u_1}{\partial t^2} - gh_0 \frac{\partial^2 u_1}{\partial x_1^2} - \frac{1}{3} h_0^2 \frac{\partial^4 u_1}{\partial t^2 \partial x_1^2} = -\frac{1}{\rho} \frac{\partial^2 p}{\partial t \partial x_1}. \tag{9.30}$$

Since this equation is valid for an inviscid fluid, the flow must be gravity and inertia dominated. Furthermore, because the water depth, h_0, is assumed finite, it must play a role. Therefore, it is logical to take ρ, g, and h_0 as the basic quantities in Buckingham's Pi theorem upon which all other quantities depend. Hence, we may introduce the following dimensionless quantities (show by a formal derivation that ρ, g, and h_0 is a dimensionally independent set or basis):

$$\bar{x}_1 = \frac{x_1}{h_0}, \quad \bar{p} = \frac{p}{\rho g h_0}, \quad \bar{u}_1 = \frac{u_1}{\sqrt{g h_0}}, \quad \bar{t} = t\sqrt{\frac{g}{h_0}} \tag{9.31}$$

Note that these can be shown by dimensionally analyzing equation (9.33). If we use equation (9.31) in equation (9.30), and note that

$$\frac{\partial}{\partial x_1} \rightarrow \frac{1}{h_0}\frac{\partial}{\partial \bar{x}_1}, \quad \frac{\partial}{\partial t} \rightarrow \sqrt{\frac{g}{h_0}}\frac{\partial}{\partial \bar{t}},$$

and so forth for the second derivatives, we obtain

$$\frac{\partial^2 \bar{u}_1}{\partial \bar{t}^2} - \frac{\partial^2 \bar{u}_1}{\partial \bar{x}_1^2} - \frac{1}{3}\frac{\partial^4 \bar{u}_1}{\partial \bar{t}^2 \partial \bar{x}_1^2} = -\frac{\partial^2 \bar{p}}{\partial \bar{t} \partial \bar{x}}. \tag{9.32}$$

Equation (9.32) can now be used in the numerical analysis of the problem. Once $\bar{u}_1$ is solved, equation (9.31) can be used to determine the dimensional velocity for a particular water depth. Note that equation (9.31) can be obtained by Buckingham's Pi theorem once we assume that

$$u_1 = f(x_1, h_0, p, \rho, g, t), \tag{9.33}$$

and take ρ, g, and h_o as a dimensionally independent set (see Prob. 9.12.26).

Let us give another example to show how nondimensionalization may be invaluable, this time in understanding the physics of a problem. Assume that we are dealing with very deep water and therefore, we do not consider the water depth h_0 as a parameter. Instead, we may, for instance, consider a characteristic body length. As a specific example, consider a ship steadily moving in very deep water and in the positive x_1 direction. We may write the velocity potential as

$$\Phi = \Phi(x_1, x_2, x_3, L, \eta, t, U, g, f), \tag{9.34}$$

where L is the length of the ship, η is the surface elevation, U is the speed of the ship, and f describes the geometry of the hull, i.e., $x_3 = f(x_1, x_2)$. None of the variables in equation (9.34) depends on the mass density of water. Therefore, we cannot choose three variables which are dimensionally independent. In fact, this is not necessary for the dimension of Φ lies in the $L-T$ plane only. Thus, we choose L and U as the basic variables that are dimensionally independent, and obtain

$$\bar{\Phi} = \frac{\Phi}{LU}, \quad \bar{\eta} = \frac{\eta}{L}, \quad (\bar{x}_1, \bar{x}_2, \bar{x}_3) = \frac{(x_1, x_2, x_3)}{L}, \quad \bar{t} = \frac{U}{L}t. \tag{9.35}$$

With these nondimensional variables, the dimensional boundary value problem given by

$$\frac{\partial^2 \Phi}{\partial x_1^2} + \frac{\partial^2 \Phi}{\partial x_2^2} + \frac{\partial^2 \Phi}{\partial x_3^2} = 0 \text{ (in the fluid)}, \tag{9.36}$$

$$\frac{\partial \eta}{\partial t} - \frac{\partial \Phi}{\partial x_2} = 0, \quad g\eta + \frac{\partial \Phi}{\partial t} = 0 \text{ (on the free surface)},$$

$$\frac{\partial \Phi}{\partial x_2} = 0 \text{ as } x_2 \rightarrow -\infty \text{ (on the sea floor)},$$

$$\frac{\partial \Phi}{\partial n} = Un_1 \text{ (on the hull surface, } x_3 = f(x_1, x_2)),$$

where n_1 is the x_1 component of the unit normal vector (dimensionless) on the hull surface pointing out of the fluid, can be written in a dimensionless form as

$$\frac{\partial^2 \bar{\Phi}}{\partial \bar{x}_1^2} + \frac{\partial^2 \bar{\Phi}}{\partial \bar{x}_2^2} + \frac{\partial^2 \bar{\Phi}}{\partial \bar{x}_3^2} = 0 \text{ (in the fluid)}, \tag{9.37}$$

$$\frac{\partial \bar{\eta}}{\partial \bar{t}} - \frac{\partial \bar{\Phi}}{\partial \bar{x}_2} = 0, \quad \frac{\bar{\eta}}{Fr^2} + \frac{\partial \bar{\Phi}}{\partial \bar{t}} = 0 \text{ (on the free surface)},$$

$$\frac{\partial \bar{\Phi}}{\partial \bar{x}_2} = 0 \text{ as } \bar{x}_2 \to -\infty \text{ (on the sea floor)},$$

$$\frac{\partial \bar{\Phi}}{\partial \bar{n}} = \bar{n}_1 \text{ (on the hull surface)}.$$

In the aforementioned equations, the normal derivative $\partial/\partial \bar{n}$ refers to the derivative in the direction of the unit normal vector on the hull surface. It is interesting to note that the length Froude number, $Fr = U/\sqrt{gL}$, appears in the dynamic free-surface condition. This is another beauty of the use of dimensionless variables, i.e., the importance of the Froude number in this case became very apparent since it is part of the dynamic free-surface condition explicitly, something that we could not have seen from equation (9.36) directly. Note that as $Fr \to 0$, $\bar{\eta} \to 0$, and thus $\partial \bar{\phi}/\partial \bar{x}_2 \to 0$, i.e., the free surface becomes a rigid lid! However, at this point, one should not worry about the physics of the problem just posed. The above two sets of dimensionless equations are simply given as examples to show how one can use dimensional analysis in a mathematical problem.

9.8   Scaling of Loads

In planning model tests, one has to decide which scaling law to be used. In most ocean engineering model tests, the Froude scaling law is used as offshore platforms mostly encounter gravity waves in nature. This means that during such experiments, one must make sure to satisfy that $Fr_m = Fr_p$. Having also decided what length scale, $S_L = L_m/L_p$, to use, one can scale the other physical quantities, accordingly. For example, if we want to scale the wave forces by use of Froude's law, so that $U_m = U_p S_L^{1/2}$, we can first determine the scaling of time and acceleration:

$$U_m = \sqrt{S_L} U_p = \frac{L_m}{t_m} = \sqrt{S_L}\frac{L_p}{t_p} \Rightarrow t_m = \sqrt{S_L}t_p, \ a_m = \frac{U_m}{t_m} = \frac{\sqrt{S_L}U_p}{\sqrt{S_L}t_p} = a_p, \tag{9.38}$$

where t is the time. That the scale for accelerations are the same in the model and prototype scales is not surprising since the gravitational acceleration in the model and prototype scales is the same, i.e., $g_m = g_p$.

We also need to scale the masses of the model and the prototype. It is

$$S_m = \frac{m_m}{m_p} = \frac{\rho_m L_m^3}{\rho_p L_p^3} = S_\rho S_L^3. \tag{9.39}$$

Therefore, the force scaling can be written as

$$S_F = \frac{m_m \dot{U}_m}{m_p \dot{U}_p} = \frac{m_m}{m_p} = S_m = S_\rho S_L^3, \tag{9.40}$$

since the acceleration scale is 1.0.

This method of obtaining the scaling for forces can be used on any other physical quantity to determine how it is scaled to the model. However, there is way of obtaining the same scaling result. Let us show this by way of an example. Consider the dimension of force and write it as $[F] = (L, M, T^{-2}) = (1, 1, -2)$, i.e., as a vector in the three-dimensional (L, M, T) space. Writing the functional form of force as $F = f(\rho, L, g)$, and taking the set (ρ, L, g) as a dimensionally independent set of quantities, one can use the Pi theorem (Buckingham 1914) to obtain the single dimensionless π :

$$[F] = (L, M, T^{-2}) = (1, 1, -2) = [\rho]^p [L]^q [g]^r = (-3, 1, 0)^p (1, 0, 0)^q (1, 0, 2)^r$$

$$\text{or } p = 1, q = 3, r = 1 \Rightarrow \frac{F}{\rho L^3 g} = \pi \Rightarrow \frac{F_m}{\rho_m L_m^3} = \frac{F_p}{\rho_p L_p^3} \Rightarrow S_F = S_\rho S_L^3.$$

$$\tag{9.41}$$

This is the same result given by equation (9.40).

In Table 9.2, we show the model scales for some of the physical quantities that would be of interest during an ocean engineering experiment (L: length, M: mass, T: time, $S_L = L_m/L_p$ is the length ratio, and $S_\rho = \rho_m/\rho_p$ is the specific gravity of salt water). Many others are given in Chakrabarti (1994), but for the special case of $S_\rho = 1.0$.

Table 9.2 Some model scaling obtained by use of Froude's scaling law.

Quantity	Dimension	Scale
Length	L	S_L
Mass	M	$S_\rho S_L^3$
Mass moment of inertia	L^2M	$S_\rho S_L^5$
Moment of inertia of area	L^4	S_L^4
Time	T	$S_L^{1/2}$
Acceleration	LT^{-2}	1
Velocity	LT^{-1}	$S_L^{1/2}$
Linear spring constant	MT^{-2}	$S_\rho S_L^2$
Axial stiffness	LMT^{-2}	$S_\rho S_L^3$
Bending stiffness	L^3MT^{-2}	$S_\rho S_L^5$
Work	L^2MT^{-2}	$S_\rho S_L^4$
Power	L^2MT^{-3}	$S_\rho S_L^{7/2}$
Energy	L^2MT^{-2}	$S_\rho S_L^4$
Force	LMT^{-2}	$S_\rho S_L^3$
Moment	L^2MT^{-2}	$S_\rho S_L^4$
Stress	$L^{-1}MT^{-2}$	$S_\rho S_L$
Pressure	$L^{-1}MT^{-2}$	$S_\rho S_L$
Modulus of elasticity	$L^{-1}MT^{-2}$	$S_\rho S_L$

9.9   Elastic Structures

Model tests of elastic structures bring additional complexities to the problem as it is not practical in many cases to correctly scale the stiffness of the structure at the model scale. Such elastic structures could be tension-leg-platform tendons, oil-production risers, catenary-mooring lines, and so on, but the structure itself may be elastic as well, especially very large floating structures (VLFS) (see, e.g., Ertekin and Kim 1999). To see this, consider a tubular, beam-like object (e.g., riser), and write the scalings, from Table 9.2, of the axial and bending stiffness, and its diameter as

$$S_{EA} = \frac{(EA)_m}{(EA)_p} = \frac{F_m}{F_p} = S_F = S_m = S_\rho S_L^3,$$

$$S_{EI} = \frac{(EI)_m}{(EI)_p} = \frac{F_m L_m^4}{L_m^2} \frac{L_p^2}{F_p L_p^4} = S_F S_L^2 = S_\rho S_L^5, \qquad \frac{D_m}{D_p} = \frac{L_m}{L_p} L_p = S_L, \tag{9.42}$$

Either the bending or shear stress, τ, is scaled as

$$S_\tau = \frac{F_m}{L_m^2} \frac{L_p^2}{F_p} = \frac{S_F}{S_L^2} = \frac{S_\rho S_L^3}{S_L^2} = S_\rho S_L. \tag{9.43}$$

And the area moment of inertia is scaled as

$$S_I = \frac{I_m}{I_p} = \frac{L_m^4}{L_p^4} = S_L^4. \tag{9.44}$$

Therefore, the Young's modulus is scaled as

$$S_E = \frac{E_m}{E_p} = \frac{S_{EI}}{S_I} = \frac{S_\rho S_L^5}{S_L^4} = S_\rho S_L. \tag{9.45}$$

There are basically two problems one encounters as a result of these scalings. One is that it is very difficult to scale the geometry because D_m is typically 50–100 times smaller than D_p. The other is that the modulus of elasticity of the material used in the experiments is also 50–100 times smaller than what is used in the prototype, e.g., steel. Some engineering solutions to these kinds of problems are necessary to conduct the experiments. For example, Dillingham (1984) suggested in modeling the tendons of a tension-leg platform that all parameters are correctly scaled for the tendon except the axial stiffness that is modeled by a spring placed at the top or bottom of the tendon to provide the correct stiffness. Even this approximation involves some errors that must be carefully assessed. The structural rigidity can sometimes be reduced by distorting the structure as discussed, e.g., by Tulin (1999).

Another type of a distorted model can be achieved in ocean model tests in rather shallow water where the horizontal length dimensions are much larger than the vertical length dimension. In such cases, two different model scale ratios are used in the experiments (see, e.g., Hughes 1993; Chakrabarti 1994). Let us again set the horizontal length scale to S_L but set the vertical length scale to $S_V = h_m/h_p$, where h_m and h_p are the water depths in the model and prototype scales, respectively, and S_V, in general, is different from S_L. Next, let us consider the linear shallow-water phase speed, $c_p = (gh)^{1/2}$, and use S_V to write the Froude scaling law to obtain

$$\frac{c_m}{\sqrt{gh_m}} = \frac{c_p}{\sqrt{gh_p}} \Rightarrow c_m = \sqrt{S_V} c_p, \tag{9.46}$$

and since $c = \lambda/T$, where λ is the wave length, scaled by S_L, i.e., $S_L = \lambda_m/\lambda_p$, we have the following scaling for the wave period:

$$\frac{T_m}{T_p} = \frac{\lambda_m}{\lambda_p} \frac{c_p}{c_m} = \frac{S_L}{\sqrt{S_V}} \Rightarrow T_m = \frac{S_L T_p}{\sqrt{S_V}}. \tag{9.47}$$

If the water is not very shallow, of course the full dispersion relation based on the linear wave theory can be used in deriving the corresponding scales that will now involve hyperbolic functions. Finally, recall that the accelerations scale as 1.0 if a single length scale is used according to the Froude scaling law. In the distorted model used, however, that will not be the case. The accelerations in a distorted model would scale as

$$a_m = \frac{U_m}{T_m} = \frac{\sqrt{S_V} U_p \sqrt{S_V}}{S_L T_p} \Rightarrow \frac{a_m}{a_p} = \frac{S_V}{S_L}. \tag{9.48}$$

9.10   Some Examples of Dimensional Analysis

In this section, several illustrative examples will be given on the application of the Pi theorem. For some other applications, see, e.g., Chakrabarti (2002).

9.10.1   Flow Through a Pipe

This problem is a special one because dimensional analysis was applied for the first time by Osborne Reynolds in the study of fluid motion in pipes in the late nineteenth century.

Consider a pipe of constant cross section, not necessarily a circular one, and length l as shown in Figure 9.2. It will be assumed that the flow is due to the pressure drop per unit length of the pipe, i.e., the pressure gradient is

$$\mathcal{P} = (P_2 - P_1)/l.$$

Furthermore, we assume that the flow is steady, i.e., the velocity profile is the same at any time and for any $x = x_0$. Compressibility of the fluid is neglected. Therefore, only inertia and viscosity are considered.

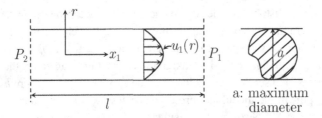

a: maximum
    diameter

Figure 9.2 A pipe of constant cross section (not necessarily a circular one).

To determine the motion of the fluid, it suffices to analyze the discharge rate, Q. Consequently, we anticipate that $Q = f(a, l, \mathcal{P}, \rho, \mu)$. The dimensions of these variables are $[Q] = L^3 T^{-1}$, $[a] = L$, $[l] = L$, $[\mathcal{P}] = L^{-2} M T^{-2}$, $[\rho] = L^{-3} M$, and $[\mu] = L^{-1} M T^{-1}$. The dynamic viscosity coefficient μ is the proportionality factor between the shear stress τ and the velocity gradient, i.e.,

$$\tau = \tau_{12} = \mu \, \partial u_1 / \partial x_2$$

in a simple two-dimensional viscous flow (see Section 10.2).

As a dimensionally independent set, we choose a, ρ, and μ (one may also choose $a, \mathcal{P}$, and ρ, for instance). We do not include Q in the set because we want to isolate it. Therefore, $[Q] = L^3 T^{-1} = [a]^p [\rho]^q [\mu]^r = L^p (L^{-3} M)^q (L^{-1} M T^{-1})^r$. Equating the exponents, we obtain $p = 1, q = -1, r = 1$. Therefore, the associated Π is

$$\Pi = \frac{Q}{a \rho^{-1} \mu} = \frac{Q \rho}{a \mu}.$$

In our formulation, $n = 5$ and $k = 3$, and therefore $n + 1 - k = 3 \Pi$s must be obtained from the Pi theorem. To obtain the second Π, i.e., Π_1, we may isolate $\mathcal{P}$, and write $[\mathcal{P}] = L^{-2} M T^{-2} = [a]^p [\rho]^q [\mu]^r$ to find $p = -3, q = -1, r = 2$. Hence,

$$\Pi_1 = \frac{\mathcal{P}}{a^{-3} \rho^{-1} \mu^2} = \frac{\mathcal{P} a^3 \rho}{\mu^2}.$$

The only variable that we have not isolated or included in the primary variable set (or dimensionally independent set) is l. As expected, $\Pi_2 = l/a$ by inspection. Finally then,

$$\frac{Q \rho}{a \mu} = F_1 \left(\frac{\mathcal{P} a^3 \rho}{\mu^2}, \frac{l}{a} \right).$$

As an exercise, one can start with $a, \mathcal{P}$, and ρ, and show that

$$\frac{Q \rho^{1/2}}{a^{5/2} \mathcal{P}^{1/2}} = F_2 \left(\frac{\mu}{\mathcal{P}^{1/2} a^{3/2} \rho^{1/2}}, \frac{l}{a} \right).$$

One can also show the relation between F_1 and F_2. To do this, let

$$\Pi_1 = \frac{\mathcal{P} a^3 \rho}{\mu^2}, \quad \Pi_1^* = \frac{\mu}{\mathcal{P}^{1/2} a^{3/2} \rho^{1/2}}, \quad \Pi = \frac{Q \rho}{a \mu}, \quad \Pi^* = \frac{Q \rho^{1/2}}{a^{5/2} \mathcal{P}^{1/2}}.$$

Noting that $\Pi_1^* = 1/\sqrt{\Pi_1}, \Pi^* = \Pi_1^* \Pi$, we have $\Pi = F_1(\Pi_1, \Pi_2)$ or

$$\frac{\Pi^*}{\Pi_1^*} = F_1 \left(\frac{1}{(\Pi_1^*)^2}, \Pi_2 \right).$$

But $\Pi^* = F_2(\Pi_1^*, \Pi_2)$. Therefore,

$$F_1(\Pi_1, \Pi_2) = F_1 \left(\frac{1}{(\Pi_1^*)^2}, \Pi_2 \right) = \frac{1}{\Pi_1^*} F_2(\Pi_1^*, \Pi_2).$$

In experiments, where pipe length l is very large, one can neglect the entrance and exit effects and discard l/a as a parameter. Assume that $\mathcal{P} = \mathcal{P}(Q, a, \rho, \mu)$ and $Q \propto U_m a^2$, where U_m is the mean velocity, and starting with a, Q, and ρ as an independent set, one can show that

$$\frac{Pa}{\rho U_m{}^2} = F_3\left(\frac{U_m a}{\nu}\right) = F_3(Re),$$

where Re is the Reynolds number. $F_3(Re)$ is called the resistance coefficient of the pipe. See Sedov (1959, p. 31) for $-F_3(Re) \equiv \psi$ as a function of Re based on the radius of the pipe. For a circular pipe, it can be shown that $-F_3(Re) = 8/Re$ when the flow is laminar (fluid particles move parallel to the pipe wall). This analytical result is due to Poiseuille. Laminar flow is valid up to about $Re = 1300$. A transition and a turbulent flow regimes follow for $Re > 1300$. In turbulent flow, fluid particles move in an erratic fashion even in a direction perpendicular to the pipe wall (see Chapter 10 for details).

We emphasize that the most important point in this example is the existence of the possibility of constructing new dimensionless variables from other dimensionless variables.

9.10.2   Water Waves

Consider a train of periodic progressive waves moving over shallow water as shown in Figure 9.3. The fluid density, ρ, is constant and the sea floor is flat. We initially assume that viscosity does not play a role, and therefore the fluid motion is inertial. Let us analyze the phase speed, c_p, of the waves. We write $c_p = f(\lambda, h_o, g, \rho, A)$ since these are the only variables that, we think, are important. We may choose (ρ, g, h_0) as the primary variables since this set is dimensionally independent. Since $n = 5$, $k = 3$, we must find $n + 1 - k = 3\Pi$s. For the first Π, we consider c_p and, as usual, write $[c_p] = [g]^p [h_0]^q [\rho]^r$ to obtain $p = q = 1/2, r = 0$. Hence,

$$\Pi = \frac{c_p}{g^{1/2} h_0{}^{1/2}} = \frac{c_p}{\sqrt{g h_0}}.$$

For the second Π, we isolate A by $[A] = [g]^p [h_0]^q [\rho]^r$, to obtain $p = r = 0, q = 1$. Hence, $\Pi_1 = A/h_0$. Similar calculations give $\Pi_2 = \lambda/h_0$ (or by inspection). Therefore,

$$\frac{c_p}{\sqrt{g h_0}} = F\left(\frac{A}{h_0}, \frac{\lambda}{h_0}\right).$$

Thus, this relation shows that the dimensionless phase speed of waves is a function of wave amplitude and wave length in general. Note that the right-hand side of the aforementioned equation can be written as $G(A/\lambda)$, which is the wave slope (clearly, G is a function different from F).

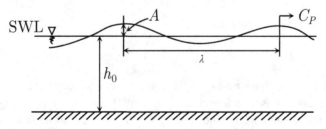

Figure 9.3 Periodic, progressive waves in shallow water.

If we have had considered the viscous effects by including μ in our list, then we would have obtained (show this)

$$\frac{c_p}{\sqrt{gh_0}} = G\left(\frac{A}{h_0}, \frac{\lambda}{h_0}, \frac{\nu}{h_0\sqrt{gh_0}}\right),$$

where the function G is different from the function F in the previous equation. Note that in an inviscid fluid, c_p does not depend on ρ, but in a viscous fluid it does ($\nu = \mu/\rho$). Thus, in the case of an inviscid fluid, we could have started without the density as a parameter. However, one must keep in mind that the density affects the forces since pressure is dependent on the density.

The last dimensionless ratio, on the right-hand side of the aforementioned equation, shows that the smaller the water depth is, the more important the viscosity would be in determining the phase speed. Obviously, one can add more variables to the list of variables that c_p can depend on. Among them are the sea-floor roughness, surface tension, and surface shear stresses on the air–water interface.

It was stated that if the result of the use of the Pi theorem contradicts with the experimental data, then one of the possibilities is that some relevant variables which should have been included in the dimensional analysis were left out by mistake. To see this by way of an example, let us assume instead that $c_p = f(\lambda, h_o, \rho, A)$, or that c_p is not a function of the gravitational acceleration. Then $n = 4$, $k = 2$, and therefore we must find 3Πs. Let us choose (ρ, h_0) as the dimensionally independent set of variables. For the first Π, we consider c_p and, as usual, write $[c_p] = [h_0]^q[\rho]^r$ to discover that we are unable to write three equations with three unknowns since $L, M,$ and T are all involved in the equation. Hence, we cannot nondimensionalize c_p with the physical quantities that it is assumed to depend on. This is the beauty of dimensional analysis; it tells us that some physical quantity is missing in our analysis.

9.10.3 Drag Force on a Sphere

We consider a sphere moving with a constant velocity U in an unbounded viscous fluid (no boundaries are present other than the boundary of the sphere itself). The drag force on the body depends on d, U, ρ, ν and e, i.e., $D = f(d, U, \rho, \nu, e)$, where d is the diameter of the sphere and e is the surface-roughness thickness. Let us choose ρ, d, U as a dimensionally independent set. Then, since $[D] = LMT^{-2}$, $[d] = L$, $[U] = LT^{-1}$, $[\rho] = L^{-3}M$, $[\nu] = L^2T^{-1}$, and $[e] = L$, we have $LMT^{-2} = [\rho]^p[d]^q[U]^r$ or $1 = -3p + q + r$; $1 = p$; $-2 = -r$. Therefore, $p = 1$, $q = r = 2$ and

$$\Pi = \frac{D}{\rho d^2 U^2}.$$

Because we have six variables and a set of three dimensionally independent variables, we must have $n + 1 - k = 3\Pi$s. To determine Π_1, we isolate ν : $[\nu] = L^2T^{-1} = [\rho]^p[d]^q[U]^r$ or $2 = -3p + q + r$, $0 = p$, $-1 = -r$. Therefore, $q = r = 1$ and $p = 0$, and

$$\Pi_1 = \frac{\nu}{dU} \qquad \text{or} \qquad \Pi_1 = \frac{Ud}{\nu}.$$

To determine Π_2, we isolate e : $[e] = L = [\rho]^p [d]^q [U]^r$ or $1 = -3p + q + r$, $0 = p$, $0 = -r$. Therefore, $p = r = 0$, $q = 1$, and $\Pi_2 = e/d$. This result could be obtained by inspection, i.e., without any calculations since e can only be nondimensionalized by d.

Finally, $\Pi = F(\Pi_1, \Pi_2)$ or

$$\frac{D}{\rho U^2 d^2} = F\left(\frac{Ud}{\nu}, \frac{e}{d}\right).$$

The aforementioned equation can be written in a more traditional form:

$$\frac{D}{\frac{1}{2}\rho U^2 S} = C_D(Re, k), \quad S = \frac{\pi d^2}{4}, \quad k = \frac{e}{d},$$

where S is the frontal (or projected) area and k is called the dimensionless roughness coefficient. The coefficient $1/2$ in the denominator is customary and has nothing to do with dimensional analysis. See Sedov (1959) who shows the variation of the drag coefficient $C_w = C_D$ ($W = D$, $v = U$) as a function of the Reynolds number, $R = Re$, for a smooth sphere, i.e., $k = 0$. For the effect of $k \neq 0$ on the drag coefficient, see, e.g., Schlichting (1968).

9.10.4   Flow Past a Flat Plate

Consider a rectangular flat plate of width B and length l moving edgewise with velocity U in an unbounded fluid. The resistance experienced by the plate in a viscous fluid can be anticipated in the following functional form; $R = f(U, l, B, \nu, \rho, e)$, where e is the roughness thickness as in Section 9.10.3. Choosing (U, l, ρ) as the dimensionally independent set of variables, and noting that $[R] = LMT^{-2}$, $[U] = LT^{-1}$, $[B] = [l] = [e] = L, [\nu] = L^2 T^{-1}$, and $[\rho] = L^{-3} M$, we have $LMT^{-2} = [U]^p [l]^q [\rho]^r$ or $1 = p + q - 3r$, $1 = r$, $-2 = -p$. Therefore, $p = q = 2$, $r = 1$, and hence $\Pi = R/U^2 l^2 \rho$. Note that $n + 1 - k = 4\Pi$s must be obtained.

Isolating ν, B, and e, separately, we can obtain the following: $\Pi_1 = Ul/\nu = Re$, $\Pi_2 = B/l, \Pi_3 = e/l \equiv k$. Therefore, $\Pi = F(\Pi_1, \Pi_2, \Pi_3)$ or

$$\frac{R}{\frac{1}{2}\rho S U^2} = G\left(Re, \frac{B}{l}, k\right),$$

where $S = lB$ is the surface area of the plate. The coefficient $1/2$ is again customary. Here, the thickness of the plate is assumed to be very small. If it is not, then the thickness should be a parameter too. The above mentioned function G is known as the frictional-drag coefficient, C_f, which may be calculated by, e.g., the Schoenherr formula:

$$\frac{0.242}{\sqrt{C_f}} = \log_{10}(Re C_f),$$

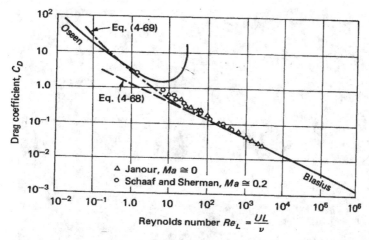

Figure 9.4 Drag coefficient of a flat plate in laminar flow (from White 1974).

which is valid if $k = e/l = 0$ (smooth plate) and $Re > 2 \times 10^6$. Note that C_f is the global (accumulated) friction coefficient, not the local one (for local one $Re_{x_1} = Ux_1/\nu$, where x_1 is measured from the upstream edge of the plate).

For $Re < 3 \times 10^5$, the flow is laminar and $C_f = 1.328/\sqrt{Re}$ according to Blasius' boundary layer theory (see Section 10.8.1). For $10^5 < Re < 2 \times 10^6$, the flow is transitional. This range may change considerably depending on the value of e. Figure 9.4 shows the experimental and theoretical variations of the drag coefficient $C_f = C_D$ for "small" Reynolds numbers, $Re = Re_L$ (from White 1974). In this figure, Eq. (4-69) refers to an interpolation formula for $1 \leq Rel_L \leq 100$, and Eq. (4-68) refers to the Blasius solution of the drag coefficient for one side of the plate (see White (1974) for more information and experimental data).

9.10.5 Ship Resistance

To determine the power requirements of a ship, experiments conducted on a small-scale ship model are necessary. In this example, we will briefly discuss the procedure to obtain the resistance of a ship (prototype) from model experiments. In other words, we will discuss how model data can be extrapolated to prototype.

Towing tanks are used around the world to measure the resistance experienced by ships and submarines, and other oceangoing objects. The towing tank of MARIN in the Netherlands is shown in Figure 9.5.

Let us suppose now that we tow a ship model in a towing tank without allowing it to trim or squat (to avoid added complexity). The towing speed, U, is constant. The resistance or the force acting on the model, in the direction of ship's movement, will be in the following form: $R = f(U, L, \rho, \mu, g)$, where L is the length of the model. Now if we choose U, L, and ρ as a dimensionally independent set of fundamental (or primary) variables, we obtain

Figure 9.5 The towing tank of MARIN in the Netherlands, © MARIN.

$$\frac{R}{\frac{1}{2}\rho U^2 L^2} = f_1\left(\frac{UL}{\nu}, \frac{U}{\sqrt{gL}}\right) = f_1(Re, Fr). \tag{9.49}$$

On the other hand, if we choose L, ρ, and g as the primary variables, we can obtain

$$\frac{R}{\rho g L^3} = f_2\left(\frac{UL}{\nu}, \frac{U}{\sqrt{gL}}\right) = f_2(Re, Fr). \tag{9.50}$$

As an exercise, one should try to obtain the two aforementioned relations.

If we denote the wetted surface of the ship by S and its volume by ∇, then the above mentioned two dimensionless resistance expressions mentioned earlier can also be written, respectively, as

$$\frac{R}{\frac{1}{2}\rho U^2 S} = f_1(Re, Fr), \quad \frac{R}{\rho g \nabla} = f_2(Re, Fr). \tag{9.51}$$

It is customary to denote f_1 and f_2 by C_T and $\bar{C}_T$, respectively:

$$C_T = f_1(Re, Fr) = \frac{R_T}{\frac{1}{2}\rho U^2 S}, \quad \bar{C}_T = f_2(Re, Fr) = \frac{R_T}{\rho g \nabla}, \tag{9.52}$$

that are clearly dimensionless. Because of the fact that both C_T and $\bar{C}_T$ (known as the total resistance coefficients) depend on Re and Fr, we still have the difficulty when we use the Froude and Reynolds scales simultaneously. At this point, it is necessary to make an assumption about the form of the function f_1 or C_T. The assumption is that the total resistance is the sum of the frictional drag, which depends on the Reynolds number, and the residual drag, which depends on the Froude number, i.e.,

$$C_T(Re, Fr) = C_f(Re) + C_R(Fr). \tag{9.53}$$

The residual drag is the sum of wave-making drag and eddy drag, and it is due to the integrated pressure (due to normal stresses) over the entire wetted surface of the

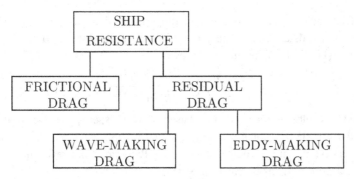

Figure 9.6 Drag components of ship resistance.

body (see Figure 9.6). Frictional drag is due to the integrated tangential stresses over the entire wetted surface of the body. The relation equation (9.53) is called Froude's hypothesis.

The basic justification behind Froude's hypothesis is based on the fact that if the Froude number of the model is the same as the Froude number of the ship, then the wave patterns generated in nature and in a towing tank will be similar, i.e., $(C_R)_m = (C_R)_p$. This means that the corresponding model speed has to be chosen such that

$$(Fr)_m = (Fr)_p = \frac{U_p}{\sqrt{gL_p}} = \frac{U_m}{\sqrt{gL_m}} \quad \text{or} \quad U_m = \sqrt{\frac{L_m}{L_p}} U_p \equiv \sqrt{S_L} U_p,$$

$$(9.54)$$

where the subscript m refers to the model and p refers to the prototype, and $S_L = L_m/L_p$ is the length scale.

Since $(C_T)_m = (C_R)_m + (C_f)_m$ and $(C_T)_p = (C_R)_p + (C_f)_p$, model test results can then be used to predict the full-scale resistance by the following relation:

$$(C_T)_p = (C_T)_m - (C_f)_m + (C_f)_p = (C_R)_m + (C_f)_p,$$

$$(9.55)$$

where we have used the basic assumption $(C_R)_m = (C_R)_p$ and $C_R = C_R(Fr)$ only (see equation (9.53)), so that $(C_R)_p = (C_T)_m - (C_f)_m$. Note that the frictional resistance coefficient C_f is for an equivalent flat plate at the corresponding Reynolds number.

As will be discussed in Chapter 10, the roughness of the surface affects the frictional forces. To account for the increase in the frictional forces due to surface roughness, it is common practice to add a constant roughness allowance, $\Delta C_f = 0.4 \times 10^{-3}$, to the total resistance coefficient $(C_T)_p$ in equation (9.55), i.e.,

$$(C_T)_p = (C_R)_m + (C_f)_p + \Delta C_f = \frac{R_{Tp}}{\frac{1}{2}\rho_p U_p^2 S_p},$$

$$(9.56)$$

where we also used equation (9.52) on the right-hand side.

The frictional resistance, C_f, is generally obtained by use of a formula based on experimental data. A commonly used formula is the International Towing Tank Conference (ITTC) 1957 formula given by

$$C_f = \frac{R_f}{\frac{1}{2}\rho U^2 S} = \frac{0.075}{(\log_{10}(Re) - 2)^2}, \tag{9.57}$$

where the Reynolds number is length based, U is the speed, S is the wetted-surface area, and R_f is the frictional resistance. When we use equation (9.57), one should consider various quantities such as the speed or Reynolds number for the model or the prototype accordingly (as in equation (9.55)).

Once the total resistance coefficient is determined from equation (9.56), one can obtain the total resistance experienced by the prototype, and hence the effective horse power (EHP) from

$$\text{EHP} = \frac{(R_T)_p \ (\text{Newton}) \ U_p \ (\text{m/s})}{745.7 \ (\text{Watts/HP})} \tag{9.58}$$

in SI units, or by

$$\text{EHP} = \frac{(R_T)_p \ (\text{lb}) \ U_p (\text{ft/ sec})}{550 \ (\text{lb} - \text{ft/sec} - \text{HP})} \tag{9.59}$$

in Imperial units, and where HP stands for horse power. For a more detailed account of the subject, see, e.g., Rawson and Tupper (1984), Vol. 2, or Comstock (1967).

Let us next give an example to how these formulas can be used in solving a practical problem.

Example

The resistance experiments of a 80 m long ship are conducted in a towing tank at a scale of 1:20. At the model speed of 3.5knot (recall that 1knot = 0.5144 m/s), the total resistance of the model is measured as 35 N (recall that 1Newton = 1 kg m/s^2 in the SI unit system). The freshwater temperature in the tank is 15°C. At this temperature, the freshwater mass density is 0.999×10^3 kg/m^3 and the kinematic viscosity is 1.1413×10^{-6} m^2/s. The prototype ship will operate in salt water of temperature 20°C. At this temperature, the mass density of salt water is 1.0247×10^3 kg/m^3 and the kinematic viscosity is 1.0565×10^{-6} m^2/s. The wetted-surface area of the prototype is $1,290$ m^2. Use the ITTC 1957 formula, equation (9.57), for the skin friction. Add roughness allowance to the ship resistance coefficient. Calculate the effective horse power of the prototype ship at the corresponding speed.

Solution:
Scales can be calculated immediately: Length scale is $S_L = L_m/L_p = 1/20$. Area scale is

$$(a) \ S_A = S_m/S_p \propto L_m^2/L_p^2 \ \Rightarrow \ S_A = S_L^2.$$

Speed can be calculated from the Froude scaling law:

$$(b) \ U_m = U_p \sqrt{S_L}.$$

Other formulas needed in the calculations are as follows (commas in subscripts should be interpreted as "or"):

$$(c) \quad (Re)_{m,p} = \frac{U_{m,p} L_{m,p}}{\nu_{m,p}}.$$

$$(d) \quad (C_f)_{m,p} = \frac{0.075}{\left(\log_{10}(Re)_{m,p} - 2\right)^2}.$$

$$(e) \quad (C_R)_m = (C_T)_m - (C_f)_m = (C_R)_p.$$

$$(f) \quad (C_T)_p = (C_R)_m + (C_f)_p + \Delta C_f.$$

$$(g) \quad (C_T)_{m,p} = \frac{(R_T)_{m,p}}{\frac{1}{2}\rho_{m,p} U_{m,p}^2 S_{m,p}}.$$

First, $(R_T)_m$ is measured and $(C_T)_m$ is thus calculated from Eq. (g). Next, we can calculate the Reynolds number of the model and prototype from Eq. (c). The residual resistance coefficient of the model can be calculated from Eq. (e) once the frictional resistance coefficient of the model is calculated from Eq. (d). The total resistance coefficient of the prototype is then calculated by using Eq. (f), where given roughness allowance is added and frictional resistance coefficient of the prototype is calculated from Eq. (d). Once the total resistance coefficient of the prototype is known from Eq. (f), one can calculate the total resistance experienced by the prototype from Eq. (g). The calculations are summarized in Tables 9.3 and 9.4.

Table 9.3 Input data (*) and calculated values for the model and prototype at 1:20 scale.

	U (knot)	U (m/s)	L (m)	S (m^2)	T (°C)	ρ (kg/m^3)	ν (m^2/s)
Model	3.50*	1.80	4.0	3.225	15	0.999×10^3*	1.141×10^{-6}*
Prototype	15.65	8.05	80*	1,290*	20	1.024×10^3*	1.057×10^{-6}*

Table 9.4 Reynolds numbers, resistance coefficients, and resistance for the model and prototype.

	Re	C_f	C_R	C_T	R_T
Model	6.31×10^6	3.26×10^{-3}	3.46×10^{-3}	6.71×10^{-3}	35
Prototype	6.10×10^8	1.63×10^{-3}	3.46×10^{-3}	5.49×10^{-3}	234,719

The effective horse power can now be obtained as

$$(EHP)_p = \frac{234719 \times 8.05}{745.7} = 2534 \text{ at } V_p = 15.65 \text{ knots.}$$

9.11   Concluding Remarks

Dimensional analysis in experimental mechanics is very important. One should always look for dimensionless variables before starting an experimental program. If certain dimensionless variables have not been established before a particular experiment, then one must carry out dimensional analysis in order to reduce the number of variables and obtain a functional relation between them. This allows one to simplify the problem considerably. Dimensional analysis is also useful in checking the correctness of differential equations and in numerical analysis.

However, one should not rely on dimensional analysis alone since it cannot provide the missing variables. Nor can it tell us which variables should be included in a dimensionally independent set. Knowledge, experience, and intuition must always accompany dimensional analysis.

9.12   Self-Assessment

9.12.1

Consider the spillway problem whose lengthwise cross section is shown in Figure 9.7. The width of the spillway is b. The fluid is viscous with a mass density of ρ and kinematic viscosity ν. The flow is steady. Gravitational acceleration is g.

1. Show that (ρ, g, h_0) is a dimensionally independent set.
2. Analyze the discharge rate (volume of flow per unit time), Q, by use of dimensional analysis and including all relevant variables. Choose (ρ, g, h_0) as a dimensionally independent set.
3. Assume next that viscosity can be ignored, i.e., the fluid is inviscid. We conduct model tests at a scale of 1:20 and for $Q_m = 3\text{ft}^3/\text{s}$. What is the discharge rate, Q_p, of the prototype? (Hint: You can use the functional relation you obtained earlier and the fact that any functional dependence on b_m/h_{0m} is equal to the functional dependence on b_p/h_{0p}.)

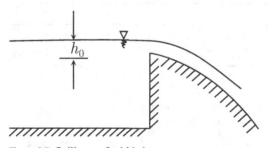

Figure 9.7 Spillway of width b.

9.12.2

1. Assume that you can use Froude's scaling law in the model experiments. Prove that no matter what length scaling factor, $S_L = L_m/L_p$, you use, the prototype accelerations will be the same as the model accelerations. For example, if the vertical acceleration on the deck of a ship is $0.8\,\text{m/s}^2$, we would measure a deck acceleration of $0.8\,\text{m/s}^2$ on the model of this ship at the corresponding conditions in the tank, regardless of what length scale we use. (Hint: First, obtain the scaling for time and then use the fact that the dimension of acceleration is the dimension of velocity over time).
2. If the frequency of vertical oscillations of a prototype ship is $\omega_p = 0.628\,\text{rad/s}$, what is ω_m if the model scale is 1:50?

9.12.3

1. A submarine is deeply submerged so that no waves are generated when it moves. If the prototype speed of the submarine is 10m/s, what would be the speed of the model if the model scale is 1:100? You can take $v_m = v_p$.
2. The same submarine now moves on the free surface with a speed of 5m/s. If viscosity can be ignored, determine the model speed.
3. If the submarine next moves with a speed of 7m/s below, but close to, the free surface so that both the inertial and the viscous forces are important, what should the viscosity of the fluid be in the model tank? Use the seawater value of v_p at 15°C.

9.12.4

Consider the conical structure shown in Figure 9.8. This structure may be used for storage, exploration, and/or production of oil in the arctic environment.

Linear (small-amplitude) waves of amplitude A and wavelength λ are impinging on the conical structure. By assuming that the wave force depends on

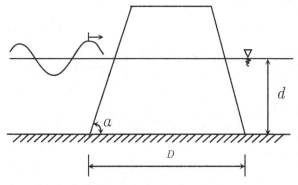

Figure 9.8 Conical oil storage structure.

$F = f(\rho, g, A, D, \lambda, d, \alpha)$, analyze the problem dimensionally to determine what dimensionless parameters the wave force depends on (choose, e.g., $(\rho,\ g,\ D)$ as a dimensionally independent set of quantities after showing that it is an acceptable set).

9.12.5

Model experiments are conducted on a 182.87 m (600 ft) long tanker. The model scale is 1:30. The below table shows various quantities specified or measured during the tests.

	Model	Prototype
Length (m)		182.87
Temperature (Celsius)	15.56	20
Water density (kg/m^3)	9.989E+02 (fresh water)	1.025E+03 (salt water)
Kinematic viscosity (m^2/s)	1.1262E-06	
Wetted surface area (m^2)		6856.4118
Roughness allowance (ΔC_f)	—	4.00E-04

The model resistance data are given in the below table.

V_m (knot)	$(R_T)_m$ (Newton)
1.00	4.00
1.49	8.89
1.75	12.00
1.99	15.24
2.24	19.33
2.49	24.00
2.69	28.00
2.79	30.44
2.90	33.33
2.92	34.22
3.00	37.33
3.04	38.89
3.10	41.20
3.15	42.22
3.20	45.11
3.30	51.33
3.40	58.00
3.49	65.56
3.60	72.89
3.70	81.56
3.84	91.24
3.99	107.55
4.15	132.44
4.30	164.67

Use the ITTC correlation line and Froude's law to calculate and plot the following, by use of a spreadsheet file (or MATLAB or a programming language such as Fortran 95) you need to create:

1. $(C_f)_m$, $(C_f)_p$, $(C_T)_m$, $(C_T)_p$ versus the Re number (be careful about the model and prototype quantities for the Re number, i.e., Re_m, Re_p), and plot them.
2. $(C_T)_m$, $(C_T)_p$, $(C_R)_m$, $(C_R)_p$ versus the Fr number, and plot them.
3. EHP$_p$ of the prototype ship at the corresponding prototype speeds, U_p, in knots, and plot it.

Note: Add roughness allowance $(\Delta C_f)_p$ to $(C_T)_p$. Show your calculations.

Ans.: EHP $= 237$ at $V_p = 5.48$ knot. Do hand calculations for $V_m = 1.0$ knot case first to make sure that your results are correct for the other calculations done through programming.

9.12.6

A production riser is attached to an offshore platform. The natural frequency of oscillation of the riser depends on $f_n = f_n(E, I, g, L, M, n)$ when the riser is assumed to be a simply supported beam (i.e., free to rotate, zero moment and no displacement) and when it is under no tension. Here, E : modulus of elasticity; I : moment of inertia of the cross-sectional area of the riser; L : length of the riser; M_L : mass per unit length of the riser; and n : an integer that refers to the mode shape of the riser.

Analyze the problem dimensionally (taking E, g, L as a dimensionally independent set of quantities) and compare your result with the exact solution of the problem (for zero tension):

$$f_n = \frac{\pi n^2}{2} \sqrt{\frac{EI}{L^4 M_L}}.$$

9.12.7

A single-degree-of-freedom system with viscous damping is undergoing forced vibrations. This system can model, e.g., the surge motions of a floating platform approximately. The excitation force is given by $F_0 \cos \omega t$, where F_0 is the amplitude of the force, ω is the angular frequency of the excitation force, and t is the time.

You will dimensionally analyze the dynamic displacement, x, that depends on

$$x = f(F_0, k, \omega, m, c),$$

where k is the spring constant, m is the total mass, and c is the damping constant. The governing differential equation of the problem is given by

$$m\frac{d^2x}{dt^2} + c\frac{dx}{dt} + kx = F_0 \cos \omega t.$$

1. Determine the dimensions of c and k such that the above mentioned equation is dimensionally homogeneous.

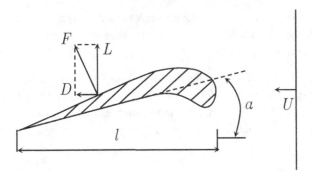

Figure 9.9 A hydrofoil of length l in uniform current.

2. Prove that (F_0, m, ω) form a dimensionally independent set.
3. Use the dimensionally independent set mentioned earlier to show that

$$\frac{x}{x_{st}} = G\left(\frac{c}{c_c}, \frac{\omega}{\omega_n}\right),$$

where $x_{st} = F_0/k$ is the static excursion, $c_c = 2\sqrt{km}$ is the critical damping coefficient, $\omega_n = \sqrt{k/m}$ is the natural (or resonant) frequency, and G is some function.

Note that the above mentioned equation cannot be obtained directly but by replacing $m\omega^2$ by k and so forth. Because of the definitions of ω_n, k, and c_c, you can obtain the above mentioned result directly by selecting F, k, m as a set.

For your information only, the exact solution of this problem is given by

$$\frac{x}{x_{st}} = \left\{ \left(\frac{2c\omega}{c_c\omega_n}\right)^2 + \left(1 - \frac{\omega^2}{\omega_n^2}\right)^2 \right\}^{-1/2}.$$

9.12.8

Consider the hydrofoil shown in Figure 9.9, where L is the lift force, D is the drag force, l is the chord length, α is the angle of attack, and U is the free stream velocity.

Dimensionally analyze this viscous flow problem to determine the functional form of the drag force by showing that:

1. (S, U, ρ) is a dimensionally independent set of quantities, where S is the planform area and ρ is the mass density of the fluid, and
2.

$$\frac{D}{\frac{1}{2}\rho S U^2} = C_F(Re, \alpha),$$

where C_F is the dimensionless drag coefficient.

9.12.9

Your goal is to conduct experiments in a model tank to simulate linear water waves at a scale of 1:40. The waves in the model tank are generated by a wave maker. The prototype data for the waves are: $T_p = 6\,\mathrm{s}$ (wave period), $A_p = 1.2\,\mathrm{m}$ (wave amplitude), and $h_p = 48\,\mathrm{m}$ (water depth can be assumed "deep").

By use of Froude's scaling law, calculate:

1. Wave period, T_m, and wave frequency, $\omega_m = 2\pi/T_m$, in the model scale.
2. Wave length in the model, λ_m, and prototype, λ_p, scales, and wave amplitude, A_m, in the model scale.
3. Wave phase speed in the model, c_m, and prototype, c_p, scales
4. Fluid particle (horizontal) acceleration (maximum) in the model, a_m, and prototype, a_p, scales from linear theory.

9.12.10

Assume that the axial thrust force, T, associated with a propeller of diameter D, that rotates with an angular velocity n (rad/s), while simultaneously moving in the axial direction with a velocity U in water of density ρ is dependent on

$$T = f(\rho, U, n, D).$$

1. Prove that (ρ, n, D) is a dimensionally independent set of quantities.
2. Dimensionally analyze (by use of the set ρ, n, D) the trust force to show that the dimensionless thrust force is some function of the advance ratio, $J = U/nD$, which is a measure of the angle of attack of the propeller blade.

9.12.11

Show if the quantities in each of the following set below are dimensionally independent or not.

1. Velocity, length, energy
2. Density, gravitational acceleration, wave length
3. Time, length
4. Force, velocity, kinematic viscosity
5. Length, acceleration, time

9.12.12

Assume that the velocity components depend on $u_i = f_i(x_1, x_2, p, \rho, g, t, h, \eta)$, $i = 1, 2$, where η is the surface elevation, p the pressure, and h the water depth (two-dimensional problem). Assume further that the evolution of long water waves is governed by the Boussinesq equations in shallow water:

$$u_t + (u \cdot \nabla) u + g\nabla\eta - \frac{h^2}{3}\nabla^2 u_t = 0, \quad \nabla^2 = \nabla \cdot \nabla, \quad \nabla = \frac{\partial}{\partial x_j} e_j, \quad j = 1,2,$$

and where the subscript t indicates differentiation with respect to time.

1. Choose a suitable,[1] dimensionally independent set of quantities.
2. Write the Boussinesq equations in indicial notation.
3. Nondimensionalize the Boussinesq equations you obtained by use of the indicial notation mentioned earlier.

9.12.13

Review the article by Schmidt and Housen (1995) and prove the following:

1. Given Eq. (1), prove that Eq. (2) is correct.
2. If $h = f(U,g)$, prove that $\Pi = hg/U^2 = $ constant (see p. 24, "Power Laws" section).

9.12.14

The resistance of a ship model is measured as $R_{Tm} = 7.4$ lb at $v_m = 3.0$ knots. The prototype ship is 590 ft and it will operate in seawater of temperature 55°F. The model scale is 1:30. Also given are: $\rho_m = 1.9385$ lb s^2/ft^4 (slug/ft^3) and $v_m = 0.12376 \times 10^{-4}$ ft^2/s. The wetted surface area of the model is $S_m = 76.43$ ft^2. The roughness allowance of $\Delta C_f = 0.4 \times 10^{-3}$ needs to be added to C_{Tp}.

Extrapolate the model data to prototype to obtain the EHP of the ship at the corresponding speed in knots.

9.12.15

Sediment transport rate may be written in a functional form:

$$Q_s = f(H,T,h,\rho,\mu,g,x_1,x_2,x_3,t,\rho_s,D),$$

where

Q_s	Sediment transport rate (mass/time)
H	Wave height
T	Wave period
h	Water depth
ρ	Water density
μ	Dynamic viscosity of water
g	Acceleration due to gravity
x_1,x_2,x_3	Spatial coordinates
t	Time
ρ_s	Density of the sediment particles
D	Diameter of the sediment particles

---

[1] Physical quantities that are constant are suitable in general, but think about the physics of the problem also.

1. Show that (H, T, ρ_s) is a dimensionally independent set of quantities.
2. Use the set (H, T, ρ_s) to dimensionally analyze the sediment transport rate, Q_s.

9.12.16

Consider a submerged breakwater (as a box sitting on the sea floor with its top surface submerged as well) in two-dimensional form, and assume that the transmitted wave height, H_T, depends on the following:

$$H_T = f(H_I, \lambda, h, d, b, \rho, g, \mu),$$

where H_I is the incoming wave height, λ the wave length, h the constant water depth, d the water depth above the submerged breakwater, b the width (in the direction of wave propagation) of the breakwater, ρ the mass density of water, g the gravitational acceleration, and μ the dynamic viscosity of water.

1. By use of the set (H_I, ρ, g), dimensionally analyze the transmitted wave height, H_T, to determine the functional dependence of the transmission coefficient $C_T = H_T/H_I$.
2. For constant values of b, d, and h, and linear waves in inviscid fluid, present an argument that the transmission coefficient is a function of the wave frequency, ω, only.

9.12.17

The speed of sound, a, in water can be written in a functional form: $a = f(E, p, \rho, \nu)$, where $E = \rho\, \partial p/\partial\rho$ is the bulk modulus of water, p the pressure, ρ the mass density, and ν the kinematic viscosity coefficient.

1. Select a dimensionally independent set of quantities among (E, p, ρ, ν).
2. By use of the set selected earlier, dimensionally analyze the problem to express the speed of sound, a, in a functional form in terms of dimensionless quantities. Note in your final result that for water $p = cE$, where c is a dimensionless constant.

9.12.18

Review the article by Ertekin and Xu (1994) and prove Eq. (11) of the article. Note: Show first that there is a maximum of two dimensionally independent set of quantities, and that the set (g, U_{10}) is one of the dimensionally independent set of quantities.

9.12.19

Review the article by Dillingham (1984) and, for the following, prove the scalings given below for a tendon of a tension-leg platform (TLP), or a drilling riser of a semi-submersible, to conduct experiments in a model tank by use of Froude's scaling law $(g = g_m = g_p)$; m : model, p : prototype.

1. For force (or weight), $[F] = (1,1,-2)$, $S_F \equiv F_m/F_p = S_\rho S_L^3$, $(S_\rho = \rho_m/\rho_p$, $S_L = L_m/L_p)$
2. For moment of inertia of area, $[I] = (4,0,0)$, $S_I \equiv I_m/I_p = S_L^4$
3. For modulus of elasticity, $[E] = (-1,1,-2)$, $S_E \equiv E_m/E_p = S_\rho S_L$
4. For axial stiffness, $[EA] = (1,1,-2)$, $S_{EA} \equiv (EA)_m/(EA)_p = S_\rho S_L^3$
5. For bending stiffness, $[EI] = (3,1,-2)$, $S_{EI} \equiv (EI)_m/(EI)_p = S_\rho S_L^5$

9.12.20

An infinitely long flat plate oscillates parallel to the x_1 axis with a given velocity $U_w(t)$. The fluid occupies the two-dimensional space $x_2 > 0$ and it is viscous. The frequency of oscillation of the plate is ω, and the amplitude of oscillation is U_0, i.e., $U_w(t) = U_0 \cos(\omega t)$. The boundary-layer thickness can be written, in functional form, as $\delta = \delta(U_0, \omega, \mu, \rho, t)$.

Show that the set (μ, ρ, t) is dimensionally independent, and then dimensionally analyze the boundary-layer thickness by use of this set.

9.12.21

H. Darcy's law of frictional loss in pipe flow is given by

$$h_L = C \frac{L}{D} \frac{U_m^2}{2g},$$

where L is the pipe length, and the dimensionless coefficient C is dependent on

$$C = C(U_m, D, \rho, \mu, e),$$

where U_m is the mean fluid velocity, D the pipe diameter, ρ the mass density of the fluid, μ the dynamic viscosity coefficient, and e is the surface-roughness thickness.

1. Prove that (ρ, U_m, D) is a dimensionally independent set of quantities.
2. Use the above mentioned set to dimensionally analyze the dimensionless coefficient C.

9.12.22

You are going to conduct model tests (in fresh water) on a wave energy conversion (WEC) device (say, a heaving buoy). You will use Froude's scaling law.

1. Prove that the wave power scales as

$$S_P = \frac{P_m}{P_p} = S_\rho S_L^{7/2},$$

where $S_\rho = \rho_m/\rho_p$, P is the wave power, and $S_L = L_m/L_p$, where L is the length scale, and the subscripts m and p refer to the model and prototype scales, respectively.

Note: In proving the above mentioned equation, you must first show how the force, mass, and velocity are scaled.

2. If you use a length scale of $S_L = 1/20$, what will the power produced by the prototype WEC in salt water be if the model WEC produces 10 Watt of power during the model tests?

9.12.23

Prove that if the Froude number similitude is valid, both the Strouhal number and the Keulegan–Carpenter number similitudes, i.e., $(St)_m = (St)_p$ and $(KC)_m = (KC)_p$, are simultaneously valid, where

$$St = \frac{U}{fL}, \qquad KC = \frac{UT}{L},$$

where f is the cyclic frequency (Hz) and T is the period. As a result, you can reach the conclusion that neither the St nor the KC similitude will hold if the Reynolds-number similitude is valid. Note that both the St and the KC numbers refer to an oscillatory (unsteady) flow, not to a steady flow.

9.12.24

1. By considering the ratios of the inertial force per unit volume to the elastic force per unit volume $(E\,A/L^3)$, where E is the Young's modulus and A is the area, derive the Cauchy number $C_y = \rho U^2/E$, generally used in hydroelastic scaling.
2. By considering the ratios of the inertial force per unit volume to the pressure force per unit volume $(p\,A/L^3)$, where p is the pressure, derive the Euler number, $Eu = p/\rho U^2$.

9.12.25

Consider

$$\frac{\rho'}{\rho} = \alpha^a \beta^b \gamma^c,$$

where the mass density ρ' is measured in the SI unit system (kg/m^3) and ρ is measured in the Imperial unit system $(slug/ft^3)$. Use the homogeneity theorem, equation (9.14), to obtain ρ given that $\rho' = 1000$ kg/m^3. Note: $l' = \alpha\,l$, $\alpha = 3.281$, and so forth.

9.12.26

Obtain equation (9.31) by assuming that equation (9.33) is true and taking ρ, g, and h_0 as a dimensionally independent set.

9.12.27

Consider the nonlinear Level I Green–Naghdi equations (see Ertekin 1984) in shallow water.

The continuity equation representing mass conservation is given by

$$\frac{\partial \eta}{\partial t} + \nabla \cdot (h + \eta)V = -\frac{\partial h}{\partial t},$$

and the combined conservation of linear and director momentum equation (when the atmospheric pressure, $\hat{p}$, is taken as zero) is given by

$$\frac{DV}{Dt} + g\nabla\eta + \frac{1}{6}\left[-\frac{D^2h}{Dt^2}\nabla(2\eta - h) + \nabla(4\eta + h)\frac{D^2\eta}{Dt^2} + (h + \eta)\nabla\left(2\frac{D^2\eta}{Dt^2} - \frac{D^2h}{Dt^2}\right)\right] = 0,$$

where $h = h(x_1, x_2, t)$ is the water depth in the absence of any waves, $V = u_1(x_1, x_2, t)$ $e_1 + u_2(x_1, x_2, t)e_2$ is the velocity vector with components u_1 and u_2 along the x_1 and x_2 directions, respectively, $\eta = \eta(x_1, x_2, t)$ is the surface elevation, g is the gravitational acceleration, D/Dt is the material derivative operator (in two-dimensional form), and D^2/Dt^2 is the second material derivative operator (in two-dimensional form).

1. Write these equations in two-dimensional form (in the $x_1 - x_3$ plane) when the water depth is constant, i.e., $h = h_0$, everywhere.
2. Nondimensionalize these two-dimensional equations you just obtained for constant water depth by use of the dimensionally independent set (ρ, g, h_0).

Part IV

Marine Applications

We here build on the fundamentals of the previous parts to model flows and bodies in flows. We study both situations where flow solutions are to be obtained far enough from the boundaries of objects (i.e., where viscous effects could be ignored relative to other flow phenomena), and situations where phenomena close to the boundary (such as turbulence) need to be understood in order to predict behavior in the large. In most of this work we assume linearity, particularly in the context of motion of bodies in waves.

10 Viscous-Fluid Flow

10.1 Introduction

Beginning with this chapter, we attempt to connect some of the general mathematical formulations of the earlier chapters with the application space commonly encountered in the marine arena. In this chapter, we will discuss the motion of a viscous fluid in the presence of solid boundaries. Viscosity plays a role in a number of marine applications, particularly where our interest is in determining the resistance force encountered by ships, towed bodies including offshore platforms, etc. As well, several motion problems involving offshore platforms as well as diverse phenomena such as sediment transport in near-shore waters require the consideration of viscous forces.

We will first discuss the concept of viscosity and the stress tensor. Following that, the conservation equations that govern the motion of a viscous, incompressible fluid will be introduced.

Because of difficulties in obtaining analytical solutions of the equations that govern the motion of a viscous fluid, only a few exact solutions exist. These solutions, which are rather easy to obtain, are generally used to test the accuracy of some numerical and perturbation methods. In other words, before fully implementing a numerical method to solve a nonlinear problem or a complicated linear problem, one would test a particular scheme used by numerically solving a problem whose solution is known. One can also find approximate solutions of some viscous-flow problems. These approximations are generally made for either low- or high-Reynolds-number flows. The latter category includes the boundary-layer approximations.

These topics will be discussed to some extent here, and we will conclude this chapter with laminar and turbulent boundary layers that exist in different flow regimes.

10.2 Viscosity and the Stress Tensor

The most illustrative way to introduce the concept of viscosity is, possibly, by considering the motion of a plate moving in the x_1 direction with a constant velocity U as shown in Figure 10.1. The space between the top (moving) plate and bottom (fixed) plate is filled with a viscous fluid of constant density ρ.

A real fluid adheres to boundaries (this condition is called the no-slip boundary condition and is valid for most fluids), $u_1(h) = U$ and $u_1(0) = 0$. Indeed, experimental data as well as theoretical prediction, see equation (10.50), would show that

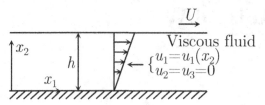

Figure 10.1 A flat plate moving with velocity U above a viscous fluid.

$$u_1(x_2) = \frac{U}{h}x_2, \quad u_2 = u_3 = 0, \tag{10.1}$$

where u_i are the components of the fluid particle velocity vector, u, i.e.,

$$u = u_1 e_1 + u_2 e_2 + u_3 e_3 = u_i e_i, \tag{10.2}$$

and e_i are the unit base vectors in the direction i. The solution equation (10.1) for the velocity is known as the Plane Couette flow (see Section 10.5.1). Furthermore, these experiments would show that the shear stress, $\tau(x_1, 0, x_3, t)$, on the fixed plate is constant and related to the particle velocity by

$$\tau = \mu \frac{du_1}{dx_2}. \tag{10.3}$$

In equation (10.3), the dynamic viscosity coefficient, μ, is simply the constant of proportionality between the shear stress and the velocity gradient. Note that $[\mu] = L^{-1}MT^{-1} = (-1, 1, -1)$. It varies from one fluid to another and also depends on temperature and pressure.

In general, the above stress component for the force is in the direction of x_1. Of course, there also are the normal components of stress, i.e., if we use the vector notation, we have

$$\tau = \left(\mu \frac{du_1}{dx_2}, -p, 0\right) \text{ (on the fixed wall)} \tag{10.4}$$

and

$$\tau = \left(-\mu \frac{du_1}{dx_2}, p, 0\right) \text{ (on the moving wall).} \tag{10.5}$$

These arguments are valid for a Newtonian fluid (stress is a linear function of strain) at a constant temperature.

In simple terms then, viscosity is the ability of a fluid to flow freely. For light fluids like water, we may sometimes ignore viscosity by saying that the fluid is inviscid, i.e., $\mu = 0$. We also define $\nu = \mu/\rho$ as the kinematic viscosity of the fluid. For salt water at 15°C, for example, $\nu = 1.19 \times 10^{-6}$ m²/s, which is very small.

The relation between the stress and strain tensors is given by the constitutive equations.

Let us now reconsider the control volume in the fluid as shown in Figure 10.2. This volume must be thought of as an infinitesimal volume upon which neighboring volumes exert stresses (see also Section 2.6). On the other three faces of the cube, shear stresses will be in opposite directions. Therefore, there are nine functions that

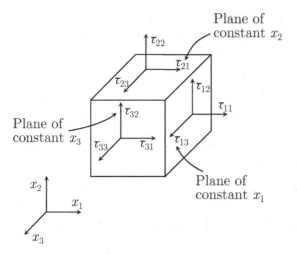

Plane of
constant x_2

τ_{22}

τ_{21}

τ_{23}

τ_{12}

τ_{32}

τ_{11}

Plane of
constant x_3

τ_{33}

τ_{31}

τ_{13}

x_2

x_1

Plane of
constant x_1

x_3

Figure 10.2 An infinitesimal control volume of viscous fluid with surface stresses.

can be combined under the stress tensor. In this case, we have a second-order tensor τ_{ij} which was given by equation (2.75) before.

If we consider a surface element with a unit normal vector $\boldsymbol{n} = (n_1, n_2, n_3)$, the stress (surface force per unit area) on this surface may be given by a vector (say in the x_1 direction) $\boldsymbol{\tau}_1 = \tau_{11} n_1 \boldsymbol{e}_1 + \tau_{12} n_2 \boldsymbol{e}_2 + \tau_{13} n_3 \boldsymbol{e}_3$. In general,

$$\tau_i = \tau_{ij} n_j, \quad i, j = 1, 2, 3. \tag{10.6}$$

In equation (10.6), it is understood that summation is over the repeated index j, and i is the free index (recall Einstein's summation convention).

By use of the conservation of angular momentum or the moment of momentum principle, one can show that $\tau_{ij} = \tau_{ji}$, i.e., the stress tensor is symmetric, see Prob. 10.11.17. We will derive next the other conservation equations that will be used.

10.3   Continuity and the Navier–Stokes Equations

Let us consider a material volume (a volume that contains the same particles), $V(t)$, that changes with time. Since the mass of the fluid is given by

$$m = \iint \int_{V(t)} \rho \, dV,$$

the conservation of mass equation can be postulated as

$$\frac{Dm}{Dt} = \frac{D}{Dt} \iint \int_{V(t)} \rho \, dV = 0, \tag{10.7}$$

where D/Dt is the material derivative or substantial derivative following the fluid particles, see Section 1.4. If we denote the surface of the volume $V(t)$ by $S(t)$, then $S(t)$ must move with the same normal velocity of the fluid particles on the boundary.

Let us next evaluate equation (10.7). Under the usual smoothness (continuous derivatives exist) assumption, one can show that, for any scalar function $F(t)$ given by

$$F(t) = \iint_{V(t)} \int f(x,t)dV,$$

we can write its material derivative as

$$\frac{DF(t)}{Dt} = \frac{D}{Dt} \iint_{V(t)} \int f(x,t)dV = \iint_{V(t)} \int \left\{ \frac{\partial f}{\partial t} + \frac{\partial}{\partial x_j}(fu_j) \right\} dV. \qquad (10.8)$$

Equation (10.8) is known as the Reynolds transport theorem (see, e.g., Fung (1977)). One-dimensional form of this theorem is known as Leibnitz's rule (see, e.g., Wylie (1975)). By applying equation (10.8) to equation (10.7), we obtain

$$\iint_{V(t)} \int \left[\frac{\partial \rho}{\partial t} + \frac{\partial}{\partial x_j}(\rho u_j) \right] dV = 0. \qquad (10.9)$$

Since the volume, $V(t)$, is arbitrary, the integrand must be zero, and therefore,

$$\frac{\partial \rho}{\partial t} + \frac{\partial}{\partial x_j}(\rho u_j) = 0 \qquad (10.10)$$

or

$$\frac{D\rho}{Dt} + \rho \frac{\partial u_i}{\partial x_i} = 0. \qquad (10.11)$$

This is the general conservation of mass statement.

On the other hand, if we have an incompressible fluid, we must require that

$$\frac{D\rho}{Dt} = \frac{\partial \rho}{\partial t} + u_j \frac{\partial \rho}{\partial x_j} = 0. \qquad (10.12)$$

Equation (10.12) also is known as the incompressibility condition, see also Section 1.4. And from equations (10.11) and (10.12), we can also call equation (10.13), the incompressibility condition. Note that if ρ is constant then $D\rho/Dt = 0$, but the converse is not necessarily true. In other words, incompressibility does not necessarily imply that the density is constant. A typical example is a stratified fluid for which $\rho = \rho(x_1,x_2,x_3)$ and $D\rho/Dt = 0$.

Then, for an incompressible fluid,

$$\frac{\partial u_j}{\partial x_j} = 0 \text{ or } u_{j,j} = 0, \quad j = 1,2,3. \qquad (10.13)$$

In vector notation, equation (10.13) is written as the divergence of u being zero, i.e.,

$$\nabla \cdot u = 0, \qquad (10.14)$$

where ∇ is the gradient vector operator:

$$\nabla = \frac{\partial}{\partial x_j}e_j, \; j = 1,2,3. \qquad (10.15)$$

All the repeated or dummy indices run from 1 to 3 unless otherwise specified. Equation (10.13) or equation (10.14) is also called the continuity equation.

Let us now discuss the conservation of momentum equation postulated by

$$\frac{D}{Dt} \iiint\limits_{V(t)} \rho u_i \, dV = \iint\limits_{S(t)} \tau_{ij} n_j \, dS + \iiint\limits_{V(t)} F_i \, dV. \tag{10.16}$$

The left-hand side of equation (10.16) represents the rate of change of momentum within the volume $V(t)$, and the first term on the right-hand side is the surface forces acting on $S(t)$ and the second term is the body forces (due to gravity) that act within $V(t)$.

Now, by use of Gauss' divergence theorem (see also Section 2.3), i.e.,

$$\iiint\limits_{V(t)} \nabla \cdot \boldsymbol{Q} \, dV = \iint\limits_{S(t)} \boldsymbol{Q} \cdot \boldsymbol{n} \, dS, \tag{10.17}$$

for any vector (or tensor) $\boldsymbol{Q}$, see, e.g., Wylie (1975), equation (10.16) can be written as

$$\frac{D}{Dt} \iiint\limits_{V(t)} \rho u_i \, dV = \iiint\limits_{V(t)} \left[\frac{\partial \tau_{ij}}{\partial x_j} + F_i \right] dV, \tag{10.18}$$

where we also used equation (10.6). In equation (10.17), $\boldsymbol{n}$ is the outward unit normal vector on the surface $S(t)$, which encloses $V(t)$.

We can now use the transport theorem, equation (10.8), in equation (10.18), to obtain

$$\iiint\limits_{V(t)} \left[\frac{\partial}{\partial t}(\rho u_i) + \frac{\partial}{\partial x_j}(\rho u_i u_j) \right] dV = \iiint\limits_{V(t)} \left[\frac{\partial \tau_{ij}}{\partial x_j} + F_i \right] dV. \tag{10.19}$$

As before, $V(t)$ is arbitrary, and therefore, the integrands on the left and right sides of equation (10.19) must be equal to each other. Hence,

$$\frac{\partial}{\partial t}(\rho u_i) + \frac{\partial}{\partial x_j}(\rho u_i u_j) = \frac{\partial \tau_{ij}}{\partial x_j} + F_i. \tag{10.20}$$

The left-hand side of equation (10.20) can be expanded to write it as

$$u_i \frac{\partial \rho}{\partial t} + \rho \frac{\partial u_i}{\partial t} + u_i u_j \frac{\partial \rho}{\partial x_j} + \rho u_i \frac{\partial u_j}{\partial x_j} + \rho u_j \frac{\partial u_i}{\partial x_j}$$

$$= u_i \left(\frac{\partial \rho}{\partial t} + u_j \frac{\partial \rho}{\partial x_j} \right) + \rho \left(\frac{\partial u_i}{\partial t} + u_j \frac{\partial u_i}{\partial x_j} \right) + \rho u_i \frac{\partial u_j}{\partial x_j}$$

$$= \rho \left(\frac{\partial u_i}{\partial t} + u_j \frac{\partial u_i}{\partial x_j} \right) = \rho \frac{Du_i}{Dt},$$

where we have used the incompressibility condition, equation (10.12), and the continuity equation, (10.13). The conservation of momentum equation thus becomes

$$\frac{Du_i}{Dt} = \frac{1}{\rho}(\tau_{ij,j} + F_i). \tag{10.21}$$

Equation (10.21), along with equation (10.13), are called Euler's equations. The unknown τ_{ij} must be expressed in terms of the pressure and velocity gradients. The constitutive equations (see Section 10.2) specify the property of a material. One already notes from equation (10.21) that we use the short-hand index notation for derivatives, i.e., $\partial\tau_{ij}/\partial x_j \equiv \tau_{ij,j}$.

For an incompressible Newtonian fluid, the constitutive equations are repeated here for convenience:

$$\tau_{ij} = -p\delta_{ij} + \mu\left(\frac{\partial u_i}{\partial x_j} + \frac{\partial u_j}{\partial x_i}\right), \tag{10.22}$$

where p is the pressure and δ_{ij} is the Kronecker delta which is a second-order tensor defined by equation (1.45).

Note that if $\mu = 0$ or if $u_{i,j} + u_{j,i} = 0$ (meaning no strain) in equation (10.22), there are only normal stresses, and therefore $\tau_{ij} = -p\delta_{ij}$. Also if $\mu \neq 0$, the deformation of a fluid particle causes additional normal stresses.

The quantity F_i in equation (10.21), which represents the conservative body force, refers to the gravitational force and can be written as

$$F_i = -\rho g\delta_{i2}, \tag{10.23}$$

where $\delta_{i2} = 1$ if $i = 2$ and $\delta_{i2} = 0$ if $i = 1$ or 3, since the gravitational acceleration vector points toward the negative x_2 direction.

Before substituting τ_{ij} and F_i in equation (10.21), note that, for a constant μ, we have

$$\begin{aligned}\tau_{ij,j} &= -\left(p\delta_{ij}\right)_{,j} + \{\mu(u_{i,j} + u_{j,i})\}_{,j} = -p_{,j}\delta_{ij} + \mu(u_{i,jj} + u_{j,ij})\\ &= -p_{,i} + \mu(u_{i,jj} + (u_{j,j})_{,i}) = -p_{,i} + \mu u_{i,jj}\end{aligned} \tag{10.24}$$

since δ_{ij} is independent of x_j, and since

$$(u_{j,j})_{,i} = \frac{\partial}{\partial x_i}\left(\frac{\partial u_1}{\partial x_1} + \frac{\partial u_2}{\partial x_2} + \frac{\partial u_3}{\partial x_3}\right) = \frac{\partial}{\partial x_i}(\nabla \cdot \boldsymbol{u}) = 0$$

from the continuity equation, (10.14), for an incompressible fluid. If we now use equations (10.23) and (10.24) in equation (10.21), we obtain

$$\frac{Du_i}{Dt} = -\frac{1}{\rho}p_{,i} - g\delta_{i2} + \nu u_{i,jj}. \tag{10.25}$$

Equation (10.25) are called the Navier–Stokes (N-S) equations. In expanded form, equation (10.25) can be written as

$$\frac{\partial u_1}{\partial t} + u_1\frac{\partial u_1}{\partial x_1} + u_2\frac{\partial u_1}{\partial x_2} + u_3\frac{\partial u_1}{\partial x_3} = -\frac{1}{\rho}\frac{\partial p}{\partial x_1} + \nu\left(\frac{\partial^2 u_1}{\partial x_1^2} + \frac{\partial^2 u_1}{\partial x_2^2} + \frac{\partial^2 u_1}{\partial x_3^2}\right),$$

$$\frac{\partial u_2}{\partial t} + u_1\frac{\partial u_2}{\partial x_1} + u_2\frac{\partial u_2}{\partial x_2} + u_3\frac{\partial u_2}{\partial x_3} = -\frac{1}{\rho}\frac{\partial p}{\partial x_2} - g + \nu\left(\frac{\partial^2 u_2}{\partial x_1^2} + \frac{\partial^2 u_2}{\partial x_2^2} + \frac{\partial^2 u_2}{\partial x_3^2}\right),$$

$$\tag{10.26}$$

$$\frac{\partial u_3}{\partial t} + u_1\frac{\partial u_3}{\partial x_1} + u_2\frac{\partial u_3}{\partial x_2} + u_3\frac{\partial u_3}{\partial x_3} = -\frac{1}{\rho}\frac{\partial p}{\partial x_3} + \nu\left(\frac{\partial^2 u_3}{\partial x_1^2} + \frac{\partial^2 u_3}{\partial x_2^2} + \frac{\partial^2 u_3}{\partial x_3^2}\right).$$

Note that the last terms in equations (10.26) are $\nu\Delta u_1, \nu\Delta u_2$, and $\nu\Delta u_3$, where Δ is the Laplacian operator given by

$$\Delta(\;) \equiv \frac{\partial^2(\;)}{\partial x_1{}^2} + \frac{\partial^2(\;)}{\partial x_2{}^2} + \frac{\partial^2(\;)}{\partial x^3} \equiv \frac{\partial^2(\;)}{\partial x_j \partial x_j} = (\;)_{,jj}. \tag{10.27}$$

Some authors prefer the ∇^2 (del-squared or Nabla-squared) notation to the Δ notation for the Laplacian. Recall that ∇ is given by equation (10.15), and $\nabla^2 = \nabla \cdot \nabla$.

Equations (10.13) and (10.25) govern the motion of an incompressible and viscous fluid. They represent four unknowns, u_i and p, and four equations. It is important to emphasize that the N-S equations are nonlinear (contain the products of the unknowns, e.g., $u_1 \partial u_1 / \partial x_1$), whereas the continuity equation is linear. The biggest difficulty in solving the N-S equations is due to the nonlinear convective acceleration terms, i.e., $(u \cdot \nabla)u$, shown in the vector form of the following N-S equations:

$$\frac{\partial u}{\partial t} + (u \cdot \nabla)u = -\frac{1}{\rho}\nabla p + \nu\Delta u + \frac{1}{\rho}F, \tag{10.28}$$

where the body force F was given by equation (10.23), or by $F = -\rho g e_2$.

The other conservation equation, namely the conservation of energy of fluid mechanics, is not considered here because we assumed that the temperature does not play a role in many problems of concern to us. Note that if the temperature is constant, then the conservation of energy statement becomes conservation of mechanical energy only, and thus, it is equivalent to the conservation of linear momentum statement which already led to the derivation of the N-S equations (see Prob. 10.11.20). Finally, the conservation of angular momentum is equivalent to $\tau_{ij} = \tau_{ji}$ as discussed before (see Section 10.2), and therefore it is automatically satisfied (also see Prob. 10.11.17) in fluid mechanics.

To solve the N-S equations and the continuity equation, one needs to satisfy the boundary conditions relevant to the physical problem in question. In the following section, we will discuss these conditions, and in subsequent sections, some of these conditions will be used in solving certain boundary-value problems encountered in hydrodynamics.

10.4   Boundary Conditions

Solid boundaries are the most common boundaries in marine hydrodynamics. On solid boundaries that are impervious, the fluid velocity must be equal to the velocity of the boundary. In a viscous fluid, both the tangential and normal velocity components of fluid particles and the solid boundary must be the same at each point of contact. In other words,

$$u = q \quad \text{or} \quad u_i = q_i, \tag{10.29}$$

where u is the fluid velocity vector and q is the solid-boundary velocity vector. Equation (10.29) is called the no-slip boundary condition. Clearly, $u_i = 0$ on the boundary if the solid boundary is fixed, i.e., $q_i = 0$.

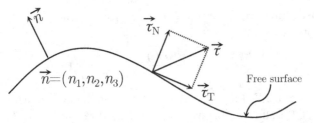

Figure 10.3 Stresses on a free surface with an outward unit normal vector *n*.

In the case of a free surface, which exists when there is a large density difference between two neighboring fluids, such as air and water, the tangential stresses are generally negligible so that the only stress acting on the surface is the normal one. Consider Figure 10.3, and note that $\tau_i = \tau_{ij} n_j$ and $\boldsymbol{\tau}_N = (\boldsymbol{\tau} \cdot \boldsymbol{n})\boldsymbol{n}$, or

$$\tau_{N_i} = \tau_k n_k n_i = \tau_{kj} n_j n_k n_i. \tag{10.30}$$

Since

$$\boldsymbol{\tau} - \boldsymbol{\tau}_N = \boldsymbol{\tau}_T, \tag{10.31}$$

where $\boldsymbol{\tau}_N$ is the normal and $\boldsymbol{\tau}_T$ is the tangential force vectors, we have $(\boldsymbol{\tau}_T \cong 0)$

$$\tau_{ij} n_j - \tau_{kj} n_j n_k n_i = 0. \tag{10.32}$$

We can also use equation (10.22) and note that $-p\delta_{ij} n_j = -p n_i$ and $-p\delta_{kj} n_j = -p n_k$ and $\tau_{ij} = -p\delta_{ij} + \mu(u_{i,j} + u_{j,i})$, we have

$$\mu[(u_{i,j} + u_{j,i})n_j - (u_{k,j} + u_{j,k})n_j n_k n_i] = 0, \tag{10.33}$$

on the free surface and for $i, j, k = 1,2,3$. If there is no surface tension, the dynamic condition on the free surface can be stated as "the normal stress on the surface must be constant," i.e., from equation (10.30), τ_N is constant and equals to $\tau_k n_k = \tau_{kj} n_j n_k$, or by use of equations (10.22) and (10.33), $\{-p\delta_{kj} + \mu(u_{k,j} + u_{j,k})\}n_j n_k$ or

$$-p + \mu(u_{k,j} + u_{j,k})n_j n_k = \text{constant}, \tag{10.34}$$

since $(\delta_{kj} n_j)n_k = n_k n_k = n_1^2 + n_2^2 + n_3^2 = |\boldsymbol{n}|^2 = 1$. Equation (10.34) is the dynamic free-surface boundary condition for a *viscous fluid*. For applications of this boundary condition, see, e.g., Ertekin and Sundararaghavan (1995) and Sundararaghavan and Ertekin (1997). Note that if the fluid is inviscid, $\mu = 0$, equation (10.34) becomes $p = \text{constant} = p_{\text{atm}}$ on the free surface.

The additional condition on the free surface is the kinematic condition that must be satisfied. This condition is also the kinematic boundary condition in ideal flow on any surface, including the rigid and flexible surfaces that are impermeable. It can be stated as the normal component of the velocity of the fluid particles, $\boldsymbol{u} \cdot \boldsymbol{n}$, on the surface must be the same as the normal velocity of the surface, $\boldsymbol{q} \cdot \boldsymbol{n}$, i.e.,

$$\boldsymbol{u} \cdot \boldsymbol{n} = \boldsymbol{q} \cdot \boldsymbol{n}, \tag{10.35}$$

so that the fluid particles remain on the surface. If the equation of the free surface is given by $F(x_1,x_2,x_3,t) = 0$, e.g., $F = \eta(x_1,x_3,t) - x_2$, where η is the free-surface elevation, then the unit normal vector to this surface is given by

$$\boldsymbol{n} = \frac{\boldsymbol{\nabla}F}{||\boldsymbol{\nabla}F||} \quad \text{or} \quad n_i = \frac{F_{,i}}{\sqrt{F_{,j}F_{,j}}}. \tag{10.36}$$

Since the total differential of F (in Lagrangian coordinates in which $x_i = x_i(t)$) is

$$dF = \frac{\partial F}{\partial t} + \frac{dx_i}{dt}\frac{\partial F}{\partial x_i} = \frac{\partial F}{\partial t} + q_i F_{,i},$$

we must have

$$\boldsymbol{q}\cdot\boldsymbol{n} = \boldsymbol{u}\cdot\boldsymbol{n} = -\frac{1}{||\boldsymbol{\nabla}F||}\frac{\partial F}{\partial t}. \tag{10.37}$$

Therefore, the kinematic free-surface condition is

$$\frac{DF}{Dt} = \frac{\partial F}{\partial t} + u_1\frac{\partial F}{\partial x_1} + u_2\frac{\partial F}{\partial x_2} + u_3\frac{\partial F}{\partial x_3} = 0, \tag{10.38}$$

where $F = \eta(x_1,x_3,t) - x_2$. This result could directly be obtained by setting $DF/Dt = 0$ since the material derivative of a function on a material surface, a surface that contains the same particles (such as the free surface of the ocean, except spray), must vanish to conserve mass. Also note that equation (10.35) is true for any geometric surface, not only for the free surface, i.e., $\boldsymbol{q}$ here can be the velocity of any geometric surface, free or not; for example, it could be the velocity vector of the sea floor (e.g., during a seaquake!).

Note that if we substitute $F(x_1,x_2,x_3,t) = x_2 - \eta(x_1,x_3,t) = 0$ in equation (10.38), then equation (10.38) will become the kinematic free surface boundary condition as will be discussed in Section 12.2.

10.5   Some Exact Solutions

As mentioned before, the basic difficulty in solving the N-S equations is with the nonlinear convective acceleration terms. If the physics of the problem is such that these nonlinear terms drop out, then it may be possible to obtain a closed-form or analytical solution to the problem. Such analytical solutions, although simple in nature, can be used to verify the accuracy of numerical methods used before solving more complex problems. In this section, we will look into some exact solutions of the N-S equations.

10.5.1   Couette, Plane Couette, and General Couette Flows

One of the viscous flow problems whose exact solution is known is the flow between two parallel walls or (plates) or Couette flow, see Figure 10.4. In Couette flow, the deriving force may be due to the pressure differential when both walls are fixed, or it may be due to the movement of one wall relative to the other.

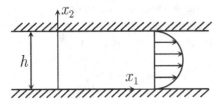

Figure 10.4 Viscous flow (due to a pressure gradient) between two parallel plates.

Consider the flow due to a pressure gradient as shown in Figure 10.4. In both the x_1 and x_3 directions, the plate dimensions are assumed to be much larger than h so that the flow is in the x_1–x_2 plane (two-dimensional), and entrance/exit effects are neglected. Also, the flow is steady by assumption ($\partial/\partial t = 0$) and is the same for all x_1. Therefore, $u_1, u_2 \neq f(x_1)$, and also $u_3 = 0$ by assumption (2-D flow). However, we must have, by assumption, $p = p(x_1)$. Note that both u_1 and u_2 are zero on the top and bottom plates (because of the no-slip condition, equation (10.29)).

With the above remarks in mind, one of the governing equations of the problem, namely the continuity equation for an incompressible fluid given by equation (10.13), reduces to

$$\frac{du_2}{dx_2} = 0. \tag{10.39}$$

Because $u_2 \neq f(x_2)$ from equation (10.39), and since $u_2 = 0$ at $x_2 = 0$ and at $x_2 = h$, u_2 must be zero everywhere. Therefore, the N-S equations given by equation (10.25) or equation (10.26) reduce to a single equation:

$$0 = -\frac{1}{\rho}\frac{dp}{dx_1} + \nu\frac{d^2u_1}{dx_2^2}, \tag{10.40}$$

that is the N-S equation in the x_1 direction only. The N-S equation in the x_2 direction is automatically satisfied. The continuity equation is also satisfied. And therefore, only the N-S equation in the x_1 direction must be solved. Note that the original partial differential equations (10.13) and (10.26) are now reduced to an ordinary differential equation, equation (10.40).

The single N-S equation (10.40) shows the balance between the pressure forces and viscous forces, since the inertia forces due to local and convective acceleration terms have dropped out and gravity does not play a role because of the absence of the free surface and the fact that the plates are horizontal.

We can now integrate the reduced N-S equation, equation (10.40), twice in the x_2 direction to obtain

$$u_1(x_2) = \frac{1}{\mu}\frac{dp}{dx_1}\left(\frac{x_2^2}{2} + c_1 x_2 + c_2\right), \tag{10.41}$$

where c_1 and c_2 are the integration constants. To determine these constants, we must impose the no-slip boundary condition equation (10.29), i.e.,

$$u_1(0) = u_1(h) = 0. \tag{10.42}$$

The constants c_1 and c_2 are obtained by substituting the above conditions in equation (10.41). As a result, one obtains $c_1 = -h/2$, $c_2 = 0$. Therefore,

$$u_1(x_2) = \frac{1}{2\mu}\frac{dp}{dx_1}x_2(x_2 - h).$$ (10.43)

Now let us define a dimensionless pressure gradient by

$$P = -\frac{h^2}{2\mu U_m}\frac{dp}{dx_1},$$ (10.44)

where U_m is the mean flow velocity defined by

$$U_m = \frac{1}{h}\int_0^h u_1(x_2)dx_2.$$ (10.45)

Then, from equations (10.43) through (10.45), we have

$$\frac{u_1(x_2)}{U_m} = P\frac{x_2}{h}\left(1 - \frac{x_2}{h}\right),$$ (10.46)

which shows that $u_1(x_2) > 0$ if $P > 0$, i.e., the flow is in the direction of negative pressure gradient, or from high pressure to low pressure, as expected. The velocity distribution $u_1(x_2)$ is parabolic and symmetric with respect to $x_2 = h/2$ at which point $du_1/dx_2 = 0$. The maximum value of the velocity is

$$u_1(x_2 = h/2) = \frac{PU_m}{4}.$$ (10.47)

In the Couette flow discussed earlier, the deriving force is the pressure gradient or differential. Another deriving force could be obtained by moving one of the walls or plates at a constant velocity U. We have briefly discussed this problem earlier (see Figure 10.1). Let us next obtain the solution to this problem which is called the Plane Couette flow.

In the Plane Couette flow, it is clear that $u_3 = 0$ because the problem is two dimensional. Then the continuity equation (10.13) gives

$$\frac{\partial u_1}{\partial x_1} + \frac{\partial u_2}{\partial x_2} = 0.$$ (10.48)

Since at every section of the plate the velocity profile is assumed to be the same, we must have $u_1 = u_1(x_2)$ only (thus $u_{1,1} = 0$). Therefore, from equation (10.48), u_2 cannot be a function of x_2 (because $u_{2,2} = 0$). Furthermore, u_2 is also not a function of t or x_1 because of the assumptions of steadiness and the neglect of entrance/exit effects, respectively. Thus, $u_2 = 0$ identically everywhere since u_2 must be zero on the solid, fixed boundaries.

Having these facts, we know that the continuity equation is satisfied exactly, and the N-S equations (equation (10.26)) simply become (assuming that there is no pressure gradient)

$$0 = \mu\frac{d^2u_1}{dx_2^2}.$$ (10.49)

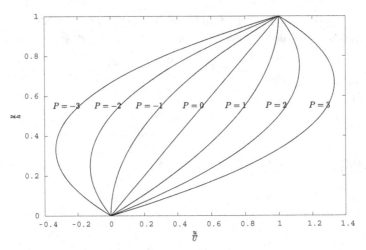

Figure 10.5 Velocity profiles in general Couette flow.

Integrating equation (10.49), we have $u_1(x_2) = c_1 x_2 + c_2$. And by use of the no-slip boundary condition, $u_1(0) = 0, u_1(h) = U$, we obtain the integration constants c_1 and c_2 as $c_1 = U/h$, and $c_2 = 0$. Therefore,

$$u_1(x_2) = \frac{U}{h} x_2, \tag{10.50}$$

which shows a linear variation in the x_2 direction (see also equation (10.1)). This flow is the Plane Couette flow.

It is possible to have a pressure gradient and a moving wall simultaneously as the sources of the driving force. Noting that the governing equations of both cases discussed above are linear, we can simply superimpose the two solutions to obtain

$$\frac{u_1(x_2)}{U} = \frac{x_2}{h}\left\{1 + P(1 - \frac{x_2}{h})\right\} \tag{10.51}$$

as the solution of the general Couette flow, where U is understood to be U_m with regard to the pressure-gradient part of the solution. The velocity profiles are shown in Figure 10.5 as functions of the dimensionless pressure gradient P. Note that reverse flow occurs when the pressure gradient resists the motion induced by the moving wall.

10.5.2   Poiseuille and Hagen–Poiseuille Flows

The problem that we will discuss next is called the Poiseuille flow, which refers to a steady flow in a conduit or duck with an arbitrary but constant cross section. As discussed before, the entrance/exit effects are neglected (the conduit is assumed to be very long), and therefore, the axial velocity components due to the initial disturbance are assumed to have died out.

Let us consider Figure 10.6. The transverse and axial velocity components are zero, i.e., $u_2 = u_3 = 0$. Then the continuity equation (that becomes $\partial u_1/\partial x_1 = 0$) shows the possible dependence of u_1 as $u_1 = f(x_2, x_3)$. We could have, of course, stated at

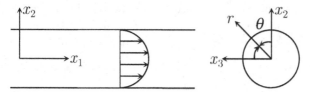

Figure 10.6 Steady flow in a conduit of arbitrary (but constant) cross section.

the outset that $u_1 \neq f(x_1)$, and thus that the continuity equation is satisfied. The N-S equations, (10.8), then become

$$0 = -\frac{1}{\rho}\frac{\partial p}{\partial x_1} + v\left(\frac{\partial^2 u_1}{\partial x_2^2} + \frac{\partial^2 u_1}{\partial x_3^2}\right), \qquad 0 = -\frac{1}{\rho}\frac{\partial p}{\partial x_2}, \qquad 0 = -\frac{1}{\rho}\frac{\partial p}{\partial x_3}. \qquad (10.52)$$

Equation (10.52) shows that $p \neq f(x_2, x_3)$. Thus,

$$\frac{1}{\mu}\frac{dp}{dx_1} = \frac{\partial^2 u_1}{\partial x_2^2} + \frac{\partial^2 u_1}{\partial x_3^2}. \qquad (10.53)$$

Since $u_1 \neq f(x_1)$ from the continuity equation, the right-hand side of equation (10.53) cannot be a function of x_1. On the other hand, because $p \neq f(x_2, x_3)$, the left-hand side of equation (10.53) can only be a function of x_1. Therefore, both sides must be equal to the same constant. Equation (10.53) is called the Poisson equation.[1]

If the cross section of the conduit is circular, elliptical, or rectangular (and some other simple shapes), one can find a closed-form solution of the Poisson equation subject to the no-slip boundary condition. The case that corresponds to the flow in a circular cylinder (or pipe), which is called the Hagen–Poiseuille flow, will be discussed next.

The use of the cylindrical coordinate system shown in Figure 10.6 is convenient especially when the cross section is circular. Noting that $u_1 \neq f(\theta)$, the Poisson equation (10.53) in this coordinate system becomes

$$\frac{1}{r}\frac{d}{dr}\left(r\frac{du_1}{dr}\right) = \frac{1}{\mu}\frac{dp}{dx_1}. \qquad (10.54)$$

Because $p \neq f(r)$, the above equation can be integrated with respect to r twice to obtain

$$u_1(r) = \frac{r^2}{4\mu}\frac{dp}{dx_1} + c_1 \log r + c_2,$$

where c_1 and c_2 are the integration constants that can be obtained by imposing the following conditions: $u_1(0)$ must be finite (physical reasoning), and $u_1(r = a) = 0$ (no-slip condition). Therefore, $c_1 = 0$ and $c_2 = -(1/4\mu)(dp/dx_1)a^2$. As a result,

$$u_1(r) = \frac{1}{4\mu}\frac{dp}{dx_1}(r^2 - a^2). \qquad (10.55)$$

---

[1] It also is the governing equation of some wave-mechanics problems in hydrodynamics and torsional-stress problems in elasticity.

Similar to the Couette problem discussed before, the velocity distribution, equation (10.55), is parabolic, and $dp/dx_1 \neq 0$ is a requirement for u_1 to be nonzero.

The total volume rate or discharge rate can now be calculated by

$$Q = \int_A u_1(r)\,dA, \tag{10.56}$$

where $dA = 2\pi r\,dr$ is the differential element of the flow cross-sectional area. Thus, by use of equation (10.55), we obtain

$$Q = -\frac{\pi}{8\mu}\frac{dp}{dx_1}a^4, \quad \frac{dp}{dx_1} < 0. \tag{10.57}$$

The mean velocity U_m is then given by

$$U_m = \frac{Q}{A} = \frac{Q}{\pi a^2} = -\frac{a^2}{8\mu}\frac{dp}{dx_1} \equiv -\frac{a^2}{8\mu}P, \tag{10.58}$$

where P is the pressure differential. Because $(u_1)_{\max} = u_1(r = 0) = -(a^2/4\mu)(dp/dx_1)$, we have

$$U_m = \frac{1}{2}(u_1)_{\max}. \tag{10.59}$$

The shear stress on the wall is given by

$$\tau(r = a) = -\mu\frac{du_1(a)}{dr} = 4\mu\frac{U_m}{a} = \text{constant.} \tag{10.60}$$

This problem was dimensionally analyzed in Section 9.10.1, where we have defined $P \equiv dp/dx_1$. From the relation between U_m and P given by equation (10.58), we can write

$$P = -\frac{8\mu U_m}{a^2} \tag{10.61}$$

or

$$\frac{Pa}{\rho U_m^2} = -\frac{8\nu}{aU_m} = -\frac{8}{Re} = F_3(Re), \tag{10.62}$$

where $F_3(Re)$, the pipe resistance coefficient, was discussed in Section 9.10.1. Also from the relation between Q and P obtained earlier, i.e.,

$$Q = -\frac{\pi}{8\mu}Pa^4, \tag{10.63}$$

we have

$$\frac{Q\rho}{a\mu} = -\frac{\pi}{8}\frac{Pa^3\rho}{\mu^2} = F_1\left(\frac{Pa^3\rho}{\mu^2}\right), \tag{10.64}$$

where F_1 was obtained (by ignoring the entrance/exit effects) in Section 9.10.1 by dimensional analysis. As mentioned in that section, the experiments show that the analytical results obtained here are valid only when the Reynolds number is low ($Re < 1000$) so that the flow is laminar. This will be discussed in subsequent sections.

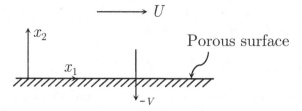

Figure 10.7 Steady viscous flow over a porous, flat surface.

10.5.3   Flow over a Porous Wall

Consider Figure 10.7, which shows a two-dimensional, steady ($\partial/\partial t = 0$), and uniform flow (or current), i.e., $u_1 \to U$ as $x_2 \to \infty$, over a porous, flat surface. We assume that the porous surface provides suction with a vertical velocity $u_2(x_1,0) = -V$, which is constant. Moreover, we assume that $u_1 \neq f(x_1)$ so that $u_1 = f(x_2)$ only, and the pressure is hydrostatic. Since the velocity u_1 is the same at any x_1, the continuity and the N-S equations for the 2-D flow become

$$\frac{\partial u_2}{\partial x_2} = 0,$$

$$u_2 \frac{du_1}{dx_2} = v \frac{d^2 u_1}{dx_2^2},$$

(10.65)

$$u_1 \frac{\partial u_2}{\partial x_1} + u_2 \frac{\partial u_2}{\partial x_2} = -\frac{1}{\rho}\frac{dp}{dx_2} - g + v\left(\frac{\partial^2 u_2}{\partial x_1^2} + \frac{\partial^2 u_2}{\partial x_2^2}\right).$$

Also, since $u_2 \neq f(x_2)$ by the continuity equation, it can only be a function of x_1. But noting that $u_2(x_1,0) = -V$ is constant, u_2 must be equal to $-V$ for all x_1 and x_2 (clearly, this is an idealized situation). Hence, the N-S equation in the x_2 direction (last equation of equation (10.65)) is satisfied, giving the hydrostatic pressure only. Therefore, we are left with the second equation in (10.65), which we can integrate, i.e.,

$$-V\frac{du_1}{dx_2} = v\frac{d^2 u_1}{dx_2^2},$$

(10.66)

to obtain

$$\frac{du_1}{dx_2} + \frac{V}{v}u_1 = C_1\frac{V}{v}.$$

(10.67)

This is a first-order ordinary differential equation whose particular solution is $u_1 = C_1 = $ constant. To determine the homogeneous solution, let $u_1 = C_2\exp(C_3 x_2)$, where C_2, C_3 are constants, and substitute it in the homogeneous differential equation ($du_1/dx_2 + Vu_1/v = 0$) to obtain

$$C_2 C_3 e^{C_3 x_2} + \frac{V}{v}C_2 e^{C_3 x_2} = 0 \quad \text{or} \quad C_3 = -\frac{V}{v}.$$

Then,

$$u_1(x_2) = C_1 + C_2 e^{-\frac{V}{v}x_2}.$$

(10.68)

The boundary and physical conditions, respectively, require $u_1(0) = 0$, $u_1(\infty) = U$. As a result, $C_1 = U$ and $C_2 = -U$, and therefore,

$$u_1(x_2) = U\left(1 - e^{-\frac{V}{\nu}x_2}\right). \tag{10.69}$$

If we assume that the thickness of the boundary layer (in which the particles slow down in comparison with particle velocities outside this layer because of the no-slip boundary condition for u_i) on the porous surface at $x_2 = 0$ (denoted by δ; and this will be discussed later in this chapter) is attained at a vertical elevation which corresponds to $u_1(x_2 = \delta) = 0.99U$, then

$$\frac{u_1(\delta)}{U} = 0.99 = 1 - e^{-\frac{V}{\nu}\delta} \quad \text{or} \quad \delta = 4.61\frac{\nu}{V}. \tag{10.70}$$

As seen in equation (10.70), the boundary-layer thickness is inversely proportional to the suction velocity and, therefore, as V increases δ will decrease. Conversely, as ν increases δ will increase; and for a small viscosity fluid, such as water, δ will be "small." This is the reason why the separation of boundary layer, which will also be discussed later in this chapter, can be minimized if one can provide suction. Again, from equation (10.70), we see that for a constant ν, as $V \to 0$, $\delta \to \infty$, since the porous flat surface is infinitely far away from any free surface by assumption. Therefore, the entire fluid domain is the boundary layer! On the other hand, note that if $V = 0$, $u_1 = 0$ for all x_2 from equation (10.69), and therefore, the solution breaks down. The same is true when $V < 0$. The basic reason for the difficulty with the $V = 0$ case is that in a viscous fluid, u_1 has to be a function of the x_1 coordinate in this problem if there is no suction (we will see this in sections 10.7 and 10.8). Recall that we have assumed $u_1 \neq f(x_1)$. The case when $V < 0$ poses another type of difficulty related to the imbalance between the convection and diffusion of vorticity (see, e.g., Currie (1974, p. 249)).

10.6   Low-Reynolds-Number Flows

A few solutions that we have given in the last section are valid, at least in the theoretical sense, for any Reynolds number. For some problems, however, an exact solution may not be possible. If this is the case, one may attempt to obtain solutions of the approximate governing equations. One particular approximation is possible when the Reynolds number (that represents the ratio of the inertia forces to viscous forces) is small. This is known as Stokes' approximation or the creeping-flow assumption. Sometimes, it also is known as the lubrication theory.

If the Reynolds number is very small, then the convective acceleration terms in the N-S equations can totally be neglected. To see this, consider the following dimensionless variables indicated by overbars:

$$u_i = U\bar{u}_i, \quad p = \frac{\rho\nu U}{l}\bar{p}, \quad x_i = l\bar{x}_i, \quad t = \frac{l^2}{\nu}\bar{t}, \tag{10.71}$$

where U and l are the characteristic velocity and length, respectively, in a given problem. Let us substitute these dimensionless variables in the N-S equations given by equation (10.25) to obtain

$$\frac{\nu U}{l^2}\frac{\partial \bar{u}_i}{\partial \bar{t}} + \frac{U^2}{l}\bar{u}_k\bar{\nabla}_k\bar{u}_i = -\frac{\nu U}{l^2}\bar{\nabla}_i\bar{p} + \frac{\nu U}{l^2}\bar{\nabla}_k\bar{\nabla}_k\bar{u}_i, \tag{10.72}$$

where $i, k = 1, 2, 3$, and the dimensionless gradient operator is

$$\bar{\nabla}_k = \frac{\partial}{\partial \bar{x}_k}. \tag{10.73}$$

Now multiply equation (10.72) by $l^2/(U\nu)$ to obtain

$$\frac{\partial \bar{u}_i}{\partial \bar{t}} + Re\,\bar{u}_k\bar{\nabla}_k\bar{u}_i = -\bar{\nabla}_i\bar{p} + \bar{\nabla}_k\bar{\nabla}_k\bar{u}_i, \tag{10.74}$$

where $Re = Ul/\nu$ is the Reynolds number.

If the Re number is small, then we can write the following approximation (since $Re \to 0$ in the limit $U \to 0$):

$$\frac{\partial \bar{u}_i}{\partial t} = -\bar{\nabla}_i\bar{p} + \bar{\nabla}_k\bar{\nabla}_k\bar{u}_i, \tag{10.75}$$

or in vector notation

$$\frac{\partial \bar{u}}{\partial \bar{t}} = -\bar{\nabla}\bar{p} + \bar{\Delta}\bar{u}. \tag{10.76}$$

If we now return to the dimensional variables given by equation (10.71), equation (10.76) becomes

$$\frac{\partial u}{\partial t} = -\frac{1}{\rho}\nabla p + \nu\Delta u, \quad \text{or} \quad \frac{\partial u_i}{\partial t} = -\frac{1}{\rho}p_{,i} + \nu\,u_{i,jj}. \tag{10.77}$$

This equation and the continuity equation, $\nabla \cdot u = 0$, are collectively called the Stokes equations, and they can be viewed as the zeroth-order problem in a perturbation expansion of the following functions:

$$\bar{u} = \bar{u}_0 + \bar{u}_1 Re + \bar{u}_2(Re)^2 + \cdots, \quad \bar{p} = \bar{p}_0 + \bar{p}_1 Re + \bar{p}_2(Re)^2 + \cdots, \tag{10.78}$$

where the small parameter used for expansion around $\bar{u}_0$ and $\bar{p}_0$ is the Re number.

If the flow is steady, the momentum equation, (10.77), becomes

$$\nabla p = \mu\nabla^2 u. \tag{10.79}$$

(A solution of this equation will be given later by equation (10.91).)

If we now take the curl of the curl of equation (10.79) and note the vector identities:

$$\nabla \times \nabla p = 0, \quad \nabla \times (\nabla \times u) = \nabla(\nabla \cdot u) - \nabla^2 u, \tag{10.80}$$

where p is a scalar function, and moreover, using the continuity equation, we can obtain

$$\nabla^4 u = 0. \tag{10.81}$$

This is called the biharmonic equation. The same equation also appears in plate theory in structural mechanics, where u refers to displacements.

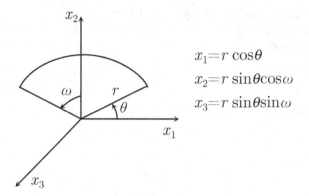

Figure 10.8 Spherical polar coordinates.

In an axisymmetric flow, the velocity is independent of the azimuthal angle ω and the azimuthal component of velocity is zero everywhere. Consider, for instance, the spherical polar coordinates (r, θ, ω) shown in Figure 10.8. Then

$$\frac{\partial \boldsymbol{u}}{\partial \omega} = 0, \quad \boldsymbol{e}_\omega \cdot \boldsymbol{u} = 0. \tag{10.82}$$

The continuity equation in spherical coordinates is written as

$$\frac{1}{r^2} \frac{\partial}{\partial r}(r^2 u_r) + \frac{1}{r \sin \theta} \frac{\partial}{\partial \theta}(u_\theta \sin \theta) = 0 \tag{10.83}$$

for an axisymmetric flow. The velocity components that satisfy the above equation are given by

$$u_r = \frac{1}{r^2 \sin \theta} \frac{\partial \psi}{\partial \theta}, \quad u_\theta = -\frac{1}{r \sin \theta} \frac{\partial \psi}{\partial r}, \tag{10.84}$$

where ψ is called the Stokes' stream function. With the radial and angular velocity components given by equation (10.84), the biharmonic equation (10.81) can be written in terms of the stream function (see, e.g., Happel and Brenner (1965) and Yih (1977)) as

$$E^4 \psi = 0, \tag{10.85}$$

where the operator $E^4 = E^2 E^2$, and the operator E^2 is given by

$$E^2 = \frac{\partial^2}{\partial r^2} + \frac{\sin \theta}{r^2} \frac{\partial}{\partial \theta} \left(\frac{1}{\sin \theta} \frac{\partial}{\partial \theta} \right). \tag{10.86}$$

Equation (10.85) for the stream function can be solved in some cases (see, e.g., Section 10.6.1) next, subject to the boundary conditions of the particular problem being investigated.

10.6.1   Fixed Sphere in Uniform Flow

As an example to a low-Reynolds-number flow, let us consider a sphere placed in an uniform flow U far from the body.

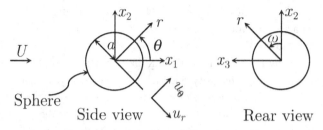

Sphere

Side view

Rear view

Figure 10.9 A fixed sphere in uniform viscous flow.

The no-slip boundary condition $u_r = u_\theta = 0$ on the cylinder dictates through equation (10.84) that

$$\frac{\partial \psi}{\partial r} = \frac{\partial \psi}{\partial \theta} = 0 \quad \text{on } r = a, \tag{10.87}$$

where a is the radius of the sphere. The coordinate system is shown in Figure 10.9. Because of axisymmetry, $\psi \neq f(\omega)$.

Far away from the sphere, u_r and u_θ are given by $u_r = U \cos \theta$, $u_\theta = -U \sin \theta$. Therefore, as $r \to \infty$,

$$u_r = \frac{1}{r^2 \sin \theta} \frac{\partial \psi}{\partial \theta} = U \cos \theta, \quad u_\theta = -\frac{1}{r \sin \theta} \frac{\partial \psi}{\partial r} = -U \sin \theta. \tag{10.88}$$

Thus, we must have

$$\psi = \frac{1}{2} U r^2 \sin^2 \theta + \text{constant as } r \to \infty. \tag{10.89}$$

This far-field boundary condition must also be satisfied.

One can use the separation of variables technique to solve the biharmonic equation (10.85) by setting $\psi(r,\theta) = F(r)G(\theta)$ and substituting it into $E^4 \psi = 0$, and imposing the boundary conditions. The solution, which was obtained by Stokes (1851), is

$$\psi = \frac{1}{4} U a^2 \sin^2 \theta \left(\frac{a}{r} - \frac{3r}{a} + \frac{2r^2}{a^2} \right),$$

$$u_r = U \cos \theta \left(1 + \frac{a^3}{2r^3} - \frac{3a}{2r} \right), \tag{10.90}$$

$$u_\theta = U \sin \theta \left(-1 + \frac{a^3}{4r^3} + \frac{3a}{4r} \right).$$

In the expression for ψ above, the first term corresponds to a doublet, the second to a Stokeslet, and the third to a uniform stream or flow (see Chapter 11). In fact, because the governing equation and the boundary conditions are linear, one can obtain the solutions of ψ separately in three different problems and then superimpose the solutions. These three problems are as follows:

1. Uniform flow (no sphere)
2. Doublet flow (constant pressure, inviscid)
3. Stokeslet flow (variable pressure, viscous)

The above approach was followed by Currie (1974) in obtaining the solution of the uniform flow past a sphere problem. A third alternative method of solution was given by Prandtl and Durand (1935).

Several remarks may be useful in analyzing the solution. First of all, the velocity components do not depend on viscosity, and therefore, on the Reynolds number (since, by assumption, $Re \to 0$.) The presence of the sphere is felt at very large distances. For example, at $r = 10a$, u_r and u_θ are only 90% of U. Another point that should be made is that the velocity components are never greater than U. In potential flow (inviscid fluid and irrotational flow), we will see later on in Chapter 11 that the magnitudes of the velocity components are close to the uniform velocity at a much smaller distance away from the sphere. Also, in potential flow, it is possible that u_r, u_θ are greater than U at certain locations. In Stokes flow, the tangential component of velocity, u_θ, is zero on the sphere because of the no-slip condition. In an inviscid flow, this is not the case as we will see later on. And finally, there is perfect fore and aft symmetry for the streamlines, exactly as in the inviscid flow case. However, it is known that at high-Reynolds-number flows there is a wake region behind a blunt body, like the sphere under discussion, in which vortices are carried away from the body. This fact alone shows that the present solution will break down at high Reynolds numbers. As a result, the streamlines will not be fore and aft symmetric.

In the present case, even though the streamlines are symmetric with respect to $\theta = \pi/2$, the pressure force or form drag is non zero, unlike in the potential-flow case. The reason for this is due to the neglect of convective acceleration terms in the momentum equation.

The pressure is governed by equation (10.79), i.e., $\nabla p = \mu \nabla^2 u$, as derived before. By use of the spherical polar coordinates and integrating the pressure equation, it can be shown that (verify this) the solution of the pressure equation (10.79) is

$$p(r,\theta) = p_\infty - \frac{3\mu a U}{2r^2} \cos\theta, \tag{10.91}$$

which is antisymmetric with respect to $\theta = \pi/2$, and thus the force due to pressure is nonzero as we will see next.

Of course, there is also the frictional drag due to surface shear stresses. The shear part of τ_{ij}, $i \ne j$, was given in equation (10.22) in Cartesian coordinates. Noting that the shear part of the stress is nothing but the rate of deformation or strain tensor, i.e.,

$$\tau_{ij} = -p\delta_{ij} + 2\mu\epsilon_{ij}, \tag{10.92}$$

where

$$\epsilon_{ij} = \frac{1}{2}(u_{i,j} + u_{j,i}), \tag{10.93}$$

one can obtain ϵ_{ij} in any other curvilinear orthogonal coordinate system (see, e.g., Appendix 2 of Yih (1977)), and show that the shear stress is

$$\tau_{r\theta} = \mu\left(\frac{1}{r}\frac{\partial u_r}{\partial\theta} + \frac{\partial u_\theta}{\partial r}\right) = -\frac{U\mu\sin\theta}{r}\left(1 - \frac{3a}{4r} + \frac{5a^3}{4r^3}\right). \tag{10.94}$$

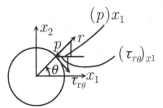

Figure 10.10 Shear- and pressure-force components on a sphere in the x_1 direction.

From Figure 10.10, $(\tau_{r\theta})_{x_1} = \tau_{r\theta} \sin \theta$, $(p)_{x_1} = p \cos \theta$, where $(\tau_{r\theta})_{x_1}$ and $(p)_{x_1}$ refer, respectively, to the shear- and pressure-force components in the x_1 direction. Therefore, the total force acting on the sphere is

$$(F)_{x_1} = \int_0^{2\pi} \tau_{ij}\, n_j \, dS = -\int_0^{\pi} \tau_{r\theta}(a,\theta)\sin\theta\, dA - \int_0^{\pi} p(a,\theta)\cos\theta\, dA, \qquad (10.95)$$

where $dS = r\, d\theta$, $\boldsymbol{n} = \cos\theta\, \boldsymbol{e}_1 + \sin\theta\, \boldsymbol{e}_2$, and $dA = 2\pi a^2 \sin\theta d\theta$. Thus, by use of equations (10.91) and (10.94) and integrating equation (10.95), we have

$$(F)_{x_1} = 4\pi\mu U a + 2\pi\mu U a = 6\pi\mu U a. \qquad (10.96)$$

Therefore, two-thirds of the force is due to friction and one-third is due to the normal pressure.[2]

By use of the traditional definition of the drag coefficient, we then have Stokes' solution:

$$C_D = \frac{F}{\frac{1}{2}\rho U^2 A} = \frac{24}{Re}, \qquad (10.97)$$

where A is the projected area of the sphere ($= \pi a^2$) and Re is the diameter-based Reynolds number, i.e.,

$$Re = \frac{2aU}{\nu}. \qquad (10.98)$$

Figure 10.11 shows the comparison among experiments, theory, and an empirical formula for a sphere. Stokes' solution, equation (10.97) here (which is the same as Eq. (3-258) in Figure 10.11), is valid for $Re < 2$ as expected since we took the limit of the N-S equations as the Reynolds number goes to zero. Oseen's approximation includes a linear convective term in the momentum equation (see, e.g., White (1974)). We give Oseen's approximation below although it is not necessarily better than Stokes' formula:

$$C_D = \frac{24}{Re}\left\{1 + \frac{3Re}{16} + \frac{9}{160}Re^2 \ln(Re)\right\}. \qquad (10.99)$$

The empirical formula obtained by White (1974), who used a curve-fitting method, seems to be valid up to $Re = 2 \times 10^5$ as seen (as in Eq. (3-265)) in Figure 10.11 (from White (1974)). This formula is given by

---

[2] One should not generalize these ratios to other problems.

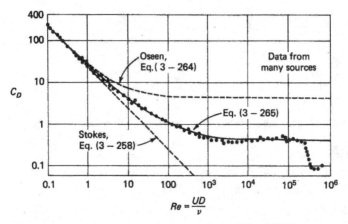

Figure 10.11 Drag coefficients of a sphere (from White (1974)).

$$C_D = \frac{24}{Re} + \frac{6}{1 + \sqrt{Re}} + 0.4. \tag{10.100}$$

10.6.2 Sphere Falling Down with Constant Velocity

As a final application of Stokes' solution for a sphere, consider a sphere (of uniform mass density) that is falling down under the action of gravity, and assume that the velocity is equal to a "small" constant number, i.e., non-accelerating sphere (also see Section 9.10.3). If we write the velocity as

$$U = f(D, \rho_s, \rho_f, \mu, g), \tag{10.101}$$

where D is the diameter of the sphere, and ρ_s and ρ_f are the density of the sphere and fluid, respectively, it can be shown by dimensional analysis (show this) that

$$U = \frac{gD^2}{\mu} G(1)(\rho_s - \rho_f), \tag{10.102}$$

where $G(1)$ is a constant. To obtain equation (10.102), one can assume that $U = f(D, \gamma, \mu)$, where $\gamma = (\rho_s - \rho_f)g$, and choose (D, γ, μ) as a dimensionally independent set.[3]

For low Re numbers, $G(1)$ can be determined explicitly from Stokes' solution as follows. Since, from equation (10.96), $F = 3\pi\mu D U$ is the total hydrodynamic force ($D = 2a$ is the diameter of the sphere), F must be balanced by the weight of the sphere (downward force), W, minus the buoyancy force (upward force), B, because of Archimedes' principle; otherwise the sphere will accelerate. Hence,

$$F = W - B = \pi D^3 (\rho_s - \rho_f)g/6 = 3\pi\mu D U$$

---

[3] Recall that if $n + 1 - k = 1$, as it is here, then Π must be equal to a constant.

or

$$U = \frac{gD^2}{18\mu}(\rho_s - \rho_f), \tag{10.103}$$

where we have used equation (10.96). The velocity U is called the terminal velocity. The comparison of equations (10.102) and (10.103) shows that $G(1) = 1/18$ (a constant, provided that $Re \to 0$).

10.6.3   Circular Cylinder in Two Dimensions

In the case of two-dimensional problems, such as a uniform flow past a circular cylinder whose axis is perpendicular to the incoming flow, it is not possible to satisfy both the no-slip and far-field (uniform flow) boundary conditions when solving the biharmonic equation for the stream function. This is called Stokes' paradox. The reason for this is that in two-dimensional problems, the inertia terms (convective accelerations) cannot be neglected entirely. For a detailed discussion on the subject, see, e.g., Currie (1974, Art. 8.8). To circumvent the problem, Oseen suggested to linearize the inertia terms in the N-S equations. Since $Re \to 0$, the inertia terms must drop out; the solution of the N-S equations (and the continuity equation) with linear inertia terms is valid only in the far field, not near the body. Then, one may seek two solutions: one is exact in the near field and the other is exact in the far field. These solutions are matched in an intermediate region. This method is generally called the matched asymptotic expansion (see, for instance, Van Dyke (1975)). The Oseen–Lamb solution (see Lamb (1932, Art. 343)) is shown in Figure 10.12 (from White (1974)) and is given by

$$C_D = \frac{8\pi}{Re\{0.5 - \gamma + \ln(8/Re)\}}, \tag{10.104}$$

where $\gamma = 0.577216\ldots$ is the Euler's constant (see, e.g., Abramowitz and Stegun (1972)) and $Re = UD/\nu$ ($D = 2a$ = diameter of the circular cylinder). This solution is valid up to $Re = 1.0$.

There is no closed-form solution of the full mass and momentum equations at higher Re numbers except in a few cases for which numerical solutions exist. White (1974) fitted a curve to the experimental data that is shown in Figure 10.12 and valid for $1 \le Re \le 2 \times 10^5$. This curve is given by the following empirical formula:

$$C_D = 1.0 + 10(Re)^{-2/3}. \tag{10.105}$$

For the drag coefficients C_D, of some other 2-D shapes, see Figure 13.19.

The drag force for the above two drag coefficients in equations (10.104) and (10.105) is defined by

$$\frac{\text{Total force}}{\text{Cylinder length}} = \frac{1}{2}\rho C_D U^2 D. \tag{10.106}$$

Note that, in this steady flow problem, the drag force is proportional to U^2. However, if the flow is unsteady (could be time harmonic for instance), then U^2 must be replaced

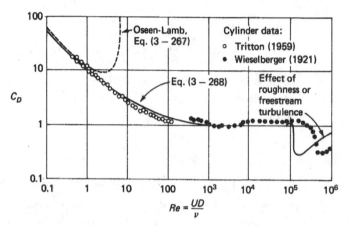

Figure 10.12 Drag coefficients for a circular cylinder (from White (1974)).

by $U|U|$ to account for the proper sign of the velocity, i.e., particle velocity vector may change sign in an unsteady flow, such as in periodic wave motion.

10.7   Boundary Layer and Separation

The exact and approximate solutions of the N-S equations discussed in the last two sections are accurate as long as the conditions, which hold by assumption, continue to exist. The same is true for the laminar boundary-layer theory which is an approximation. In this section, we will discuss the concept of the "boundary layer" around fully submerged bodies in a viscous fluid with "low viscosity," such as water. We used the terminology low viscosity because we are interested in viscous flows confined to a thin layer surrounding a body (recall equation (10.70) for the boundary-layer thickness). This is the case for many of the applications in ocean engineering.

The nature of the flow within the boundary layer is dependent on the shape of the body, roughness of the surface, the smoothness of the incoming flow (or the uniformity of the flow), and the magnitude of the flow velocity. There are several other factors such as the temperature variations and steadiness of the incoming flow (these two are not considered here) which would also affect the flow within the boundary layer. In an inviscid fluid, the no-slip boundary condition on a solid-fluid-interface boundary is relaxed so that the tangential component of the particle velocity on the boundary is not restricted, while the normal component must be equal to the normal velocity of the solid boundary. Whereas in a viscous, or real, fluid, both the tangential and normal components of particle velocity must be equal to the solid-boundary velocity. As a result, in a layer called the boundary layer around the body, fluid particles do slow down to the velocity of the body, if any. Depending on the Reynolds number, the thickness of the boundary layer may vary. At high Reynolds numbers, the particle velocities can reach the value of the free-stream velocity at very small distances away from the boundary from its zero value (if the body is fixed) on the boundary, i.e., the

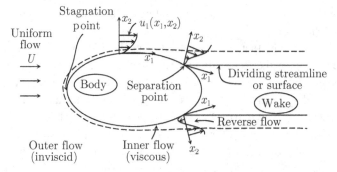

Figure 10.13 Viscous flow about an arbitrarily shaped three-dimensional body.

boundary layer is quite thin at high Re numbers. On the other extreme, the boundary layer would be very thick if the Re number is very small as in Stokes' (or creeping) flow discussed in Section 10.6.

Let us now consider viscous flow about an arbitrarily shaped body as shown in Figure 10.13. Outside the boundary layer, shown by dotted lines, viscous effects can be neglected because the velocity gradients are small. In this outer flow region, the fluid can be assumed inviscid. And if, in addition, the flow far from the body is irrotational initially, then it will remain so forever by Kelvin's theorem (see, e.g., Currie (1974)). This outer flow will be discussed in Chapter 11.

Within the boundary layer, velocity gradients are expected to be large, as shown in Figure 10.13. Outside the boundary layer, the particle velocity is a function of the x_i coordinate at a given time, and this is the free-stream velocity modified by the presence of the body, *but not viscosity*. Within the boundary layer, the particle velocity is a function of the local (x_1, x_2) coordinates (assuming we have a two-dimensional body), and it becomes zero on the solid surface. At one point in the downstream region, the velocity profile may become tangent to the surface normal because of an increase in the pressure in the outer flow which causes the particles inside the boundary layer to further slow down. This point is called the separation point. The increase in the pressure in the outer flow may be due to a rather abrupt change in the body geometry, such as sharp corners, e.g., rectangular nose of a truck or the tail of a hydrofoil. This, in turn, causes the pressure gradients across a streamline to become unreasonably high, and in fact, to become infinite as the radius of the body curvature goes to zero as in the case of a sharp corner. To see this theoretically, one can write the momentum equations in streamline or natural coordinates and obtain $\partial p/\partial n = \rho U^2/R$, for a steady, inviscid, and incompressible flow, where n represents the normal direction to streamlines and R is the radius of curvature of the streamline which coincides with the body surface in an inviscid fluid (see, e.g., Potter and Foss (1982, Art. 8.2)). As $R \to 0, \partial p/\partial n \to \infty$ when $|U| > 0$.

Within the boundary layer (forward of separation point), the pressure is not a function of the vertical coordinate x_2 as we will see later on. Therefore, the pressure can be obtained from the outer-flow solution. This means that one has to first solve the ideal fluid problem in the outer region. Note that separation may be delayed or avoided by streamlining the body so that the pressure within the boundary layer will be constant in the normal direction.

We have seen earlier that the drag coefficient depends on the Re number. The dependence of C_D on the Reynolds number showed a single curve behavior, however, with some distinct features. Since these features are very much related to the concept of boundary layer, let us discuss the C_D-Re curve for a circular cylinder in two-dimensions (see Figure 10.12). As discussed before, for about $Re < 2$, C_D is almost a straight line (creeping flow). We have also mentioned before that for these creeping flows, the presence of the body is felt at very large distances from the body and that the viscous forces are dominant everywhere. This can be thought of as having a boundary layer that covers almost the entire fluid domain.

As the Re number increases, less and less amount of particles are slowed down around the body causing the boundary layer to become thinner because the inertia forces due to convective accelerations become larger. At approximately $Re = 5$, separation occurs for the circular cylinder (see Figure 10.14, from Batchelor (1967)). This is because the flow forms its own boundary such that the unreasonably high pressure gradients become physically reasonable. At about $Re = 40$, the von Karman vortex street is observed in the wake region. Vortex street is a result of periodic vortex shedding. The vorticity is generated along the surface of the body due to shear stresses, which in turn is a result of the no-slip body boundary condition. Then these vortices are diffused along the boundary and convected toward the wake (if it exists) (see Figure 10.15).

Between $Re = 5$ and about $Re = 50$ the vortices are symmetric. But if the Reynolds number is further increased, the flow becomes unstable due to weak damping action of viscosity. This instability causes oscillations of the wake, and therefore, vortices (lumps of vortex) form an oscillating row (see Figure 10.15, from Batchelor (1967), that shows the streaklines[4]). For more information on the instability of boundary layers and Karman vortex street formation, see, e.g., Lamb (1932, Art. 156).

The oscillations of the vortex street can be represented by the Strouhal number which basically is the ratio of the local accelerations ($\partial/\partial t$ terms) to convective accelerations ($u_j\partial/\partial x_j$ terms), i.e.,

$$St \propto \frac{\text{local accelerations}}{\text{convective accelerations}} = \frac{\partial u_1/\partial t}{u_1\partial u_1/\partial x_1} \Rightarrow St = \frac{fL}{U}, \tag{10.107}$$

where f is the cyclic frequency, $f = 1/T$ Hertz (Hz), T(sec) is the period of oscillation of the vortex street, and L and U are the characteristic length and velocity, respectively. In the circular cylinder case, the St number[5] generally varies between 0.1 and 0.2. The oscillation of the vortex street causes oscillatory lift forces in the transverse direction, e.g., strumming of a power cable exposed to wind or of an offshore platform mooring-line in a current field (see, e.g., Blevins (1977)).

---

[4] A streakline is a line that is traced out by a marker fluid (such as a dye) injected into the fluid at a fixed point in space.

[5] Note that $St_m = St_p$ if $Fr_m = Fr_p$, see Prob. 9.12.23.

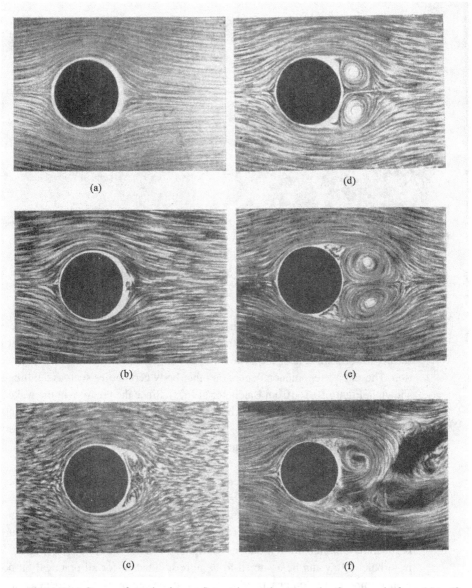

(a)

(d)

(b)

(e)

(c)

(f)

Figure 10.14 Stages of steady viscous flow at increasing oncoming flow speeds, from (a) to (f) (from Batchelor (1967)).

At about $Re = 5,000$, the wake becomes turbulent, i.e., fluid particles move unsteadily in an erratic fashion. This is unlike the case in laminar flow. After the turbulence is observed, the vortex street slowly diminishes due to turbulent eddy diffusion. However, during all these stages, the boundary layer forward of the separation point remains laminar. When the Re number reaches about 200,000, eddies that are formed in the wake penetrate into the boundary layer because of very high magnitude

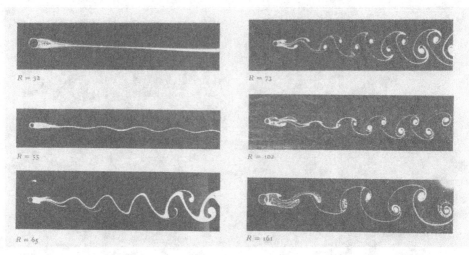

Figure 10.15 Oscillations of the wake, Karman vortex street, behind a circular cylinder as a function of the Reynolds number. The observed lines are streak lines (from Batchelor (1967)).

of vorticity which is a result of high velocities in the outer flow. As a result, the boundary layer becomes turbulent, and thus, thicker. However, the separation point moves further toward the rear of the cylinder because of an increase in momentum convection. The same phenomenon occurs on other body geometries such as a falling sphere shown in Figure 10.16 (from Batchelor (1967)). Since the pressure in the wake is high compared with the pressure in the boundary layer, and the pressure is very high at the forward stagnation point, this movement of the separation point in the downstream direction causes the drag coefficient to reduce in a dramatic way as can also be seen in Figures. 10.11 and 10.12. This Re number range is called the transition flow.

After the boundary layer becomes completely turbulent, the drag coefficient remains almost constant for $Re > 5 \times 10^6$, see Schlichting (1968). This is a very important observation since the Reynolds number is very high in the oceans due to the low kinematic viscosity coefficient for water, and therefore, the flow around an object is generally turbulent even if the oncoming flow is not. As a result, the drag coefficient in turbulent flow can be assumed to be almost constant for all practical purposes. It is also important to emphasize that the transition region is highly dependent on the roughness of the surface. The rougher the surface is, the earlier, in general, will boundary layer become turbulent.

10.8   Laminar Boundary Layer

In laminar flow, fluid particles move in an orderly manner in which the time mean of a flow quantity is equal to itself. Although this is possible in theory for any Re number, it is not always the case in nature. When the Re number reaches a critical value, say Re_{crit}, the neglect of high frequency fluctuations in time is not a good assumption

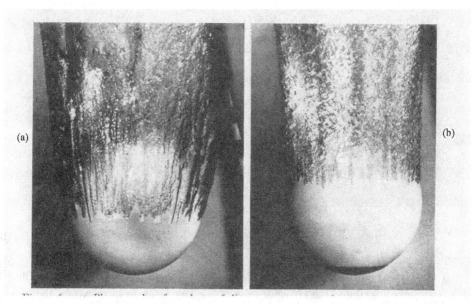

Figure 10.16 Boundary layer around a falling sphere of diameter 22 cm entering water at 64 cm/s. (a) Smooth surface; (b) near the nose, a rough patch of sand is used to make the flow near the nose turbulent. As a result, the boundary layer separation is delayed (from Batchelor (1967)).

anymore. This means that the flow becomes unstable in time, and therefore, small disturbances in the flow field do not die out, but rather grow in size. As it will be discussed later in this chapter, these disturbances cause the flow to become turbulent.

As an example to Re_{crit}, consider the Hagen–Poiseuille flow in a pipe. The Hagen–Poiseuille solution discussed in Section 10.5.2 is valid, in theory, for any Re number. However, if one conducts pipe-flow experiments, one will see that $Re_{crit} = 1000$ approximately. Here the Re number is based on the radius, a, of the pipe and the mean flow velocity U_m, i.e., $Re = U_m a / \nu$. For $Re < Re_{crit}$, one observes laminar flow in the experiments, i.e., disturbances die out and the flow returns to the Hagen–Poiseuille flow after a period of time. But if $Re > Re_{crit}$, the flow will become turbulent. Note, however, that experiments can be conducted in very carefully prepared conditions so that laminar flow in a pipe can be achieved for Re numbers up to about 50000!

Let us next give various definitions of boundary-layer thickness (these definitions, in general, are also used for turbulent boundary layer discussed in Section 10.9, but in the time-averaged sense). We have already seen the most commonly used definition of the boundary-layer thickness in Section 10.5.3, i.e., the value of $x_2 = \delta$ at which $u_1 = 0.99U$, where U is the uniform flow velocity at $x_2 = \infty$. Obviously, it is possible to define boundary-layer thickness more precisely, as we will do next.

One definition is the displacement thickness given by

$$\int_0^{\delta_1(x_1)} u_1(x_1, x_2) dx_2 = \int_{\delta_1(x_1)}^{\infty} (U - u_1) dx_2$$

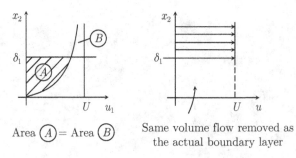

Area $\textcircled{A}$ = Area $\textcircled{B}$ Same volume flow removed as
                                                       the actual boundary layer

Figure 10.17 The decrease in volume flow that defines the displacement thickness.

or

$$\delta_1(x_1) = \int_0^{\infty} \left(1 - \frac{u_1(x_1, x_2)}{U}\right) dx_2. \tag{10.108}$$

If one multiplies δ_1 by U, then $U\delta_1$ is the volume (discharge rate) of the fluid that is absent owing to the presence of the boundary layer. In other words, the decrease in volume flow is given by (see Figure 10.17)

$$U\delta_1 = \int_0^{\infty} (U - u_1)dx_2 = \int_0^{\delta_1(x_1)} (U - u_1)dx_2 + \int_{\delta_1(x_1)}^{\infty} (U - u_1)dx_2$$

$$= \int_0^{\delta_1(x_1)} (U - u_1)dx_2 + \int_0^{\delta_1(x_1)} u_1 dx_2 = \int_0^{\delta_1(x_1)} U dx_2,$$

where δ_1 also represents the amount the streamlines that are displaced from the boundary. Also, $u_1(x_2 > \delta_1(x_1)) = U$.

Another definition of boundary-layer thickness is the momentum thickness given by

$$\rho U^2 \delta_2 = \rho \int_0^{\infty} u_1(U - u_1)dx_2$$

or

$$\delta_2 = \int_0^{\infty} \frac{u_1}{U}\left(1 - \frac{u_1}{U}\right) dx_2, \tag{10.109}$$

where $\rho U^2 \delta_2$ represents the total loss of momentum (momentum defect), and δ_2 is the thickness of the layer whose total momentum, without loss, equals the total lost momentum. Note that $\rho(U - u_1)$ represents the lost momentum per unit volume and $u_1 dx_2$ represents the rate of flow. Recall that momentum = mass * velocity.

The last definition of boundary-layer thickness that we will give is the energy thickness given by

$$\rho U^3 \delta_3 = \rho \int_0^{\infty} u_1(U^2 - u_1{}^2)dx_2$$

or

$$\delta_3 = \int_0^\infty \frac{u_1}{U}\left(1 - \frac{u_1^2}{U^2}\right) dx_2, \tag{10.110}$$

where $\rho U^3 \delta_3$ represents the mechanical energy dissipated due to the presence of the boundary layer, and δ_3 is the thickness of the layer whose total energy, without loss, equals the total lost mechanical energy. Note that $\rho(U^2 - u_1^2)$ represents the lost energy per unit volume. Recall that work = force * distance.

It is important to emphasize that the total volume flow, $U\delta_1$, momentum, $\rho U^2 \delta_2$, and energy, $\rho U^3 \delta_3$, losses are compared with those given by the inviscid-fluid theory $(\mu = 0)$.[6]

When we have an object whose shape is arbitrary, rather than a flat plate, one may ask the question on how δ is defined. This is a good question because in the case of an arbitrary body geometry, the velocities outside the boundary layer are not necessarily equal to the free-stream velocity. In such cases, one can compare the velocities in inviscid and viscous flows and then decide where the edge of the boundary layer is by finding the locations at which the velocities in viscous flow are equal to, for example, 99% of the velocities in potential flow.

10.8.1 Laminar Boundary Layer on a Flat Plate

Let us now consider the steady flow past a flat plate problem in an unbounded flow and derive Prandtl's laminar-boundary-layer equations[7] (see Figure 10.18). By assuming that the flow is steady, i.e., no time dependence, and two-dimensional, i.e., no x_3 dependence, the N-S equations, equation (10.25), (without the body-force (or gravity) term in an unbounded fluid), become

$$\frac{\partial u_1}{\partial x_1} + \frac{\partial u_2}{\partial x_2} = 0,$$

$$u_1 \frac{\partial u_1}{\partial x_1} + u_2 \frac{\partial u_1}{\partial x_2} = -\frac{1}{\rho}\frac{\partial p}{\partial x_1} + \nu\left(\frac{\partial^2 u_1}{\partial x_1^2} + \frac{\partial^2 u_1}{\partial x_2^2}\right), \tag{10.111}$$

$$u_1 \frac{\partial u_2}{\partial x_1} + u_2 \frac{\partial u_2}{\partial x_2} = -\frac{1}{\rho}\frac{\partial p}{\partial x_2} + \nu\left(\frac{\partial^2 u_2}{\partial x_1^2} + \frac{\partial^2 u_2}{\partial x_2^2}\right).$$

These equations need to be solved by satisfying the following no-slip boundary conditions and the outer flow matching (physical) conditions:

$$u_1(x_1,0) = u_2(x_1,0) = 0, \; u_1(x_1,x_2 = \infty) = U, \; u_2(x_1,x_2 = \infty) = 0. \tag{10.112}$$

We will now have order estimates for various quantities and operators. What we mean by the word "Order" is that a function, say $f(\epsilon)$, is $O(g(\epsilon))$ when the "small" parameter ϵ goes to zero and if in the limit $f(\epsilon)/g(\epsilon) = M$, where M is a finite, positive real

[6] It is also noted that if $u_1 = U$ for $x_2 \geq \delta$, then the upper limit of the integrals can be replaced by δ.
[7] The same equations can be derived by letting $Re \gg 1$ as will be discussed in Section 10.8.2.

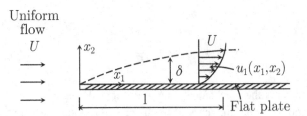

Figure 10.18 Uniform viscous (laminar) flow around a flat plate in an unbounded fluid.

number, and $g(\epsilon)$ is a function that depends on ϵ. The function $g(\epsilon)$ is called the gauge function. For example, $f(\epsilon) = \sin \epsilon = O(\epsilon)$ since in the limit $\epsilon \to 0$, $\sin \epsilon / \epsilon \to 1$. Here $g(\epsilon) = \epsilon$. Also $f(\epsilon) = \cos(\epsilon) = O(1)$ since as $\epsilon \to 0$, $\cos(\epsilon)/\epsilon^0 \to 1$ (see Section 12.3 for more order estimates).

Based on these definitions, the order estimates for our problem are

$$u_1 = O(U), \quad \frac{\partial}{\partial x_1} = O\left(\frac{1}{l}\right), \quad \frac{\partial}{\partial x_2} = O\left(\frac{1}{\delta}\right), \quad \frac{\delta}{l} \ll 1, \quad \frac{\partial}{\partial x_2} \gg \frac{\partial}{\partial x_1},$$

$$\frac{\partial^2}{\partial x_1^2} = O\left(\frac{1}{l^2}\right), \quad \frac{\partial^2}{\partial x_2^2} = O\left(\frac{1}{\delta^2}\right), \quad \frac{\partial u_1}{\partial x_2} = O\left(\frac{U}{\delta}\right), \quad \delta \ll 1, \quad \frac{\partial u_1}{\partial x_1} = O\left(\frac{U}{l}\right),$$

$$\tag{10.113}$$

where δ is the thickness of the boundary layer, l is the distance measured from the edge of the plate, and "$\ll$" means "much less than." Therefore, we can show, by use of equation (10.113), that

$$\frac{\partial u_1}{\partial x_2} \gg \frac{\partial u_1}{\partial x_1}, \quad \frac{\partial^2 u_1}{\partial x_2^2} \gg \frac{\partial^2 u_1}{\partial x_1^2}. \tag{10.114}$$

We can obtain, from the continuity equation in equation (10.111), that

$$\frac{\partial u_2}{\partial x_2} = O\left(\frac{U}{l}\right). \tag{10.115}$$

Note that equation (10.113) implies that $Re \gg 1$ as we will show shortly. Let us next show that $u_1 \gg u_2$. Since,

$$\frac{\partial u_1}{\partial x_2} = O\left(\frac{U}{\delta}\right) \gg \frac{\partial u_2}{\partial x_2} = O\left(\frac{U}{l}\right),$$

and $u_2(x_1,0) = u_2(x_1,\delta) = 0$, we must therefore have

$$u_1 \gg u_2. \tag{10.116}$$

Indeed, from equation (10.115), we have

$$u_2 = O\left(\frac{U\delta}{l}\right) \tag{10.117}$$

since $x_2 = O(\delta)$, and this shows that u_2 is small.

Next, let us consider the boundary-layer thickness δ, and show that the order estimates, equation (10.113), lead to the conclusion that $Re \gg 1$. To do that, consider the friction force, τ, per unit volume:

$$\frac{\partial \tau}{\partial x_2} = \mu \frac{\partial^2 u_1}{\partial x_2^2} \quad \text{or} \quad \frac{\partial \tau}{\partial x_2} \propto \mu \frac{U}{\delta^2}, \tag{10.118}$$

where we used equation (10.113). We know that the inertia force (due to convective accelerations) per unit volume is proportional to $\rho U^2/l$ (see equation (9.2)). These must be balanced within the boundary layer; otherwise the convective accelerations will drop out and, as a result, we will end up with the case of creeping flow (Section 10.6). Therefore,

$$\mu \frac{U}{\delta^2} \propto \rho \frac{U^2}{l} \Rightarrow \delta \propto \sqrt{\frac{\nu l}{U}}. \tag{10.119}$$

As a result, for $\delta/l \ll 1$, equation (10.119) implies that

$$\frac{\delta^2}{l^2} \propto \frac{\nu}{Ul} = \frac{1}{Re} \ll 1 \Rightarrow Re \gg 1, \tag{10.120}$$

i.e., the Reynolds number must be very large when equation (10.113) is valid.

By use of these order estimates and our initial assumptions, and the fact that $u_1 \gg u_2$, the second and third equations in equation (10.111) become

$$u_1 \frac{\partial u_1}{\partial x_1} + u_2 \frac{\partial u_1}{\partial x_2} = -\frac{1}{\rho} \frac{\partial p}{\partial x_1} + \nu \frac{\partial^2 u_1}{\partial x_2^2}, \quad 0 = \frac{1}{\rho} \frac{\partial p}{\partial x_2}. \tag{10.121}$$

The second equation in (10.121) shows that $p \neq f(x_2)$. This is an important result that shows that the solution of the inviscid-flow problem outside the boundary layer can be used to determine the pressure inside the boundary layer! In other words, the inviscid solution outside the boundary layer can be obtained first, and then the solution inside the boundary layer can be obtained second by solving the viscous-flow problem within the boundary layer.

Let us next consider the pressure gradient term in equation (10.121). Because the pressure is not a function of the vertical coordinate, and since U is constant (by assumption) in the outer region, the pressure is constant everywhere (from Bernoulli's equation, see Section 11.2), i.e.,

$$p + \frac{1}{2} \rho U^2 = \text{constant.} \tag{10.122}$$

And naturally, $\partial p/\partial x_1 = 0$ in equation (10.121). Note that this is true only for a flat plate where $u_1 = U$ everywhere outside the boundary layer.

Finally then, we have the laminar boundary-layer equations for a flat plate for $p \neq f(x_1)$:

$$\frac{\partial u_1}{\partial x_1} + \frac{\partial u_2}{\partial x_2} = 0, \quad u_1 \frac{\partial u_1}{\partial x_1} + u_2 \frac{\partial u_1}{\partial x_2} = \nu \frac{\partial^2 u_1}{\partial x_2^2}, \tag{10.123}$$

which are called Prandtl's boundary-layer equations for a flat plate.

Instead of the original N-S equations that are elliptic, we now have a parabolic partial differential equation as the linear momentum equation. Recall that a second-order partial differential equation is given in the following form:

$$Au_{\alpha\alpha} + 2Bu_{\alpha\beta} + Cu_{\beta\beta} = Du_\alpha + Eu_\beta + Fu + G,$$

where α and β are independent variables and subscripts indicate differentiation, and where A through G are functions that are independent of u, is elliptic if $AC > B^2$, parabolic if $AC = B^2$, and hyperbolic if $AC < B^2$.

If $U = f(x_1)$ (which would be the case for a body with curvature), then an additional equation, namely,

$$-\frac{1}{\rho}\frac{\partial p}{\partial x_1} = U\frac{dU}{dx_1}, \tag{10.124}$$

must also be satisfied (see equation (10.122)).

Blasius Solution

To solve Prandtl's boundary-layer equations, (10.123), subject to the boundary conditions given by equation (10.112), we can introduce a stream function such that

$$u_1 = \frac{\partial \psi}{\partial x_2}, \; u_2 = -\frac{\partial \psi}{\partial x_1}, \tag{10.125}$$

since equation (10.123) does not contain the third dimension x_3. Equations (10.123) then become a single equation in terms of ψ :

$$\frac{\partial \psi}{\partial x_2}\frac{\partial^2 \psi}{\partial x_1 \partial x_2} - \frac{\partial \psi}{\partial x_1}\frac{\partial^2 \psi}{\partial x_2^2} = \nu\frac{\partial^3 \psi}{\partial x_2^3}. \tag{10.126}$$

Note that no length scale is preferred in the above equation (i.e., the coefficients do not contain the length variable), and therefore, it is reasonable to expect that the velocity profiles will be similar along the x_1 direction. To this effect, one can introduce a similarity variable, say η, by

$$\eta = \frac{x_2}{\delta(x_1)}, \tag{10.127}$$

or since δ is inversely proportional to the square root of the local Reynolds number (see equation (10.130)) by equation (10.119), we can also write the similarity variable as

$$\eta = x_2\sqrt{\frac{U}{\nu x_1}}. \tag{10.128}$$

One can also use dimensional analysis to obtain this similarity variable η. To see this, assume that $u_1 = u_1(x_1, x_2, U, \nu, \rho)$, and choose (x_1, U, ρ) as a dimensionally independent set to obtain

$$\frac{u_1}{U} = G\left(\sqrt{\frac{Ux_1}{\nu}}, \frac{x_2}{x_1}\right) = H\left(x_2\sqrt{\frac{U}{\nu x_1}}, \frac{x_2}{x_1}\right), \tag{10.129}$$

where dimensional analysis gives us the freedom to write the second equality on the right-hand side by multiplying the first dimensionless quantity by x_2/x_1. The first dimensionless variable of the function H is recognized as η. Obviously, η can also be written as $\eta = (x_2/x_1)\sqrt{Re_{x_1}}$, where the local Reynolds number (which varies with the horizontal coordinate) is defined by

$$Re_{x_1} = \frac{Ux_1}{\nu}. \tag{10.130}$$

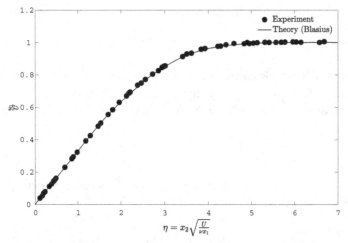

Figure 10.19 Blasius solution and experimental data for the velocity profile inside the laminar boundary layer around a flat plate.

Assuming now that ψ depends on this similarity variable η given by equation (10.128) in the following form

$$\psi(x_1, x_2) = \sqrt{\nu U x_1} F(\eta), \tag{10.131}$$

where F is a function that depends on η only, and carrying out the derivatives, equation (10.126) becomes

$$F'''(\eta) + \frac{1}{2} F(\eta) F''(\eta) = 0, \tag{10.132}$$

where primes denote ordinary differentiation with respect to η (see, e.g., Richardson (1989)). The boundary conditions, equation (10.112), become

$$F(0) = F'(0) = 0, \quad F'(\eta) \to 1 \text{ as } \eta \to \infty. \tag{10.133}$$

Equations (10.132) and (10.133) represent a nonlinear boundary-value problem in one variable, η. However, we now have a single ordinary differential equation to be solved instead of the original partial differential equations. The solution can be obtained numerically and is known as Blasius' solution. Numerical values are given in Table 10.1 (originally from Howarth (1938)) and the calculated and experimentally (Nikuradse 1942) obtained velocity distributions in the boundary layer are shown in Figure 10.19; the agreement is excellent.

The skin friction (or the shear stress) on the surface of the plate can be obtained from

$$\tau_{12}(x_1) = \mu \frac{\partial u_1}{\partial x_2}(x_1, 0) = \mu \sqrt{\frac{U^3}{\nu x_1}} F''(0), \tag{10.134}$$

or, after we use the values in Table 10.1:

$$\tau_{12} = 0.332 \rho U^2 Re_{x_1}^{-1/2}, \tag{10.135}$$

where the local Reynolds number was given by equation (10.130). Equation (10.135) shows that τ_{12} decreases by $\sqrt{x_1}$ along the surface of the flat plate.

The total drag force (per unit width and up to the point x_1), D, is obtained from

$$D = \int \tau_{12}(x_1)dx_1 = 0.664 \rho U^2 \sqrt{\frac{\nu x_1}{U}}. \tag{10.136}$$

Therefore, the frictional-drag coefficient, C_F, and the local skin-friction coefficient, C_f, are given, respectively, by

$$C_F = \left(\frac{D}{x_1}\right) \frac{1}{\frac{1}{2}\rho U^2} = \frac{1.328}{\sqrt{Re}}, \quad Re = \frac{Ul}{\nu}, \quad C_f(x_1) = \frac{\tau_{12}}{\frac{1}{2}\rho U^2} = \frac{0.664}{\sqrt{Re_{x_1}}}, \quad Re_{x_1} = \frac{Ux_1}{\nu}. \tag{10.137}$$

The coefficients in equation (10.137) are for one side of the plate (see, e.g., Newman (1978 p. 17), for the values of C_F as a function of Re. Note that C_f is not valid near the leading edge since δ/x_1 is not $\ll 1$ there, and therefore, the boundary-layer assumptions break down. Also, C_F is valid for up to $Re \cong 10^5$, see Newman (1978, Figure 2.3).

Also, note that

$$C_F = \frac{1}{L} \int_0^L C_f(x_1)dx_1,$$

so that $C_F = 2C_f$ (see, e.g., White (1974, Art. 4.1)) if $Re = Re_{x_1}$.

Various boundary-layer thicknesses, discussed in Section 10.8, can now be given (by use of Table 10.1 and numerical evaluation of the integrals in equation (10.108) through equation (10.110):

$$\delta(x_1) = 5.0 \sqrt{\frac{\nu x_1}{U}} = 5.0 x_1 Re_{x_1}^{-1/2}, \tag{10.138}$$

$$\delta_1(x_1) = 1.722 \sqrt{\frac{\nu x_1}{U}} = 1.722 x_1 Re_{x_1}^{-1/2} = 0.344\,\delta, \tag{10.139}$$

$$\delta_2(x_1) = 0.663 \sqrt{\frac{\nu x_1}{U}} = 0.664 x_1 Re_{x_1}^{-1/2} = 0.1326\,\delta. \tag{10.140}$$

$$\delta_3(x_1) = 1.043 \sqrt{\frac{\nu x_1}{U}} = 1.043 x_1 Re_{x_1}^{-1/2} = 0.2086\,\delta. \tag{10.141}$$

Note that $D = \rho U^2 \delta_2$ from equations (10.136) and (10.140), and $C_f(x_1) = \delta_2/x_1$. The variations of the horizontal and vertical components of the particle velocity in the boundary layer are shown in Figure 10.20.

Table 10.1 The function $F(\eta)$ for the boundary layer along a flat plate at zero incidence

$\eta = x_2\frac{\sqrt{U}}{\nu x_1}$	F	$F' = \frac{u_1}{U}$	F''
0.00000	0.00000	0.00000	0.33206
0.20000	0.00664	0.06641	0.33199
0.40000	0.02656	0.13277	0.33147
0.60000	0.05974	0.19894	0.33008
0.80000	0.10611	0.26471	0.32739
1.00000	0.16557	0.32979	0.32301
1.20000	0.23795	0.39378	0.31659
1.40000	0.32298	0.45627	0.30787
1.60000	0.42032	0.51676	0.29667
1.80000	0.52952	0.57477	0.28293
2.00000	0.65003	0.62977	0.26675
2.20000	0.78120	0.68132	0.24835
2.40000	0.92230	0.72899	0.22809
2.60000	1.07252	0.77246	0.20646
2.80000	1.23099	0.81152	0.18401
3.00000	1.39682	0.84605	0.16136
3.20000	1.56911	0.87609	0.13913
3.40000	1.74696	0.90177	0.11788
3.60000	1.92954	0.92333	0.09809
3.80000	2.11605	0.94112	0.08013
4.00000	2.30576	0.95552	0.06424
4.20000	2.49806	0.96696	0.05052
4.40000	2.69238	0.97587	0.03897
4.60000	2.88826	0.98269	0.02948
4.80000	3.08534	0.98779	0.02187
5.00000	3.28329	0.99155	0.01591
5.20000	3.48189	0.99425	0.01134
5.40000	3.68094	0.99616	0.00793
5.60000	3.88031	0.99748	0.00543
5.80000	4.07990	0.99838	0.00240
6.00000	4.27964	0.99898	0.00240
6.20000	4.47948	0.99937	0.00155
6.40000	4.67938	0.99961	0.00098
6.60000	4.87931	0.99977	0.00061
6.80000	5.07928	0.99987	0.00037
7.00000	5.27926	0.99992	0.00022
7.20000	5.47925	0.99996	0.00013
7.40000	5.67924	0.99998	0.00007
7.60000	5.87924	0.99999	0.00004
7.80000	6.07923	1.00000	0.00002
8.00000	6.27923	1.00000	0.00001
8.20000	6.47923	1.00000	0.00001
8.40000	6.67923	1.00000	0.00000
8.60000	6.87923	1.00000	0.00000
8.80000	7.07923	1.00000	0.00000

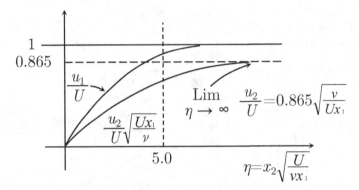

Figure 10.20 Variations of horizontal and vertical components of particle velocities inside the laminar boundary layer.

10.8.2   Further Remarks on Laminar Boundary Layer

1. In the beginning of the last section, it was mentioned that Prandtl's laminar boundary-layer equations can also be obtained by nondimensionalizing the N-S equations and assuming that the Re number is large. Let us show this next. For a two-dimensional fluid motion, we have the continuity and the N-S equations:

$$\frac{\partial u_1}{\partial x_1} + \frac{\partial u_2}{\partial x_2} = 0, \tag{10.142}$$

$$\frac{\partial u_1}{\partial t} + u_1 \frac{\partial u_1}{\partial x_1} + u_2 \frac{\partial u_1}{\partial x_2} = -\frac{1}{\rho}\frac{\partial p}{\partial x_1} + \nu \left(\frac{\partial^2 u_1}{\partial x_1^2} + \frac{\partial^2 u_1}{\partial x_2^2} \right), \tag{10.143}$$

$$\frac{\partial u_2}{\partial t} + u_1 \frac{\partial u_2}{\partial x_1} + u_2 \frac{\partial u_2}{\partial x_2} = -\frac{1}{\rho}\frac{\partial p}{\partial x_2} + \nu \left(\frac{\partial^2 u_2}{\partial x_1^2} + \frac{\partial^2 u_2}{\partial x_2^2} \right).$$

In equation (10.143), we have neglected the body (or gravitational) force without loss in generality. Therefore, the pressure in these equations is the dynamic pressure.

Next, we define the following dimensionless variables (that are of order unity):

$$\bar{x}_1 = \frac{x_1}{L}, \; \bar{x}_2 = \frac{x_2}{B}, \; \bar{t} = \frac{Ut}{L}, \; \bar{u}_1 = \frac{u_1}{U}, \; \bar{u}_2 = \frac{u_2}{V}, \; \bar{p} = \frac{p - p_0}{\rho U^2}, \tag{10.144}$$

where $L, B, U, V,$ and p_0 are the characteristic values of $x_1, x_2, u_1, u_2,$ and p, respectively. Note that we will be able to eliminate B and V seen in equation (10.144).

If we now substitute these dimensionless variables in equation (10.142), we have

$$\frac{U}{L}\frac{\partial \bar{u}_1}{\partial \bar{x}_1} + \frac{V}{B}\frac{\partial \bar{u}_2}{\partial \bar{x}_2} = 0, \tag{10.145}$$

from which we conclude that

$$\frac{V}{B} = O\left(\frac{U}{L}\right). \tag{10.146}$$

But since we have assumed that $Re = UL/\nu \gg 1$, and require that the inertia forces balance viscous forces (within the boundary layer) in the momentum equation, we have, for instance,

$$u_1 \frac{\partial u_1}{\partial x_1} = O\left(\nu \frac{\partial^2 u_1}{\partial x_2^2}\right) \Rightarrow \frac{U^2}{L}\overline{u_1}\frac{\partial \overline{u_1}}{\partial \overline{x_1}} = O\left(\nu \frac{U}{B^2}\frac{\partial^2 \overline{u_1}}{\partial \overline{x_2^2}}\right)$$

or

$$\frac{U^2}{L} = O\left(\nu \frac{U}{B^2}\right) \Rightarrow B = O\left(\frac{L}{\sqrt{Re}}\right). \tag{10.147}$$

By use of equation (10.147) in equation (10.146), we have

$$V = O\left(\frac{BU}{L}\right) = O\left(\frac{U}{\sqrt{Re}}\right). \tag{10.148}$$

We can now eliminate B and V in equation (10.144), by use of equations (10.147) and (10.148), to obtain

$$\overline{x_2} = \frac{x_2}{B} = \frac{x_2}{L}\sqrt{Re}, \quad \overline{u_2} = \frac{u_2}{V} = \frac{u_2}{U}\sqrt{Re}. \tag{10.149}$$

If we now use equation (10.144) (with $\overline{x_2}$ and $\overline{u_2}$ replaced by equation (10.149)) in equation (10.143), we obtain

$$\frac{U^2}{L}\frac{\partial \overline{u_1}}{\partial \overline{t}} + \frac{U^2\overline{u_1}}{L}\frac{\partial \overline{u_1}}{\partial \overline{x_1}} + \frac{U^2\overline{u_2}}{\sqrt{Re}}\frac{\partial \overline{u_1}}{\partial \overline{x_2}}\frac{\sqrt{Re}}{L} = -\frac{1}{\rho}\frac{\rho U^2}{L}\frac{\partial \overline{p}}{\partial \overline{x_1}} + \nu\left(\frac{U}{L^2}\frac{\partial^2 \overline{u_1}}{\partial \overline{x_1}^2} + \frac{U}{L^2}\frac{\partial^2 \overline{u_1}}{\partial \overline{x_2}^2}Re\right),$$

$$\frac{U^2}{L\sqrt{Re}}\frac{\partial \overline{u_2}}{\partial \overline{t}} + \frac{U^2\overline{u_1}}{L}\frac{1}{\sqrt{Re}}\frac{\partial \overline{u_2}}{\partial \overline{x_1}} + \frac{U^2\overline{u_2}}{L\sqrt{Re}}\frac{\sqrt{Re}}{\sqrt{Re}}\frac{\partial \overline{u_2}}{\partial \overline{x_2}}$$

$$= -\frac{\sqrt{Re}}{\rho}\frac{\rho U^2}{L}\frac{\partial \overline{p}}{\partial \overline{x_2}} + \nu\left(\frac{U}{\sqrt{Re}L^2}\frac{\partial^2 \overline{u_2}}{\partial \overline{x_1}^2} + \frac{U}{\sqrt{Re}}\frac{Re}{L^2}\frac{\partial^2 \overline{u_2}}{\partial \overline{x_2}^2}\right),$$

which, of course, are equal to

$$\frac{\partial \overline{u_1}}{\partial \overline{t}} + \overline{u_1}\frac{\partial \overline{u_1}}{\partial \overline{x_1}} + \overline{u_2}\frac{\partial \overline{u_1}}{\partial \overline{x_2}} = -\frac{\partial \overline{p}}{\partial \overline{x_1}} + \frac{1}{Re}\frac{\partial^2 \overline{u_1}}{\partial x_1^2} + \frac{\partial^2 \overline{u_1}}{\partial x_2^2}, \tag{10.150}$$

$$\frac{\partial \overline{u_2}}{\partial \overline{t}} + \overline{u_1}\frac{\partial \overline{u_2}}{\partial \overline{x_1}} + \overline{u_2}\frac{\partial \overline{u_2}}{\partial \overline{x_2}} = Re\frac{\partial \overline{p}}{\partial \overline{x_2}} + \frac{1}{Re^2}\frac{\partial^2 \overline{u_2}}{\partial x_1^2} + \frac{\partial^2 \overline{u_2}}{\partial x_2^2}.$$

Moreover, if we impose the condition $Re \gg 1$ on equation (10.150), we obtain

$$\frac{\partial \overline{u_1}}{\partial \overline{t}} + \overline{u_1}\frac{\partial \overline{u_1}}{\partial \overline{x_1}} + \overline{u_2}\frac{\partial \overline{u_1}}{\partial \overline{x_2}} = -\frac{\partial \overline{p}}{\partial \overline{x_1}} + \frac{\partial^2 \overline{u_1}}{\partial \overline{x_2}^2}, \quad 0 = -\frac{\partial \overline{p}}{\partial \overline{x_2}}. \tag{10.151}$$

The dimensional form of equation (10.151) is precisely the same as equation (10.121), where the $\partial/\partial t$ term was set to zero because of steadiness.

2. Because of the no-slip boundary condition, we have $u_1(x_1,0) = u_2(x_1,0) = 0$. Therefore, the first equation of equation (10.121) evaluated at $x_2 = 0$ becomes

$$\nu \frac{\partial^2 u_1}{\partial x_2^2} = \frac{1}{\rho}\frac{\partial p}{\partial x_1}. \tag{10.152}$$

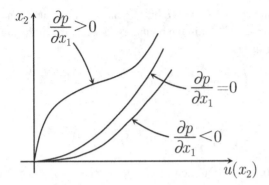

Figure 10.21 Typical (horizontal) component of particle velocity for different pressure gradients inside a boundary layer.

The pressure gradient $\partial p/\partial x_1 > 0$ is called the unfavorable pressure gradient or adverse pressure gradient, and $\partial p/\partial x_1 < 0$ is called the favorable pressure gradient. Beyond a certain point on the body surface, the particle velocity may be reduced. This reduction in velocity depends on the body geometry as well as the Reynolds number. For example, the tangential velocity at the edge of the boundary layer in the rear of a circular cylinder is less than the one in the front of it. As a result, the pressure in the rear of the circle will increase (from Bernoulli's equation), and therefore, the pressure gradient will be adverse, causing the particles to further slow down. In fact, the adverse pressure gradient may force the particles to move backward (reverse flow) after the convective acceleration $\partial u_1/\partial x_2$ becomes zero on the body surface, i.e., the velocity profile becomes tangent to the body normal direction. The point on the body surface at which this phenomenon occurs is called the separation point. Since the shear stress is equal to $\mu \partial u_1/\partial x_2$ on $x_2 = 0$, the shear stress vanishes at this point. Hence, the point of zero shear stress is also the point of separation. Typical velocity profiles for different pressure gradients are shown in Figure 10.21. Just before, at, and after the separation point, the velocity profiles look something like in Figure 10.22.

3. Because we have assumed that the outer flow is constant in the case of a flat plate, we have found that the pressure is constant from Bernoulli's equation. However, if $\partial p/\partial x_1$ does not vanish (or if the flow is nonuniform), one can expand the velocity in a power series in terms of the similarity variable η. This method is known as the Pohlhausen method (see, e.g., Newman (1978, Art. 3.15).

10.9   Turbulent Boundary Layer

In laminar flow, the velocities are assumed to be independent of time; fluid particles move in an orderly fashion. For example, in the case of laminar flow in pipes, we know that particles move parallel to pipe walls. This is not always the case we encounter in nature. After increasing the velocity beyond a "critical" Reynolds number, the flow

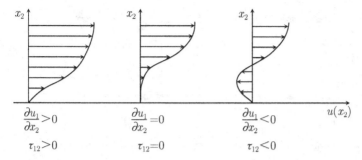

$$\frac{\partial u_1}{\partial x_2} > 0 \qquad \frac{\partial u_1}{\partial x_2} = 0 \qquad \frac{\partial u_1}{\partial x_2} < 0 \qquad u(x_2)$$

$$\tau_{12} > 0 \qquad\qquad \tau_{12} = 0 \qquad\qquad \tau_{12} < 0$$

Figure 10.22 Horizontal velocity profiles inside a boundary layer before, at, and after the separation point.

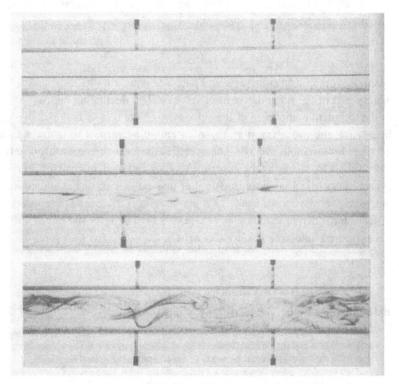

Figure 10.23 Laminar (top), transition (middle), and turbulent (bottom) flows in a laboratory tube, simulating blood flow in an artery (from Caro et al. (2012)).

may become disturbed and stay that way, i.e., a small disturbance introduced in a steady flow may not die out. This unsteady and erratic motion of particles is known as turbulence. Between the laminar and turbulent flow regimes, there is a transition regime where certain features of both regimes can be observed. These three flow characteristics are shown in Figure 10.23 taken from Caro et al. (2012).

Due to the time dependency of particle velocities in a turbulent flow, velocity profiles at different instances of time in a circular pipe, for instance, may look something like the one in Figure 10.24 (on the right).

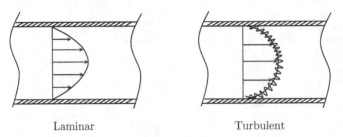

Laminar   Turbulent

Figure 10.24 Laminar (left) and turbulent (right) velocity profiles in a pipe.

The most common method in dealing with a very confused flow is to define the velocity components by

$$u_i = \bar{u}_i + u_i', \tag{10.153}$$

where $\bar{u}_i$ is the mean (time average) part (see equation (10.156)) and u_i' is the fluctuating part of the ith component of velocity. The fluctuating part, u_i', is also known as the turbulent component of the velocity u_i. Note that $\bar{u}_i$ may still depend on time. In general, the averaging may be an ensemble average, a time average, or a space average. In ensemble average, one averages the measurements taken at the same x_i and t values by repeating the experiments over and over again. This averaging is necessary in statistical analysis of data. Let us next discuss the averaging properties of functions.

If we have two functions $f = f(x_1, x_2, x_3, t) \equiv f(x_i, t)$ and $g(x_i, t)$, then, for both ensemble and time averages, we can write

$$\overline{f + g} = \bar{f} + \bar{g}, \quad \overline{\frac{\partial f}{\partial x_i}} = \frac{\partial \bar{f}}{\partial x_i}, \tag{10.154}$$

and similarly for other derivatives, including the time derivative. Also, if the duration, T, of the experiments is large, we can write

$$\overline{cf} = c\bar{f} \text{ if } c \neq f(t), \quad \overline{\bar{f}g} = \overline{\bar{f}\bar{g}} = \bar{f}\bar{g}, \quad \overline{\bar{f}} = \bar{f},$$
$$\overline{\int f\, dx_i} = \int \bar{f}\, dx_i, \quad \overline{f'\bar{g}} = 0, \tag{10.155}$$

where the time average is defined by

$$\bar{f}(x_i, t) = \lim_{T \to \infty} \frac{1}{T} \int_{t-\frac{T}{2}}^{t+\frac{T}{2}} f(x_i, \hat{t})\, d\hat{t}, \tag{10.156}$$

where $\hat{t}$ is the dummy integration variable. Note that $\bar{f}$ is not a function of $\hat{t}$ so that it can be moved out of the integral sign in equation (10.156) if f is replaced by $\bar{f}$ to show that $\overline{\bar{f}} = \bar{f}$.

If we write $f = \bar{f} + f'$ then $\overline{f'} = 0$ if T is large, i.e., $T \gg 1$; in fact, $T \to \infty$. On the other hand, it can be shown, by use of equation (10.156), that

$$\overline{f\,g} = \bar{f}\,\bar{g} + \overline{f'g'}, \tag{10.157}$$

see Prob. 10.11.8.

Next, we would like to express the continuity and the N-S equations in terms of the average velocities. To do this, we introduce equation (10.153) into equation (10.13) to obtain

$$\frac{\partial \bar{u}_i}{\partial x_i} + \frac{\partial u_i'}{\partial x_i} = 0. \tag{10.158}$$

By taking time average of the entire equation (10.158), and noting that $\overline{u_i'}$ is zero if $T \gg 1$, and by use of the first equation in (10.154), we must have

$$\frac{\partial \bar{u}_1}{\partial x_1} + \frac{\partial \bar{u}_2}{\partial x_2} + \frac{\partial \bar{u}_3}{\partial x_3} = 0, \quad \frac{\partial u_1'}{\partial x_1} + \frac{\partial u_2'}{\partial x_2} + \frac{\partial u_3'}{\partial x_3} = 0. \tag{10.159}$$

Following the same procedure, the N-S equations, (10.26) or (10.27), become

$$\frac{\partial \bar{u}_i}{\partial t} + \bar{u}_j \frac{\partial \bar{u}_i}{\partial x_j} + \overline{u_j' \frac{\partial u_i'}{\partial x_j}} = -\frac{1}{\rho} \frac{\partial \bar{p}}{\partial x_i} - g\delta_{i2} + v\overline{u_{i,jj}}, \quad i,j = 1,2,3, \tag{10.160}$$

where we have used some of equation (10.154), equation (10.155), and equation (10.157). However, we can see, for example, that

$$\overline{u_j' \frac{\partial u_1'}{\partial x_j}} = \overline{u_1' \frac{\partial u_1'}{\partial x_1}} + \overline{u_2' \frac{\partial u_1'}{\partial x_2}} + \overline{u_3' \frac{\partial u_1'}{\partial x_3}} = \frac{\partial}{\partial x_1}\overline{(u_1'^2)} + \frac{\partial}{\partial x_2}\overline{(u_1'u_2')} + \frac{\partial}{\partial x_3}\overline{(u_1'u_3')}$$

$$- \overline{u_1'\left(\frac{\partial u_1'}{\partial x_1} + \frac{\partial u_2'}{\partial x_2} + \frac{\partial u_3'}{\partial x_3}\right)}. \tag{10.161}$$

And the last term in parentheses on the right-hand side of equation (10.161) is zero from the second equation of (10.159).

Two more equations similar to equation (10.161) can easily be obtained. We therefore have

$$\overline{u_j' \frac{\partial u_i'}{\partial x_j}} = \frac{\partial}{\partial x_j}\overline{(u_i'u_j')}, \quad i,j = 1,2,3. \tag{10.162}$$

If we now use equation (10.162) in equation (10.160), and move it to the right-hand side of the equation (for traditional reasons), we obtain the turbulent-flow Reynolds equations whose mean is unsteady:

$$\frac{\partial \bar{u}_i}{\partial t} + \bar{u}_j \frac{\partial \bar{u}_i}{\partial x_j} = -\frac{1}{\rho} \frac{\partial \bar{p}}{\partial x_i} - g\delta_{i2} + v\Delta\bar{u}_i - \frac{\partial}{\partial x_j}\overline{(u_i'u_j')}, \quad i,j = 1,2,3. \tag{10.163}$$

Except the last term in equation (10.163), these are the time average of the original N-S equations, (10.27). If we have had modified the stress tensor, given by equation (10.22), in an average sense, by writing it as

$$\overline{\tau}_{ij} = -\bar{p}\delta_{ij} + \mu\left(\frac{\partial \bar{u}_i}{\partial x_j} + \frac{\partial \bar{u}_j}{\partial x_i}\right) - \overline{\rho u_i'u_j'}, \tag{10.164}$$

then we would have obtained equation (10.163) directly from the conservation of momentum equations, but we had no way of knowing this. Equations (10.163) are also known as the unsteady Reynolds-averaged Navier–Stokes (URANS) equations. If $\overline{u_i} \neq f(t)$, then the local time derivative in equation (10.163) is zero. The steady form of equation (10.163) is sometimes known as the RANS equations.

In equation (10.164), the first term on the right-hand side represents the normal stresses, the second, laminar stresses, and the third, turbulent stresses. The turbulent stress components are called the Reynold's stresses. We now have 10 unknowns (the components of the matrix contain 6 unknowns because of symmetry) and 4 equations. The unknown quantities are $\overline{p}$, $\overline{u}_1$, $\overline{u}_2$, $\overline{u}_3$, $\overline{u_1'u_1'}$, $\overline{u_1'u_2'}$, $\overline{u_1'u_3'}$, $\overline{u_2'u_2'}$, $\overline{u_2'u_3'}$, and $\overline{u_3'u_3'}$. Therefore, the solution of these equations, without additional equations, is not possible. This is called the "closure problem," which has been investigated rather unsuccessfully since the times of Prandtl. As a result, one often resort to experimental techniques to determine these new components of the stress tensor, but there are a number of closure models as well, see, e.g., Wilcox (1993). Also, it is important to note that the Reynolds stresses are inertial in nature. See Ertekin and Rodenbusch (2016) and Wehausen, Webster, and Yeung (2016) for a discussion on some computational fluid dynamics (CFD) tools available currently to overcome the closure problem in an empirical way.

The addition of the Reynolds stresses to the stress tensor increases the viscous drag on a body. This means that the velocity gradients in a turbulent boundary layer are larger than what they are in a laminar boundary layer. This results in an increase in the boundary-layer thickness in a turbulent flow compared with a laminar flow.

10.9.1   Turbulent Boundary Layer on a Flat Plate

We now consider a two-dimensional flat plate submerged in a fluid of infinite extent (unbounded fluid) for simplicity. We assume that the oncoming velocity is uniform and parallel to the plate. Thus, we expect that, at the cusped leading edge of the plate, fluid particles move in an orderly fashion, i.e., the boundary layer is laminar. Further away from the leading edge, in the downstream region, the fluid-particle motions will become unstable at high Reynolds numbers, and thereafter particles will follow an erratic trajectory as shown in Figure 10.25. The local Reynolds number at point B is about $2 \times 10^5 < Re_{x1} < 8 \times 10^5$ and at point C it is about $3 \times 10^5 < Re_{x1} < 3 \times 10^6$, where the local Reynolds number was given by equation (10.130) (see White (1974,

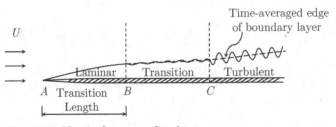

Figure 10.25 Viscous flow past a flat plate.

Fig. 6.14)). We gave a range of Re_{x_1} for the points B and C because these points may vary due to several factors, including how smooth the plate is or how uniform the oncoming flow is. In this section, we will only consider turbulent flow inside the boundary layer. Recall that for laminar flow over a flat plate C_F can be obtained by Blasius solution. For turbulent flow, it can be obtained to a great degree of accuracy by Schoenherr's formula for a flat plate, equation (10.194) as we will see later.

We expect that the functional form of the longitudinal, mean-velocity component is given by

$$\overline{u_1} = f(x_1, x_2, \mu, \rho, U) \tag{10.165}$$

in a two-dimensional, steady flow. Note that we used the mean velocity since we will average the values of various flow quantities. The wall stress is then given by

$$\overline{\tau_{12}}(x_1, 0) = \mu \frac{d\overline{f}}{dx_2}(x_1, 0, \mu, \rho, U). \tag{10.166}$$

And the boundary-layer thickness is defined by

$$0.99U = \overline{u}_1 (x_1, \delta, \mu, \rho, U), \tag{10.167}$$

so that, in reality, $\overline{\tau_{12}} = \overline{\tau_{12}}(x_1, U)$ and $\delta = \delta(x_1, U)$, when μ and ρ are treated as constants (recall that temperature and compressibility variations are neglected here).

For a fixed x_1 and U, there should only be one $\overline{\tau_{12}}$ and one δ; therefore, we can use $\overline{\tau_{12}}$ and δ instead of U and x_1, respectively, in equation (10.165), as the independent variables, as long as the Jacobian of the transformation, J, given by

$$J = \begin{vmatrix} \dfrac{\partial \delta}{\partial x_1} & \dfrac{\partial \delta}{\partial U} \\[2mm] \dfrac{\partial \overline{\tau_{12}}}{\partial x_1} & \dfrac{\partial \overline{\tau_{12}}}{\partial U} \end{vmatrix}$$

does not vanish, i.e., $J \neq 0$. This condition ensures that the derivatives exist, and the transformation from (x_1, U) to $(\overline{\tau}_{12}, \delta)$ is one to one and onto so that we can go back to (x_1, U) from $(\overline{\tau}_{12}, \delta)$. This transformation is done because the wall shear stress can be measured in experiments, and also we can look inside the boundary layer closely for different x_2/δ ratios as we will see shortly.

Hence, $\overline{u}_1$ now depends on a new function, say G, i.e.,

$$\overline{u_1} = G(\delta, x_2, \mu, \rho, \overline{\tau_{12}}) \tag{10.168}$$

such that

$$0.99U = G(\delta, \mu, \rho, \overline{\tau_{12}}), \tag{10.169}$$

where we have used equation (10.167). From equations (10.166) and (10.168), we then have

$$\overline{\tau_{12}} = \mu \frac{dG}{dx_2}(\delta, 0, \mu, \rho, \overline{\tau_{12}}). \tag{10.170}$$

Let us now analyze equation (10.168) dimensionally by taking $(\mu, \rho, \overline{\tau_{12}})$ as a dimensionally independent set of variables. We can then obtain (show this)

$$\Pi = \frac{\overline{u_1}}{\sqrt{\overline{\tau_{12}}/\rho}}, \quad \Pi_1 = \frac{x_2 \sqrt{\overline{\tau_{12}}/\rho}}{\nu}, \quad \Pi_2 = \frac{x_2}{\delta}, \text{ and therefore } \Pi = F(\Pi_1, \Pi_2).$$

$$(10.171)$$

Therefore, by defining a new " velocity"

$$\overline{u_\tau} = \sqrt{\frac{\overline{\tau_{12}}}{\rho}},$$

$$(10.172)$$

called the friction velocity, we can write the following from equation (10.171):

$$\frac{\overline{u_1}}{\overline{u_\tau}} = F\left(\frac{x_2 \overline{u_\tau}}{\nu}, \frac{x_2}{\delta}\right).$$

$$(10.173)$$

We next consider three layers within the boundary layer:

1. Viscous sublayer or laminar sublayer, $x_2/\delta \ll 1$: In this layer, viscous (laminar) stresses are dominant and the Reynolds stresses are negligible since turbulent fluctuations are very small because the particles are very close to the wall, and also because of the wall-boundary condition. The flow in the vicinity of the wall behaves as laminar, and $\overline{u_1}$ is determined by the wall stress alone.
2. Outer layer, $x_2/\delta = O(1)$: In this layer, the Reynolds stresses dominate the viscous (laminar) stresses because the higher the Reynolds number is, the smaller the viscous terms are. Recall that the Reynolds stresses are inertial in nature.
3. Overlap layer, $\nu/\overline{u_\tau} \ll x_2 \ll \delta$: The above two layers overlap in a region (layer) called the overlap region where both the viscous and Reynolds stresses are important when $\nu/\overline{u_\tau} \ll x_2 \ll \delta$ (see equation (10.184). In other words, both laminar and turbulent flows are present in this layer.

Let us discuss these layers in some detail:

1. In the viscous sublayer, we can take $x_2/\delta \cong 0$, so that from equation (10.173) we have

$$\frac{\overline{u_1}}{\overline{u_\tau}} = F\left(\frac{x_2 \overline{u_\tau}}{\nu}, 0\right) = F_1\left(\frac{x_2 \overline{u_\tau}}{\nu}\right) \equiv F_1(x_2{}^*) \text{ (inner law)}, \quad (10.174)$$

where we have defined $x_2^* = x_2 \overline{u_\tau}/\nu$. Equation (10.174) is called the inner law or the law of the wall. To determine the functional form of $F_1(x_2^*)$, consider the shear stress on the wall ($x_2 = 0$) and equation (10.174):

$$\overline{\tau}_{12} = \mu \frac{d\overline{u_1}}{dx_2} = \mu \overline{u_\tau} \frac{dF_1}{dx_2} = \mu \overline{u_\tau} \frac{\overline{u_\tau}}{\nu} \frac{dF_1}{dx_2{}^*}, \quad (10.175)$$

since $dx_2^*/dx_2 = \overline{u_\tau}/\nu$. Therefore, at $x_2^* = 0$, we must have (recall the definition of $\overline{u_\tau}$ through equation (10.172))

$$\frac{dF_1(0)}{dx_2{}^*} = 1.$$

$$(10.176)$$

Equation (10.176) and the fact that $x_2^* \ll 1$ (very close to the wall) show that F_1 is a linear function of $x_2{}^*$, and therefore, we can write

$$F_1\left(\frac{x_2\bar{u}_\tau}{\nu}\right) \cong \frac{x_2\bar{u}_\tau}{\nu} = x_2{}^* \quad \Rightarrow \quad \bar{u}_1 = \bar{u}_\tau \, x_2^* = \bar{u}_\tau{}^2 x_2/\nu. \tag{10.177}$$

2. In the outer layer, x_2/δ must be more important in equation (10.173) since we are away from the wall. But we cannot ignore the viscosity-related terms at the outset. Hence, we can write, from equation (10.173) and by following Landweber (1957),

$$\frac{\bar{u}_1}{\bar{u}_\tau} = F(x_2{}^*,\eta), \quad \eta \equiv \frac{x_2}{\delta}, \quad \text{or}$$

$$\frac{\bar{u}_1}{\bar{u}_\tau} = F(\delta^*,1) + [F(x_2{}^*,\eta) - F(\delta^*,1)] \equiv F(\delta^*,1) - F_2(\delta^*,\eta), \tag{10.178}$$

where η is the same similarity variable defined by equation (10.127), and where

$$\delta^* \equiv \frac{\delta\bar{u}_\tau}{\nu} = \frac{x_2^*}{\eta}, \quad F(\delta^*,1) = \frac{U}{\bar{u}_\tau}, \quad \eta = \frac{x_2}{\delta} = \frac{x_2^*\nu}{\bar{u}_\tau\,\delta} = \frac{x_2^*}{\delta^*}. \tag{10.179}$$

The second equation in (10.179) is a result of the first equation of (10.133). Equation (10.178) can then be written as

$$\frac{U - \bar{u}_1}{\bar{u}_\tau} = F_2(\delta^*,\eta) \quad \text{(velocity defect law)}, \tag{10.180}$$

where $F_2(\delta^*,\eta) \equiv F(\delta^*,1) - F(x_2^*,\eta)$, and it was further assumed (Townsend (1954)) that

$$-F_2(\delta^*,\eta) = \left(F(x_2{}^*,\eta) - F(\delta^*,1)\right) \neq f(x_2{}^*). \tag{10.181}$$

Obviously, this is an empirical "law." We remark, from equation (10.180), that

$$\lim_{x_2 \to \delta} \bar{u}_1 = U \quad \Rightarrow \quad F_2(\delta^*,1) = 0. \tag{10.182}$$

Equation (10.180), which is valid in the outer region, is called the velocity defect law because of the relative velocity used instead of the absolute velocity. The form of $F_2(\delta^*,\eta)$ will be obtained in connection with the overlap layer discussed next, see equation (10.192).

3. There must be a region where both the inner law equation (10.174) and the velocity defect law equation (10.180) are both valid. In this overlap layer, we have, from equations (10.177) and (10.178),

$$F_1(x_2{}^*) = \frac{\bar{u}_1}{\bar{u}_\tau} = F(\delta^*,1) - F_2(\delta^*,\eta). \tag{10.183}$$

A rigorous proof of equation (10.183) can be obtained by use of the matched asymptotic-expansion method. In order that equation (10.183) is valid, $x_2/\delta \ll 1$ and $x_2^* = \bar{u}_\tau x_2/\nu \gg 1$ must be satisfied simultaneously. Then the overlap layer will lie between

$$\frac{\nu}{\bar{u}_\tau} \ll x_2 \ll \delta. \tag{10.184}$$

Next, let us take the partial derivatives of equation (10.183) with respect to η and δ^* (note that x_2^* and $\overline{u_\tau}$ are assumed to be independent of η). We should then have

$$\frac{\partial}{\partial \eta} : \quad \delta^* \frac{dF_1}{dx_2^*} = -\frac{dF_2}{d\eta},$$

$$\frac{\partial^2}{\partial \delta^* \partial \eta} : \quad \frac{dF_1}{dx_2^*} + x_2^* \frac{d^2 F_1}{dx_2^{*2}} = -\frac{\partial^2 F_2}{\partial \eta^2}, \qquad (10.185)$$

$$\frac{\partial^2}{\partial \eta^2} : \quad \delta^{*2} \frac{d^2 F_1}{dx_2^{*2}} = -\frac{\partial^2 F_2}{\partial \eta^2}.$$

After some algebra, equation (10.185) can be combined to give

$$\frac{x_2^* \frac{d^2 F_1}{dx_2^{*2}}}{\frac{dF_1}{dx_2^*}} = \frac{\eta \frac{\partial^2 F_2}{\partial \eta^2}}{\frac{dF_2}{d\eta}} = \frac{\delta^* \frac{\partial^2 F_2}{\partial \delta^* \partial \eta}}{\frac{\partial F_2}{\partial \eta}} - 1. \qquad (10.186)$$

One can now further restrict the form of F_2, by following Landweber (1957), to write $F_2(\delta^*, \eta) = H(\delta^*) G(\eta)$, i.e., F_2 is assumed to be separable. Then equation (10.186) becomes

$$\frac{x_2^* \frac{d^2 F_1}{dx_2^{*2}}}{\frac{dF_1}{dx_2^*}} = \frac{\eta \frac{d^2 G}{d\eta^2}}{\frac{dG}{d\eta}} = \frac{\delta^* \frac{dH}{d\delta^*}}{H} - 1 = n - 1, \qquad (10.187)$$

where n is some constant, since η and x_2^* are independent variables. We note, for later reference, that experimental data indicate $0 \leq n \leq 0.2$.

If one integrates equation (10.187), there results

$$\frac{dF_1}{dx_2^*} = Ax_2^{*n-1}, \quad \frac{dG}{d\eta} = B\eta^{n-1}, \quad H = D\delta^{*n}, \qquad (10.188)$$

where A, B, and D are integration constants.

If we take $n = 0$ and integrate the first two equations in equation (10.188), we obtain

$$F_1(x_2^*) = A \log\left(\frac{x_2 \overline{u_\tau}}{\nu}\right) + C_1, \quad G(\eta) = B \log(\eta) + \overline{C_2}, \quad H(\delta^*) = D, \qquad (10.189)$$

where C_1 and $\overline{C_2}$ are the new integration constants. Now, from equation (10.183) and the separable form of F_2, we write

$$F_1(x_2^*) = F(\delta^*, 1) - G(\eta) H(\delta^*), \qquad (10.190)$$

and recalling that $x_2^* = \eta \delta^*$, we must have

$$F(\delta^*, 1) = A \log(\delta^*) + C_1 + D\overline{C_2}, \qquad (10.191)$$

where we have used equation (10.189). Note that in obtaining equation (10.191), we have discovered that A must be equal to $-BD$ since $F(\delta^*, 1)$ should be a function of δ^* only. Redefining the constant as $C_2 \equiv D\overline{C_2}$, we have

$$F_2(\delta^*, \eta) = -A \log(\eta) + C_2. \qquad (10.192)$$

Substituting equations (10.189) and (10.192) in the inner law, equation (10.174), and the velocity defect law, equation (10.180), respectively, we therefore have the logarithmic laws in the overlap layer, i.e.,

$$\frac{\overline{u_1}}{u_\tau} = A \log \left(\frac{u_\tau x_2}{\nu} \right) + C_1, \quad \frac{U - \overline{u_1}}{u_\tau} = -A \log \left(\frac{x_2}{\delta} \right) + C_2 \quad \text{(logarithmic laws)}.$$

$$(10.193)$$

See Prob. 10.11.22 as an application of the logarithmic law.

The constants A, C_1, and C_2 must be determined from experiments. Schoenherr's flat plate frictional drag formula, for example, for a plate of length l and of unit width, is given by

$$C_f^{-1/2} = 4.13 \log_{10}(C_f Re_l), \quad C_f = \frac{Force}{\frac{1}{2}\rho U^2 l}, \quad (10.194)$$

where $Re_l = Ul/\nu$, l is the plate length and C_f is the total (accumulated) drag coefficient, is based on equation (10.193) (see also Section 9.10.4), and the force is per unit width. Equation (10.194) is frequently used in determining the frictional drag of the wetted surface of ships for example. This formula is very satisfactory for a broad range of Reynolds numbers (see Newman (1978)). Another formula, namely the ITTC 1957 formula, discussed before (see equation (9.57)), is also based on the logarithmic laws.

If we had chosen the constant $n \neq 0$, then we would have, for instance,

$$F_1(x_2{}^*) = A_1 \, x_2{}^{*n} + \overline{c_1}, \quad (10.195)$$

where A_1 and $\overline{c}_1$ are the integration constants. These type of equations are known as the power laws and we shall look into one such approximation ($n = 1/7$) to determine shear stresses on a wall.

The experimentally obtained constants

$$A = 2.5, \; C_1 = 5.1 \; \text{if} \; \frac{u_\tau x_2}{\nu} > 30, \; \text{and} \; C_2 = 2.35 \; \text{if} \; \frac{x_2}{\delta} < 0.15, \quad (10.196)$$

are widely used to determine the logarithmic velocity distribution in equation (10.193). However, because of the fact that $\overline{\tau_{12}}$ and δ were chosen as independent variables, it is not possible to obtain the shear stress and the boundary-layer thickness without further assumption (see Newman (1978), Section 3.18). Instead, one may use the 1/7-power approximation for the velocity distribution inside the boundary layer.

The 1/7-power approximation, which corresponds to $n = 1/7$, $A = 8.74$, $C_1 = 0$, (based on experimental data, see Schlichting (1968, Chapters XX and XXI)), is given by

$$\frac{\overline{u_1}}{u_\tau} = 8.74 \left(\frac{u_\tau x_2}{\nu} \right)^{1/7} \quad (10.197)$$

or

$$u_\tau = 0.15 \, \overline{u_1}^{7/8} \left(\frac{\nu}{x_2} \right)^{1/8}. \quad (10.198)$$

Hence,

$$\overline{\tau_{12}} = \rho\,\overline{u_\tau}^2 = 0.0225\rho\overline{u_1}^{7/4}\left(\frac{\nu}{x_2}\right)^{1/4}.$$ (10.199)

By use of the definition of the boundary-layer thickness, i.e., $\overline{u_1} = 0.99U$ $(x_2 = \delta)$, in equation (10.199), we obtain

$$\overline{\tau_{12}} = 0.0221\rho U^2 Re_\delta^{-1/4},$$ (10.200)

where $Re_\delta = U\delta/\nu$ is the Reynolds number based on the boundary-layer thickness, and $3000 \le Re_\delta \le 70000$. This $\overline{\tau_{12}}$ will be written in terms of Re_{x_1} later in Section 10.9.1 (see equation (10.214)).

We can now use the definition of the momentum thickness equation (10.109), where the upper limit of integration is replaced by $\overline{\delta}$ because of the matching condition at the edge of the boundary layer, to obtain (show this)

$$\overline{\delta_2} = \frac{7}{72}\overline{\delta},$$ (10.201)

because (see, e.g., Schlichting (1968, Chapter XX, Arts. a and b))

$$\frac{\overline{u_1}}{U} = \left(\frac{x_2}{\overline{\delta}}\right)^{1/7}.$$ (10.202)

See Prob. 10.11.21 as an application of the 1/7-power law. Equation (10.202), instead of equation (10.197), can be the starting point for the 1/7-power approximation. Note that equations (10.200) and (10.201) will be used in the next section to determine $\overline{\tau}_{12}$ and $\overline{\delta}_2$ in terms of Re_{x_1}.

Von Karman's Momentum Integral Relation

To obtain the displacement thickness, δ_1, and the boundary-layer thickness, δ, explicitly in terms of the local Reynolds number, we will use von Karman's momentum-integral relation. Consider a 2-D control volume (see Figure 10.26) at a section of the boundary layer between x_1 and $x_1 + \Delta x_1$, and calculate the force acting on the control volume which must be equal to the rate of decrease of momentum within the control volume.

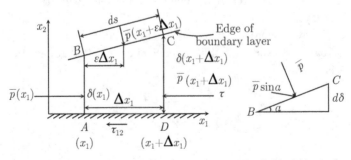

Figure 10.26 A two-dimensional control volume of the fluid within the boundary layer.

The total horizontal force acting on the face BC is

$$\sum F_{BC} = \bar{p} \sin \alpha \, ds = \bar{p} \, d\bar{\delta} = \bar{p} \frac{d\bar{\delta}}{dx_1} \Delta x_1, \tag{10.203}$$

where $\bar{p}$ is the pressure. Therefore, the total force acting on the control volume (actually, on the surface in two dimensions) $ABCD$ is

$$- \bar{p}(x_1 + \Delta x_1)\bar{\delta}(x_1 + \Delta x_1) + \bar{p}(x_1)\bar{\delta}(x_1)$$

$$+ \bar{p}(x_1 + \epsilon\Delta x_1)\frac{d\bar{\delta}(x_1 + \epsilon\Delta x_1)}{dx_1}\Delta x_1 - \overline{\tau_{12}}(x_1 + \epsilon\Delta x_1)\Delta x_1$$

$$= -\frac{\partial \bar{p}}{\partial x_1}\bar{\delta}(x_1)\Delta x_1 - \overline{\tau_{12}}\Delta x_1. \tag{10.204}$$

The right-hand side of equation (10.204) is obtained because of the fact that $d\delta/dx_1 \ll \delta$ and

$$\bar{p}(x_1 + \Delta x_1)\bar{\delta}(x_1 + \Delta x_1) = \bar{p}(x_1)\bar{\delta}(x_1) + \Delta x_1\frac{d}{dx_1}\{\bar{p}(x_1)\delta(x_1)\} + O(\Delta x_1{}^2)$$

by the Taylor series expansion in which $d\delta/dx_1$ and higher-order terms are ignored.

The rate of change (decrease) of momentum can be calculated by

$$\rho \int_0^{\bar{\delta}(x_1+\Delta x_1)} \overline{u_1}^2(x_1 + \Delta x_1, x_2)dx_2 - \rho \int_0^{\bar{\delta}(x_1)} \overline{u_1}^2(x_1, x_2)dx_2$$

$$- \rho U(x_1 + \epsilon\Delta x_1)\left\{ \int_0^{\bar{\delta}(x_1+\Delta x_1)} \overline{u_1}(x_1 + \Delta x_1, x_2)dx_2 - \int_0^{\bar{\delta}(x_1)} \overline{u_1}(x_1, x_2)dx_2 \right\}$$

$$= \rho\Delta x_1\left\{ \frac{\partial}{\partial x_1}\int_0^{\bar{\delta}(x_1)} \overline{u_1}^2 dx_2 \right\} - \rho U\Delta x_1\frac{\partial}{\partial x_1}\left\{ \int_0^{\bar{\delta}(x_1)} \overline{u_1}dx_2 \right\}, \tag{10.205}$$

where again we have used the Taylor series expansion of the momentum terms. By equating equations (10.204) and (10.205), one obtains

$$\rho\frac{\partial}{\partial x_1}\left\{ U^2\int_0^{\bar{\delta}(x_1)} \left(\frac{\overline{u_1}}{U}\right)^2 dx_2 \right\} - \rho U\frac{\partial}{\partial x_1}\left\{ \int_0^{\bar{\delta}(x_1)} \overline{u_1}dx_2 \right\} = -\frac{d\bar{p}}{dx_1}\bar{\delta}(x_1) - \overline{\tau_{12}}.$$
$$\tag{10.206}$$

However, by Bernoulli's theorem, equation (10.122), we have

$$\frac{d\bar{p}}{dx_1} = -\rho U\frac{dU}{dx_1}, \quad U = f(x_1). \tag{10.207}$$

Therefore, equation (10.206) becomes

$$\frac{\partial}{\partial x_1}(U^2\overline{\delta_2}) + U\overline{\delta_1}\frac{dU}{dx_1} = \frac{\overline{\tau_{12}}}{\rho}, \tag{10.208}$$

where $\overline{\delta_1}$ and $\overline{\delta_2}$ are the displacement and momentum thickness, respectively (see equations (10.108) and (10.109)). If U is constant, then equation (10.208) becomes

$$\overline{\tau_{12}} = \rho U^2 \frac{d\overline{\delta_2}}{dx_1}. \tag{10.209}$$

Equation (10.208), or equation (10.209) when U is constant, is known as von Karman's momentum-integral relation.[8] The drag per unit width of the plate is

$$D/b = \int_L \overline{\tau_{12}} \, dx_1 = \rho U^2 \overline{\delta_2}. \tag{10.210}$$

We can now integrate equation (10.209) and use equations (10.200) and (10.201) to obtain the boundary-layer thickness:

$$\overline{\delta} = 0.3654 x_1 Re_{x_1}^{-1/5}, \tag{10.211}$$

where Re_{x_1} is based on the leading edge distance. Here, we have assumed that $\delta(x_1 \to 0) \to 0$, and therefore, the integration constant, which should have been included in equation (10.211), has been set to zero. The momentum thickness can now be obtained by use of equations (10.201) and (10.211):

$$\overline{\delta_2} = \frac{7}{72}\overline{\delta} = 0.0355 x_1 Re_{x_1}^{-1/5}. \tag{10.212}$$

We also make use of the definition of the displacement thickness, and also equations (10.201), (10.202), and (10.211), to obtain

$$\overline{\delta_1} = \frac{\delta}{8} = 0.0457 x_1 Re_{x_1}^{-1/5}. \tag{10.213}$$

Finally, equation (10.209), which is equal to equation (10.200), becomes

$$\overline{\tau_{12}} = 0.0284 \rho U^2 Re_{x_1}^{-1/5}. \tag{10.214}$$

Inspection of equation (10.211) shows that the boundary-layer thickness in turbulent flow grows as $x_1^{4/5}$. Recall that in laminar flow δ grows as $x_1^{1/2}$ (see Section 10.8.1). Also, equation (10.202) now becomes

$$\frac{\overline{u_1}}{U} = \frac{x_2}{0.3654 \, x_1 \, Re_{x_1}^{-1/5}}, \quad x_1 > 0. \tag{10.215}$$

10.10   Concluding Remarks

In this chapter, the flow of a viscous fluid is discussed. The conservation equations for mass and momentum are derived from their integral forms. Conservation of energy

---

[8] We must emphasize that although we are discussing von Karman's momentum-integral relation here in conjunction with a turbulent boundary layer, there has been no assumption of turbulence in its derivation, and therefore, it can also be used in laminar boundary-layer calculations; see, e.g., Prob. 10.11.16 or Prob. 10.11.18.

is not considered since in this course we ignore the temperature variations and the compressibility of the fluid. Following the derivation of the continuity and the N-S equations, the boundary conditions in a viscous fluid are introduced.

It is not only instructive to deal with some exact or approximate solutions of the N-S equations but also quite useful in a comparative study where one must be sure that the solution of a particular problem, whether obtained numerically or analytically, reduces to a known solution when the same conditions are used. For example, suppose that we have obtained the numerical solution of the problem of viscous uniform flow about a sphere placed under a free surface. To check the accuracy of our numerical solution, at least partially, we can place the sphere far away from the free surface (deeply submerged) so that the effect of the free surface is minimal, if not totally negligible. The resulting numerical solution of the problem under this condition can then be compared with the Stokes solution for a creeping flow if the Reynolds number is low. As a result, this comparative study can reveal the accuracy of our numerical solution, at least when $Re \ll 1$.

Following the exact and approximate solutions of some viscous-flow problems, the concepts of boundary layer and separation are discussed. Several definitions of the boundary-layer thickness are introduced. One of the most classical problems of viscous flow, namely the laminar flow around a flat plate, is discussed, and Prandtl's laminar boundary-layer equations are derived by two different but closely related methods. These equations are solved semi-analytically to obtain Blasius' solution, which not only gives the velocity distribution in the boundary layer but also the shear stress on the plate, as well as the boundary-layer thickness. An order estimate of the quantities in the governing equations has shown that the pressure inside the boundary layer is independent of the normal coordinate, and therefore, it can be obtained from the inviscid solution outside the boundary layer. In other words, one can have the luxury of solving the inviscid problem first and then attempt to solve the viscous-flow problem which is confined to a thin layer around the plate. This, of course, is true if the fluid viscosity is not very high since the boundary-layer thickness is proportional to the square of the kinematic-viscosity coefficient. It also is shown, for a laminar boundary layer, that the shear stress on the plate decreases and the boundary-layer thickness increases by the square root of the longitudinal coordinate.

Unlike in the laminar boundary-layer theory, the turbulent flow about a body cannot be assumed steady in general. To overcome the difficulties associated with the erratic trajectories of particle motions in a turbulent flow, the particle velocity components are decomposed into the mean and fluctuating parts. With this decomposition, the time-averaged Reynolds equations are derived. These equations revealed the presence of additional stresses due to the fluctuating part of the particle velocity. Unfortunately though, the number of unknowns of the problem becomes more than the number of equations to be solved, and thus, the equations, as they are, are declared unsolvable, forcing us to resort to, for example, experimental data to determine the new unknowns.

The turbulent boundary layer on a flat plate is analyzed with the help of the dimensional dependency of the mean component of velocity. Three layers within the turbulent boundary layer are constructed, and the forms of the functions that the velocity depends on are deducted to obtain the law of the wall or inner law near the wall, the velocity defect law near the edge of the boundary layer, and the logarithmic law that is valid in the overlap layer which lies between the inner and outer layers within the boundary layer. The unknown coefficients of the problem are then determined from experimental data. The shear stress is obtained by use of the 1/7-th power approximation, and boundary-layer thicknesses are obtained by applying von Karman's momentum-integral relation to a control volume enclosing the boundary layer. This analysis has allowed us to determine the boundary-layer thickness which is shown to be proportional to the 4/5-th power of the longitudinal coordinate, in contrast with the laminar boundary-layer thickness which is proportional to the square root of the longitudinal coordinate.

10.11   Self-Assessment

10.11.1

A homogeneous sphere of specific gravity 8.0 is dropped into a large container of oil of specific gravity 0.86 and viscosity of 0.002 slug/ft – s.

First express the terminal velocity, as a function of the drag coefficient, by equation (10.100), i.e.,

$$C_D = \frac{24}{Re} + \frac{6}{1 + \sqrt{Re}} + 0.4, \quad Re = \frac{UD}{\nu},$$

when Re is not very "small." Note that since Re is a function of U also, iteration is needed to obtain the solution for the terminal velocity. Use Stokes' solution as the initial guess for the iteration, and find the terminal velocity of the sphere for the sphere diameters of:

1. $D = 0.01$ in.,
2. $D = 0.1$ in., and
3. $D = 1.0$ in.

Also, (a) indicate which one(s) of the above is a creeping motion, and (b) compare (calculate the percentage error of) your calculations above with the experimental data.

10.11.2

Suppose that a viscous fluid of thickness h is flowing down an inclined plane due to gravity as shown in Figure 10.27. The flow is steady and two-dimensional, and the fluid is incompressible. The velocities depend on the x_2 coordinate only. The atmospheric pressure and surface tension can be neglected on the free surface. And, h is assumed to remain constant.

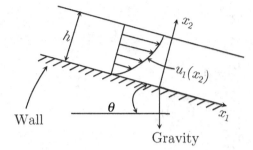

Figure 10.27 Viscous flow flowing down an inclined plane due to gravity.

1. Show that the velocity distribution is given by

$$u_1 = \rho g \frac{\sin\theta}{2\mu} x_2(2h - x_2),$$

2. and the volume flux (discharge rate) is given by

$$Q == \int_0^h u_1(x_2)\, dx_2 = \frac{\rho g h^3 \sin\theta}{3\mu}.$$

3. Also, find both the normal and shear stresses on the wall, and compare your solution to the general Couette solution between two parallel plates, $2h$ apart.
4. Discuss under what conditions the solution may break down and produce nonphysical results (hint: consider the Reynolds number and what would make it large).

10.11.3

A drill pipe is moving with constant velocity, U, inside a casing which is fixed. In other words, we have two concentric circular pipes as shown in Figure 10.28. The fluid motion is steady, and the fluid is viscous and incompressible. The radius of the casing is R_2, and the radius of the drilling pipe is R_1. The u_r and u_θ (axial-coordinate) components of the fluid velocity are zero, i.e., there is no swirl by assumption. Moreover, u_{x_3} is the same at any x_3. The hydrostatic pressure must be considered.

Determine the velocity, u_{x_3}, in the radial direction. Obtain the fluid pressure as a function of the r and x_3 coordinates. Use the equations in cylindrical coordinates for convenience. Make sure that you check if u_{x_3} satisfies the boundary condition on $r = R_1$ and $r = R_2$.

10.11.4

Air ($\nu = 1.5 \times 10^{-5} \text{m}^2/\text{s}$) blows over a flat plate at zero angle of attack and at a velocity of 3 m/s.

1. At what distance from the leading edge is the flow almost certainly turbulent?
2. What is the laminar boundary-layer thickness before this point?
3. What is the turbulent boundary-layer thickness after this point?

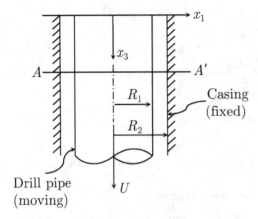

Figure 10.28 A drill pipe moving inside a casing.

10.11.5

If the boundary layer velocity profile is given by $u_1(x_2) = U \tanh(x_2/b)$, where $0 \le x_2 \le \infty$ and b is a constant,

1. Calculate $\delta, \delta_1, \delta_2$, and δ_3.
2. What is the velocity gradient at $x_2 = 0$?

10.11.6

For a slender ship-model that is 2 m long, estimate the displacement thickness at the stern by use of Blasius' result for laminar flow. The speed of the model is 1.0 m/s. Use ν for fresh water at 15°C.

10.11.7

Wind is blowing at 80 km/h over the living quarters of a ship. For air, you can take $\nu = 1.45 \times 10^{-5}\,\text{m}^2/\text{s}$.

1. At which point from the leading edge the flow will become turbulent?
2. What is the turbulent boundary-layer thickness at $x_1 = 3$ m?
3. What is the viscous sublayer thickness, $x_2 = x_2^* \nu / \bar{u}_\tau$, when $x_2^* = 5$, and $x_1 = 3$ m?
4. If the flow could be assumed laminar, what would the boundary-layer thickness be at $x_1 = 3$ m?

10.11.8

By use of the definition of the time average of a function, equation (10.156), prove that

1. $\overline{f + g} = \overline{f} + \overline{g}$
2. $\overline{\overline{f}} = \overline{f}$

3. $\overline{\overline{f} \cdot g} = \overline{f} \cdot \overline{g}$
4. $\overline{f \cdot g} = \overline{f} \cdot \overline{g} + \overline{f' \cdot g'}$

Note that $f = \overline{f} + f'$, $g = \overline{g} + g'$, $\overline{f'} = \overline{g'} = 0$.

10.11.9

If the velocity profile inside the boundary layer can be written as $u_1(x_2) = U x_2/\delta$ for $0 \le x_2 \le \delta$, and $u_1(x_2) = U$ for $\delta \le x_2 \le \infty$, where U is the constant free-stream velocity, derive

1. the displacement thickness, δ_1,
2. the momentum thickness, δ_2, and
3. the energy thickness, δ_3,

in terms of the boundary-layer thickness δ.

10.11.10

Show that viscous stress (excluding pressure) at a solid boundary wall (stationary and flat along the x_1 axis) is $\tau = \mu \omega \times n$, where ω is the vorticity vector and n is the unit normal vector on the wall. Assume that the flow is steady and two-dimensional, and the fluid is incompressible. Also, note that $\tau = \tau_{12} e_1 + \tau_{22} e_2 + \tau_{23} e_3$ and $\omega = \nabla \times u$.

10.11.11

Consider uniform flow, U, above a flat plate. The fluid is viscous. The frictional drag force (per unit width) is given by

$$D = \int \tau_{12}(x_1)\, dx_1$$

up to the point x_1. The length of the plate is 5 m and $\rho = 1,000\,\text{kg/m}^3$, $\nu = 1.14 \times 10^{-6}\,\text{m}^2/\text{s}$.

1. Calculate the drag (per unit width) on the plate if $U = 0.05$ m/s,
2. Calculate the drag (per unit width) on the plate if $U = 3.00$ m/s.

Hint: Use the equations for the shear stress in appropriate flow regime (laminar/ turbulent) to obtain the applicable drag formula.

10.11.12

A ship model is 3 m long and it is tested (by use of Froude's similitude) at a speed of $U_m = 1.0\,\text{m/s}$. The model scale is 1:50, and the model and prototype fluid kinematic viscosities are $\nu_m = 1.0 \times 10^{-6}\,\text{m}^2/\text{s}$ and $\nu_p = 1.2 \times 10^{-6}\,\text{m}^2/\text{s}$, respectively. Assume that the model is very slender, i.e., that it can be approximated as a flat plate for the purpose of this problem.

What is the boundary-layer thickness

1. at 10 cm behind the bow of the model?
2. at the stern of the model?
3. of the prototype ship at the location corresponding to $x_1 = 10$ cm in the model scale?
4. of the prototype ship at its stern?

10.11.13

Consider the Hagen–Poiseuille flow in a circular pipe. The pressure gradient is given by $dp/dx_1 = -300$ Pa/m, and the radius of the pipe is 5 cm. Fresh water flows inside the pipe at 20°C ($\nu = 1.0067 \times 10^{-6}$ m^2/s).

Calculate the:

1. Discharge rate
2. Shear stress on the pipe wall
3. Pipe resistance coefficient (how does it compare with the experimental data?)

10.11.14

Solve Prandtl's laminar boundary-layer equations for a flat plate (two-dimensional) and steady flow when $Re \gg 1$, by replacing the convective accelerations $u_1 \partial u_1/\partial x_1 + u_2 \partial u_1/\partial x_2$ by $U \partial u_1/\partial x_1$, where U is the uniform free-stream velocity. Compare your result for the skin friction coefficient with the exact Blasius solution, i.e., $c_f = 0.664/\sqrt{Re_{x1}}$.

10.11.15

Consider the steady, incompressible viscous flow along a conduit of elliptic cross section as shown in Figure 10.29. The ellipse is given by

$$\frac{x_2^2}{a^2} + \frac{x_3^2}{b^2} = 1.$$

The pressure, p, is a function of the x_1 coordinate only, and $u_1 = f(x_2, x_3)$. Assume that the pressure gradient dp/dx_1 is given in this problem, and that

$$u_1(x_2, x_3) = \eta \left\{ \frac{x_2^2}{a^2} + \frac{x_3^2}{b^2} - 1 \right\},$$

where the function η depends on $\eta = \eta(a, b, \mu, p, x_1)$.

Solve the governing equations of the problem to determine η, and thus u_1, subject to the viscous-flow boundary condition on the wall of the conduit. The solution for

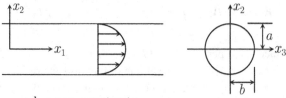

a, b, u are constants.

Figure 10.29 Steady viscous flow along a conduit of elliptic cross section.

u_1 should also be finite along the x_1-axis ($x_2 = x_3 = 0$). Note that you can use the Cartesian coordinates in solving this problem.

10.11.16

If the velocity profile in the (laminar) boundary layer over a flat plate is assumed to be given by a parabola, i.e.,

$$u_1(x_1, x_2) = A + Bx_2 + Cx_2^2 \quad \text{if} \quad x_2 < \delta \quad \text{and} \quad u_1(x_1, x_2) = U \quad \text{if} \quad x_2 \geq \delta,$$

where U is the constant free-stream velocity and A, B, C are functions of the x_1 coordinate only.

1. Determine u_1 as a function of U, x_2, and $\delta(x_1)$. Note that the vertical profile of u_1 must also be tangent to the vertical coordinate x_2 at $x_2 = \delta$, i.e., $\partial u_1/\partial x_2 = 0$ at $x_2 = \delta$, so that there is no discontinuity in u_1.
2. By use of von Karman's momentum equation, (10.208) or (10.209), which is also valid for laminar flows, show that the shear stress and the boundary-layer thickness for this velocity profile are written, respectively, as

$$\tau_{12} = 0.133 \, \rho U^2 \frac{d\delta}{dx_1}, \quad \delta(x_1) = 5.48 \, x_1 \, Re_{x_1}^{-1/2}.$$

How do these results compare with Blasius' solution?
(Hint: In obtaining the above $\delta(x_1)$, note also that $\tau_{12} = \mu \, \partial u_1/\partial x_2$ along $x_2 = 0$.)

10.11.17

Newton's second law states that the rate of change of angular momentum is equal to the total applied torque about the origin. This is known as the conservation of angular momentum or moment of momentum principle. By considering the angular momentum, i.e.,

$$\int_{V(t)} r \times \rho V \, dV = \int_{V(t)} \epsilon_{ijk} \, x_j \, \rho u_k \, dV,$$

where r is the position vector of a particle, $r = x_j e_j$, V is the particle velocity vector with components u_k, i.e., $V = u_k e_k$, and ρ is the mass density of the fluid, and postulating the conservation of angular momentum principle,

$$\frac{D}{Dt} \int_{V(t)} \boldsymbol{r} \times \rho \boldsymbol{V} \, dV = \int_{S(t)} \boldsymbol{r} \times \boldsymbol{f}_s \, dS + \int_{V(t)} \boldsymbol{r} \times \rho \boldsymbol{f}_b \, dV,$$

where $\boldsymbol{f}_s$ is the surface force per unit area ($f_{sk} = \tau_{lk} \, n_l$), $\boldsymbol{f}_b$ is the body force per unit mass, ($f_{bk} = -g \, \delta_{k2}$), and $V(t)$ is the volume of the fluid enclosed by the surface $S(t)$, prove that $\tau_{ij} = \tau_{ji}$, i.e., the stress tensor is symmetric, once the conservation of mass and momentum statements are satisfied. You must use indicial notation in your proof.

10.11.18

An approximate solution to the boundary layer on a flat plate can be obtained by use of the following horizontal component of the velocity profile in laminar flow:

$$u_1(x_1, x_2) = U \left(1 - e^{-\frac{x_2}{\delta(x_1)}}\right), \qquad 0 \le x_2 \le \infty,$$

where $\delta(x_1)$ is the boundary-layer thickness and U is the *constant*, uniform-flow velocity outside the boundary layer.

1. Verify that the choice of the velocity profile satisfies the boundary conditions on the plate, $x_2 = 0$, and as well as at $x_2 = \infty$. Is the condition at $x_2 = \delta$ satisfied?
2. Use von Karman's momentum-integral relation, equation (10.208):

$$\frac{\partial}{\partial x_1} \left(U^2 \, \delta_2\right) + U \, \delta_1 \frac{\partial U}{\partial x_1} = \frac{\tau_{12}}{\rho},$$

and the given velocity profile, and the definitions of various boundary-layer thicknesses to determine, δ_1, δ_2, and δ as functions of x_1 and $Re_{x_1} = Ux_1/\nu$.
3. Compare your approximate results for δ, δ_1, and δ_2 with Blasius' solutions for the same.

10.11.19

The following horizontal-velocity profile is assumed for the laminar boundary layer over a flat plate:

$$u_1 = \frac{3U}{2} \eta - \frac{U}{2} \eta^3, \qquad \eta = \frac{x_2}{\delta(x_1)},$$

where $\delta(x_1)$ is the boundary-layer thickness, and U is the constant free-stream velocity. The momentum thickness is given by $\delta_2 = 39 \, \delta / 280$.
   Calculate:

1. The displacement thickness, i.e., δ_1.
2. Shear stress on the plate, i.e., τ_{12} on $x_2 = 0$.
3. Boundary-layer thickness by use of von Karman's momentum-integral relation (note that U is constant).
4. The local skin-friction coefficient, i.e., $C_f(x_1)$.
   How does this compare with Blasius' solution?

10.11.20

Consider the conservation of total energy equation

$$\rho \frac{D}{Dt} \left\{ \frac{1}{2} u \cdot u + e \right\} = \nabla \cdot (u\tau - q) + f \cdot u,$$

where $f = -\rho g e_2$ is the body-force vector, e the internal energy, τ the second-order stress tensor, ρ the mass density, and q the heat-flux vector.

1. Write the indicial form of this equation.
2. Show that the conservation of thermal energy becomes

$$\rho \frac{De}{Dt} = \tau_{ij} u_{j,i} - q_{j,j},$$

if we use the conservation of momentum equation, (10.21), to drop the mechanical energy terms from the total energy equation. Thus, you can conclude that, in the absence of temperature variations, the conservation of energy is the conservation of mechanical energy which is equivalent to the conservation of momentum equation.

10.11.21

Starting with the 1/7-power law, equation (10.202),

1. show that

$$\frac{\bar{u}_1}{\bar{u}_{1Ref}} = \left(\frac{x_2}{x_{2Ref}} \right)^{1/7},$$

where $\bar{u}_1$ is the unknown wind speed at elevation x_2 and $\bar{u}_{1Ref}$ is the known wind speed at elevation x_{2Ref}, and
2. determine the wind speed at the nacelle (which is located at an elevation of 70m) of a wind turbine if the wind speed at the 10 m elevation is measured as 12 knot.

Table 10.2 Friction length for various types of surfaces

Types of surface	x_{20} (cm)
Very smooth	0.001
Calm sea	0.02
Waves on ocean surface	0.5
Rough pasture	1.0
Few trees	10.0
Forest	50.0
Downtown of a city	300.0

10.11.22

The logarithmic law, also known as Prandtl–von Karman velocity distribution, can be written for smooth or rough surfaces as

$$\frac{\bar{u}_1(x_2)}{\bar{u}_\tau} = \frac{1}{k} \ln\left(\frac{x_2}{x_{20}}\right), \qquad x_2 > x_{20},$$

where $k = 0.4$ is the von Karman constant and x_{20} is the friction length whose values are shown in Table 10.2 as a function of how rough the terrain is.

1. Show that

$$\frac{\bar{u}_1(x_2)}{\bar{u}_{1Ref}} = \frac{\ln(x_2/x_{20})}{\ln(x_{2Ref}/x_{20})}.$$

2. If the terrain is smooth, what is the wind speed at the 70m elevation if the measured wind speed at the 10m elevation is 12 knot?

11 Ideal-Fluid Flow

11.1 Introduction

There are several ocean engineering problems involving relatively large structures and relatively low speed flows (e.g., at speeds much smaller than near-acoustic velocities in water) where the overall motions and fluid forces are dominated by the fluid action away from the viscous boundary layers. Recall that the effect of viscosity is confined to a thin boundary layer around, and within a wake region behind a body (separation must occur in order that a wake region exists as discussed in Chapter 10). Therefore, for motion in the large, water can be assumed to be an inviscid fluid in most parts of the domain. Recalling that we also have assumed that the flow is incompressible (which is a very good assumption for most hydrodynamical problems of concern to us since the associated velocities are not very "high"), we have an ideal fluid to be analyzed in this chapter, where we will consider the motion of an ideal fluid of infinite extent, i.e., there is no free surface or sea floor. The presence of the free surface will be considered in Chapters 13 and 14.

11.2 Governing Equations and Boundary Conditions

By use of the incompressibility condition and the fact that viscous stresses are negligible, i.e., the fluid is inviscid, we have (from equations (10.14) and (10.25)) the governing equations of an ideal flow as follows:

$$\frac{\partial u_i}{\partial x_i} = 0, \tag{11.1}$$

$$\frac{\partial u_i}{\partial t} + u_j \frac{\partial u_i}{\partial x_j} = -\frac{1}{\rho}\frac{\partial p}{\partial x_i} + \frac{1}{\rho}F_i. \tag{11.2}$$

The above equations are called Euler's equations. The comparison of equation (11.2) with equation (10.25) shows that the stress tensor for an ideal fluid ($\mu = 0$) is given by

$$\tau_{ij} = -p\delta_{ij}. \tag{11.3}$$

As before, the vector $F = F_i e_i$ in equation (11.2) is the body force vector (see equation (10.23)) given by

$$F = (0, -\rho g, 0).\tag{11.4}$$

Additionally, we will assume that the flow not the fluid is irrotational, i.e., that the circulation around any closed contour, C, completely within the fluid domain, defined by

$$\Gamma = \oint_C u_i \, dx_i = \oint_C u \cdot dx = \oint_C (u_i e_i) \cdot (dx_j e_j)\tag{11.5}$$

is constant, or Γ = constant (note that equation (11.5) is a line integral). This can be proven by Kelvin's theorem which states that if the motion of an inviscid fluid is irrotational at any one instant of time, then it will remain so indefinitely. If we assume that at $t = -\infty$ the fluid was at rest, then the circulation $\Gamma = 0$ at $t = -\infty$, and therefore, $\Gamma = 0$ for all t.

Now we can invoke the Stokes theorem (assuming u is a continuous and differentiable function), see Section 2.4:

$$\int\int_S (\nabla \times u) \cdot dS = \oint_C u \cdot dx,\tag{11.6}$$

which is equal to equation (11.5), i.e., the circulation Γ. We can therefore show that, for an arbitrary surface S, we must have

$$\nabla \times u = 0 \quad \text{if} \quad \Gamma = 0,\tag{11.7}$$

or

$$\nabla \times u = \begin{pmatrix} e_1 & e_2 & e_3 \\ \frac{\partial}{\partial x_1} & \frac{\partial}{\partial x_2} & \frac{\partial}{\partial x_3} \\ u_1 & u_2 & u_3 \end{pmatrix} = \left(\frac{\partial u_3}{\partial x_2} - \frac{\partial u_2}{\partial x_3} \right) e_1$$

$$+ \left(\frac{\partial u_1}{\partial x_3} - \frac{\partial u_3}{\partial x_1} \right) e_2 + \left(\frac{\partial u_2}{\partial x_1} - \frac{\partial u_1}{\partial x_2} \right) e_3 = 0.\tag{11.8}$$

Equation (11.8) is equal to the vorticity vector, ω, which represents twice the rate of rotation of a fluid element with respect to its own principal axes. Therefore if $\Gamma = 0$, then for $\omega = \omega_i e_i$, we have (see also Section 1.4)

$$\omega_i = \epsilon_{ijk} u_{k,j} = 0, \quad i, j, k = 1, 2, 3,\tag{11.9}$$

where ϵ_{ijk} is the permutation symbol, equation (1.46), that we recall for convenience:

$$\epsilon_{ijk} = \begin{cases} 1 & \text{if } i, j, k \text{ are an even permutation of } 1, 2, 3, \\ -1 & \text{if } i, j, k \text{ are an odd permutation of } 1, 2, 3, \\ 0 & \text{if } i, j, k \text{ are not a permutation of } 1, 2, 3. \end{cases}\tag{11.10}$$

We say that the flow is irrotational if $\omega = 0$. This fact has very important implications. Consider now the curl of the velocity vector, i.e., $\nabla \times u = 0$. Since the magnitude of a vector product is given by

$$||\boldsymbol{\nabla} \times \boldsymbol{u}|| = ||\boldsymbol{\nabla}||\,||\boldsymbol{u}||\sin\theta|, \tag{11.11}$$

where θ is the angle between the two vectors, θ must be equal to $n\pi$, $n = 0, 1, 2, \ldots$, if $\boldsymbol{\nabla} \times \boldsymbol{u} = 0$. Therefore, $\boldsymbol{u}$ must be parallel to $\boldsymbol{\nabla}$. Of course, it can have a length different from $\boldsymbol{\nabla}$. Therefore, we can set

$$\boldsymbol{u} = \boldsymbol{\nabla}\phi, \tag{11.12}$$

where ϕ is a scalar, since $\boldsymbol{\nabla} \times \boldsymbol{\nabla}\phi = 0$ for any scalar function ϕ ($\boldsymbol{\nabla}$ is parallel to $\boldsymbol{\nabla}\phi$) Note that one could also write $\boldsymbol{u} = -\boldsymbol{\nabla}\phi$, since $\boldsymbol{u}$ is also parallel to $-\boldsymbol{\nabla}$. This form of $\boldsymbol{u}$ is not commonly used anymore even though there is nothing wrong with it.[1] We will use equation (11.12) for the definition of the velocity potential as it is commonly done in the literature. We caution that if one defines the velocity vector as the negative of the gradient of the potential, some of the foregoing equations may have different signs in front of different terms, and therefore, one must be careful about these.

Like $\boldsymbol{u}$, ϕ depends on x_i and t. ϕ is called the velocity potential, and a flow represented by ϕ is called potential flow. The biggest advantage of equation (11.12) is that instead of dealing with the three unknown velocity components, $u_1, u_2,$ and u_3, we now have one unknown, namely ϕ.

Once ϕ is determined, then all u_i can be determined from equation (11.12), i.e.,

$$u_1 = \frac{\partial\phi}{\partial x_1}, \; u_2 = \frac{\partial\phi}{\partial x_2}, \; u_3 = \frac{\partial\phi}{\partial x_3}. \tag{11.13}$$

Clearly, we could also write these as $u_1 = \phi_{,1}$, $u_2 = \phi_{,2}$, and $u_3 = \phi_{,3}$. Substituting equation (11.13) in the continuity equation (11.1), we have

$$u_{i,i} = \left(\frac{\partial\phi}{\partial x_i}\right)_{,i} = \frac{\partial^2\phi}{\partial x_i \partial x_i} = \phi_{,ii} = 0$$

or

$$\frac{\partial^2\phi}{\partial x_1^2} + \frac{\partial^2\phi}{\partial x_2^2} + \frac{\partial^2\phi}{\partial x_3^2} = \Delta\phi = 0. \tag{11.14}$$

Equation (11.14) is known as Laplace's equation (or the Laplace equation, as it is sometimes called). Note that Laplace's equation is a linear, second-order partial differential equation of the elliptic type (see the discussion below equation (10.123)).

11.2.1   Euler's Integral

Let us consider the governing equations obtained by use of the conservation of momentum principle. First, we know that $u_i = \phi_{,i}$. Let us substitute this into the N-S equations given by equation (10.25). In other words, let us not assume that the fluid is inviscid for the time being. We should obtain the following for a *viscous* fluid:

---

[1] On the contrary, the use of the negative sign in front of the gradient of the velocity potential means that the positive velocity is defined equivalently as the decreasing potential as one moves toward the positive coordinate axes, and thus, this can be viewed as a more natural definition.

$$\frac{\partial \phi_{,i}}{\partial t} + \phi_{,j}\phi_{,ij} + g\delta_{i2} + \frac{p_{,i}}{\rho} = \nu\phi_{,ijj}. \tag{11.15}$$

But,

$$\phi_{,ijj} = (\phi_{,jj})_{,i} = 0 \tag{11.16}$$

by Laplace's equation, (11.14), and the fact that we can freely change the order of differentiation. And therefore, we can write equation (11.15) as

$$\frac{\partial}{\partial x_i}\left\{ \frac{\partial \phi}{\partial t} + \frac{1}{2}\phi_{,j}\phi_{,j} + gx_2 + \frac{p}{\rho} \right\} = 0, \tag{11.17}$$

which shows that the terms inside the curly brackets, which correspond to the integral of equation (11.15), must collectively be function of time only. Hence,

$$\frac{\partial \phi}{\partial t} + \frac{1}{2}\| \nabla\phi \|^2 + \frac{p}{\rho} + gx_2 = C(t). \tag{11.18}$$

Equation (11.18) is known as Euler's integral. It does require that the flow be irrotational. But it does not require that the fluid be inviscid or the flow be steady! Note that some textbooks or articles refer to equation (11.18) as the unsteady Bernoulli theorem. Also note that the function $C(t)$ is constant everywhere at a given time, unlike the Bernoulli constant which varies from streamline to streamline as in the Bernoulli's theorem (or Bernoulli's equation), which will be discussed next.

It is worthwhile to remark that the time-dependent function $C(t)$ in equation (11.18) can be absorbed in ϕ because the velocity components u_1, u_2, and u_3 are obtained from the gradients of ϕ. Therefore, a "modified" ϕ, in which $C(t)$ is absorbed in, will not change the velocities.

11.2.2   Bernoulli's Theorem (or Equation)

We now discuss Bernoulli's theorem (or Bernoulli's equation) which requires that the flow be steady and the fluid be inviscid but does not require that the flow be irrotational (although it can also be used for steady irrotational flows in an inviscid fluid). Since we assume that the fluid is inviscid, we can start with Euler's equations, equation (11.2) (momentum equations), and set the local time acceleration, $\partial/\partial t$ term, to zero (because of the steadiness assumption) to obtain

$$u_j u_{i,j} + g\delta_{i2} + \frac{p_{,i}}{\rho} = 0. \tag{11.19}$$

Next, let us multiply the above equation by u_i to write (assume that $\rho = $ constant here, but in general this assumption can be relaxed)

$$u_i u_j u_{i,j} + u_i \frac{\partial}{\partial x_i}\left(gx_2 + \frac{p}{\rho} \right) = 0. \tag{11.20}$$

However, the first term above can be written as

$$u_i u_j u_{i,j} = u_j \frac{\partial}{\partial x_j}\left(\frac{1}{2}u_i u_i \right), \tag{11.21}$$

so that equation (11.20) becomes

$$u_j \frac{\partial}{\partial x_j}\left(\frac{1}{2}u_i u_i\right) + u_i \frac{\partial}{\partial x_i}\left(g x_2 + \frac{p}{\rho}\right) = 0. \tag{11.22}$$

Moreover, since i and j are dummy indices, we can regroup likewise terms to write

$$u_i \frac{\partial}{\partial x_i}\left\{\frac{1}{2}u_j u_j + g x_2 + \frac{p}{\rho}\right\} = 0. \tag{11.23}$$

Since $u_i \neq 0$ in general and the flow is steady by assumption, the terms inside the curly brackets in equation (11.23) must at most be constant, i.e.,

$$\frac{1}{2}u_j u_j + g x_2 + \frac{p}{\rho} = C_B = \text{constant}. \tag{11.24}$$

C_B is constant since the derivatives in equation (11.23) are taken along a material surface, which is the stream surface. Recall that if $DF/Dt = 0$ for any function F, then F must be constant on the material surface. Thus, C_B, which is called the Bernoulli constant, varies from streamline to streamline since $u_i \partial/\partial x_i\{\ \}$ in equation (11.23) are the convective acceleration terms in the material derivative taken along a path that contains the same particles (recall that we consider steady flow here). Since there is no mass flow in or out of a streamline, this path must correspond to a streamline. The different values of the Bernoulli constant on different streamlines can be shown to be related to the mass flow between the two streamlines. In the case of uniform flow at infinity for example, we can set $C_B = U_\infty^2/2 + p_\infty/\rho$, where U_∞ and p_∞ are the ambient values of the velocity and pressure, respectively.

11.2.3   Body-Boundary Condition

When we compare Euler's equations (11.2) to the N-S equations (10.25), we see that Euler's equations are one order lower because of the absence of viscous terms. As a result of this, we must relax the no-slip boundary condition given by equation (10.29), and require that the kinematic boundary condition on a boundary be stated as "the normal component of fluid velocity, $u \cdot n$, is equal to the normal component of the velocity of the boundary surface, $q \cdot n$." This condition was given by equation (10.35).

By use of equation (11.13), we then have

$$u \cdot n = (u_1 e_1 + u_2 e_2 + u_3 e_3) \cdot (n_1 e_1 + n_2 e_2 + n_3 e_3)$$
$$= \left(\frac{\partial\phi}{\partial x_1}e_1 + \frac{\partial\phi}{\partial x_2}e_2 + \frac{\partial\phi}{\partial x_3}e_3\right) \cdot \left(\frac{\partial x_1}{\partial n}e_1 + \frac{\partial x_2}{\partial n}e_2 + \frac{\partial x_3}{\partial n}e_3\right) \tag{11.25}$$
$$= \frac{\partial\phi}{\partial x_1}\frac{\partial x_1}{\partial n} + \frac{\partial\phi}{\partial x_2}\frac{\partial x_2}{\partial n} + \frac{\partial\phi}{\partial x_3}\frac{\partial x_3}{\partial n} = \frac{\partial\phi}{\partial n},$$

since (see Figure 11.1),

$$\cos\alpha_1 = \frac{\partial x_1}{\partial n} = \frac{n_1}{\|n\|} = n_1, \qquad \cos\alpha_2 = \frac{\partial x_2}{\partial n} = n_2, \qquad \cos\alpha_3 = \frac{\partial x_3}{\partial n} = n_3,$$

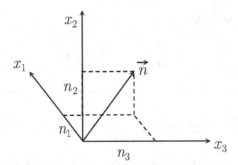

Figure 11.1 Components of the unit normal vector.

where $|\boldsymbol{n}| = 1$ is the magnitude of the unit normal vector on the body surface, pointing out of the fluid. Therefore, the body-boundary condition is given by

$$\frac{\partial \phi}{\partial n} = \boldsymbol{q} \cdot \boldsymbol{n}. \tag{11.26}$$

Of course, if the body is fixed (such as a breakwater or even the sea floor itself), then $\partial \phi / \partial n = 0$ on the body surface. It can be shown that the boundary condition equation (11.26) and the governing equation (11.14) are sufficient to uniquely determine the velocity potential ϕ up to a constant if $\boldsymbol{q}$ is known. If $\boldsymbol{q}$ is not known (which may be the case in free-surface problems for example) then we must also impose the dynamic boundary condition, i.e., that the pressure must be continuous at the free-surface interface. This is a dynamic condition since it is related to surface forces. This condition will be discussed in Chapter 12 within the context of "free surface."

11.3   Elementary Singularities

It is possible to construct some simple body geometries by superposition of elementary singularities, meaning that we use certain mathematical expressions to define a flow field by allowing the fluid velocities to become infinite at a point in the flow field! Although we do not have such infinities in nature, we eliminate the difficulties with such singularities by keeping them out of the flow field of interest as we will see later on. Let us first start the discussion by an elementary singularity, called a source (or a sink).

If the fluid mass is conserved, then the flow across the circumference of any spherical control surface whose center is at $r = 0$, must be constant if we have a source or a sink of constant strength at the origin. We can define m as the source or sink strength, which is the volume of fluid leaving the control surface S per unit time (discharge rate). Recalling that $\boldsymbol{\nabla} \cdot \boldsymbol{u}$ is the rate of change of mass in an incompressible fluid (see the derivation of the continuity equation in Section 10.3), one can define the source or sink strength by

$$m = \int_{V} \boldsymbol{\nabla} \cdot \boldsymbol{u} \, dV = \int_{S} \boldsymbol{u} \cdot \boldsymbol{n} \, dS,$$

where the surface integral on the right is written by use of Gauss' divergence theorem, equation (10.17). If we now use the spherical coordinates in evaluating the above integral, we can obtain

$$m = 4\pi r^2 u_r = 4\pi r^2 \frac{d\phi}{dr},$$

and by integrating this (and keeping in mind that $\phi = \phi(r)$), we have

$$\phi(r) = -\frac{m}{4\pi r}, \qquad \text{(3-D source)} \qquad (11.27)$$

where m is the source strength, r the radial coordinate, and u_r the radial component of the velocity vector, i.e., $u_r = \partial\phi/\partial r$. The constant of integration in equation (11.27) is zero since $\phi(r) \to 0$ as $r \to \infty$. Note that the dimension of m is $[m] = L^3/T$, i.e., m is the volume flux per unit time, or the discharge rate. Therefore, ρm is the mass per unit time pumped in (or out) of the control surfaces.

In the case of a sink, $\phi(r)$ will change sign, i.e., $\phi = m/4\pi r$. Note that equation (11.27) satisfies Laplace's equation except at the origin, where there is a singularity (the function blows up) and it represents radial streamlines originating at $r = 0$.

Since Laplace's equation that governs ϕ is linear (and there are no boundaries so that we do not have to be concerned with any nonlinear boundary conditions), we can superimpose uniform flow, given by

$$\phi = Ux_1, \quad \text{(uniform flow)} \qquad (11.28)$$

onto equation (11.27) to obtain

$$\phi = Ux_1 - \frac{m}{4\pi} \frac{1}{\sqrt{x_1^2 + x_2^2 + x_3^2}}, \qquad \text{(uniform flow + 3-D source)} \qquad (11.29)$$

which represents a three-dimensional, semi-infinitely long body placed in uniform flow as shown in Figure 11.2. Because $\sum m \neq 0$, the body is not a closed one. To close the body, one must put a sink (which has the same strength as the source) in the flow field. As an example to the case of $\Sigma m = 0$, we can write

$$\phi = Ux_1 - \frac{m}{4\pi r_1} + \frac{m}{4\pi r_2}, \qquad \text{(Rankine Ovoid)} \qquad (11.30)$$

where $r_1 = \sqrt{(x_1 - \xi_1)^2 + x_2^2 + x_3^2}$ and $r_2 = \sqrt{(x_1 - \xi_2)^2 + x_2^2 + x_3^2}$, to represent a flow around a closed body, where ξ_1 and ξ_2 are the longitudinal (x_1) coordinates of the singularities (located on $x_2 = 0$ for simplicity here), namely the source and sink. The streamlines generated by equation (11.30) define a body called Rankine Ovoid or Rankine Solid. In general, the location of a singularity has the coordinates (ξ, η, ζ), and therefore, $r = \sqrt{(x_1 - \xi)^2 + (x_2 - \eta)^2 + (x_3 - \zeta)^2}$.

In the case of a sink and a source that are infinitesimally close to each other, and are superimposed on uniform flow, one can show that

$$\phi = Ux_1 + \frac{\mu x_1}{4\pi r^3}, \quad \mu = 2ml, \qquad \text{(uniform flow + 3-D doublet)} \qquad (11.31)$$

where μ is called the dipole moment or doublet strength. The second term on the right-hand side of equation (11.31) is called a or a doublet, and $2l$ is the distance between

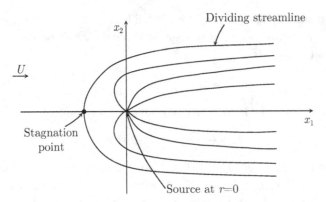

Figure 11.2 Superposition of uniform flow and a three-dimensional source.

the two singularities, each being located at a distance of l from the origin and on $x_2 = 0$. Equation (11.31) represents a sphere, whose radius is $a = (\mu/2\pi U)^{1/3}$, placed in uniform stream; this can be obtained by enforcing the body-boundary condition in an ideal fluid: $\partial\phi/\partial r = 0$, on $r = a$. If a is given, then $\mu = 2\pi U a^3$ would be the required strength of the dipole (see Prob. 11.6.1).

In the two-dimensional case, everything is assumed to be constant in the third coordinate (pointing out of the paper). Therefore, V can be considered as the volume per unit depth, and S can be considered the surface area per unit depth, so that the strength of a source can be written as

$$m = A \int_S \nabla \cdot u \, dS = A \oint_C u \cdot n \, dC,$$

where A is a constant equal to the depth of the 2-D region (which is closed) whose contour is denoted by C. We then have the source strength

$$m = \int_0^{2\pi} u_r r d\theta = 2\pi r u_r = 2\pi r \frac{\partial \phi}{\partial r}$$

or

$$\phi = \frac{m}{2\pi} \log r, \quad \text{(2-D source)} \tag{11.32}$$

where u_r again is the radial component of the velocity in polar coordinates. The dimension of m in 2-D is L^2/T, i.e., m is the discharge rate per unit depth. Here $\log r = \log_e r = \ln r$, i.e., it is the natural log.

The case of a sink can similarly be obtained, and the result is the same as equation (11.32) except that there will be a minus sign in front of m, i.e., $\phi = -m \log r/2\pi$.

Similarly, for a dipole in 2-D, see, e.g., Potter and Foss (1982), and along $x_2 = 0$, the velocity potential is given by (see also Figure 11.3)

$$\phi = \frac{\mu}{2\pi} \frac{x_1}{r^2}. \quad \text{(2-D dipole)} \tag{11.33}$$

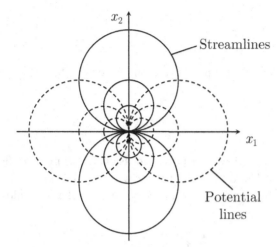

Figure 11.3 Two-dimensional dipole.

These elementary singularities can be distributed inside or on a body to exactly satisfy the body-boundary condition equation (11.26) and Laplace's equation (11.14). For example, if we superimpose uniform flow on a dipole in two dimensions, and use the polar coordinates, we obtain

$$\phi = \frac{\mu x_1}{2\pi r^2} + U x_1 = \frac{\mu \cos\theta}{2\pi r} + U r \cos\theta$$

$$= \left(\frac{\mu}{2\pi r} + U r\right) \cos\theta \quad \text{(uniform flow + 2-D doublet)}.$$

And by use of the boundary condition equation (11.26) for a fixed circle in 2-D flow, we have

$$\frac{\partial \phi}{\partial n} = -\frac{\partial \phi}{\partial r} = 0, \quad \text{on } r = a,$$

which implies that

$$\left(\frac{-\mu}{2\pi r^2} + U\right) \cos\theta = 0 \implies \mu = 2\pi a^2 U. \tag{11.34}$$

This solution represents potential flow around a circle.

Another type of singularity in two dimensions is a vortex that represents a flow of zero radial velocity everywhere, i.e.,

$$u_r = \frac{\partial \phi}{\partial r} = 0, \quad u_\theta = \frac{1}{r}\frac{\partial \phi}{\partial \theta} = \frac{c}{r}, \tag{11.35}$$

where r is the radial coordinate and c is a constant. Note that this flow field represents irrotational flow everywhere except the origin (show that this statement is true by calculating the vorticity in polar coordinates). If we denote the strength of the vortex by Γ, then the circulation that is associated with the point singularity at the origin is given by

$$\Gamma = \oint_C \boldsymbol{u} \cdot d\boldsymbol{x} = \int_0^{2\pi} u_\theta \, r \, d\theta = 2\pi c. \tag{11.36}$$

Note that positive values of c correspond to a counterclockwise flow (for details, see, e.g., Currie (1974)).

11.3.1   Complex Potential

Either in a two-dimensional flow or a three-dimensional axisymmetric flow, there exists a scalar function called the stream function $\psi(x_1, x_2, t)$, which is a collection of streamlines, $\psi = $ constant. For example, in two dimensions, the continuity equation (with $u_3 = $ constant) given by

$$\frac{\partial u_1}{\partial x_1} + \frac{\partial u_2}{\partial x_2} = 0 \tag{11.37}$$

reveals that we can write

$$u_1 = \frac{\partial \psi}{\partial x_2}, \quad u_2 = -\frac{\partial \psi}{\partial x_1}. \tag{11.38}$$

Therefore, it is possible to solve for ψ first and then determine the velocity components. The comparison of equation (11.38) with equation (11.13) in two dimensions shows that

$$\frac{\partial \phi}{\partial x_1} = \frac{\partial \psi}{\partial x_2}, \quad \frac{\partial \phi}{\partial x_2} = -\frac{\partial \psi}{\partial x_1}. \tag{11.39}$$

Equations (11.39) are called the Cauchy–Riemann equations (see also Section 3.3). It can be shown that ϕ (velocity potential) and ψ (the stream function) are the real and imaginary parts of an analytic function, $F(z)$, called the complex potential, given by

$$F(z) = \phi + i\psi, \tag{11.40}$$

where $i = \sqrt{-1}$, see Sections 3.1 and 3.7. For example, in terms of the complex potential, $F(z)$, a vortex is given by

$$F(z) = -i\frac{\Gamma}{2\pi} \log z = \phi + i\psi, \quad \phi = c\theta, \quad \psi = -c \log r, \quad z = x_1 + ix_2 = re^{i\theta}. \tag{11.41}$$

Note that $F(z)$ is an analytic function of a complex variable, z, if the derivative dF/dz exists at a point z_o and dF/dz is independent of the direction in which it is calculated. Furthermore, $F(z)$ must be continuous at $z = z_0$. Then, we can write

$$\frac{dF}{dz} = \frac{d\phi}{dz} + i\frac{d\psi}{dz} = \frac{\partial \phi}{\partial x_1} + i\frac{\partial \psi}{\partial x_1} = \frac{\partial \psi}{\partial x_2} - i\frac{\partial \phi}{\partial x_2}, \tag{11.42}$$

or

$$\frac{dF}{dz} \equiv w = u_1 - iu_2, \tag{11.43}$$

which is called the complex velocity. Note that equation (11.41) is obtained by use of equation (11.39).

In cylindrical polar coordinates, the complex velocity is given by

$$\frac{dF}{dz} \equiv w = (u_r - iu_\theta)\, e^{-i\theta}. \tag{11.44}$$

We can give some elementary 2-D flows in terms of the complex potential, F. For example, uniform flow is given by

$$F(z) = Uz, \tag{11.45}$$

where U is the uniform current velocity. The complex potential for a source of strength m located at a point z_0 is

$$F(z) = \frac{m}{2\pi} \log(z - z_0), \tag{11.46}$$

and the sink will be the negative of this equation. The complex potential for a doublet (or dipole) of strength μ located at the origin is

$$F(z) = \frac{\mu}{2\pi z}, \tag{11.47}$$

where μ is given by equation (11.34). For a vortex, the complex potential was given by equation (11.41).

In two dimensions, and for bodies in an unbounded fluid, $F(z)$ is an extremely useful tool that can be used to calculate the velocities (if, of course, the flow is irrotational and the fluid is incompressible and inviscid). Even quite complicated geometries can be handled with the use of $F(z)$ along with the conformal mapping technique. For more information on the application of conformal mapping techniques, see, e.g., Milne-Thompson (1968).

Many of these and other elementary singularities can be calculated by various methods, including conformal mapping, and the results can be visualized easily, see, e.g., Ertekin and Padmanabhan (1994).

11.4   Distribution of Singularities and the Green Function Method

We have just seen that it is possible to construct various (simple) body geometries by placing elementary singularities in the flow field. It also is possible to obtain arbitrary body geometries by continuously distributing singularities on or inside a body. This method is called the Green function method. Clearly, it is much easier to place singularities on a surface that we know, e.g., the hull surface of a ship, rather than guessing locations of singularities when they are placed inside the body.

A Green function is also called a source function, and it is one of the bases for finding solutions to diffraction problems in coastal and offshore engineering. A numerical method, called the boundary-element method (BEM), uses this approach. As we will see later on, the Green function method requires the discretization of the boundaries on which there are distributed singularities with unknown strengths. This method is in contrast with the finite-element method (FEM) which requires that the entire fluid domain and its boundaries be discretized to solve for the velocity potential, and therefore, the velocities. For more information on the BEM, see for instance, Brebbia (1978). We shall first discuss some preliminaries before going into the Green function method.

If S is a smooth closed surface enclosing the volume V, and $\boldsymbol{u}$ is any vector whose components have continuous first partials in the domain V enclosed by S, then

$$\iiint_V (\nabla \cdot \boldsymbol{u})\, dV = \iint_S (\boldsymbol{u} \cdot \boldsymbol{n})\, dS, \tag{11.48}$$

where $\boldsymbol{n}$ is the outer unit normal vector on the surface, S, i.e., positive $\boldsymbol{n}$ points out of the fluid (or into the body). Equation (11.48) is known as the divergence theorem or the Gauss theorem (see, e.g., Sokolnikoff (1966)) for $\boldsymbol{u}$, and was given by equation (10.17) before. Now if $\boldsymbol{u} = \nabla\phi$, i.e., the flow is irrotational, then equation (11.48) becomes (after we use equation (11.25))

$$\iiint_V \Delta\phi\, dV = \iint_S \frac{\partial\phi}{\partial n}\, dS. \tag{11.49}$$

The above integral equation is known as Green's formula. Next, assume that $\boldsymbol{p} = f\nabla g$ and $\boldsymbol{q} = g\nabla f$ are any two continuous, vector-valued functions. We can write (noting that f and g are scalar functions)

$$\nabla \cdot \boldsymbol{p} = \nabla \cdot (f\nabla g) = \nabla f \cdot \nabla g + f\Delta g, \qquad \nabla \cdot \boldsymbol{q} = \nabla \cdot (g\nabla f) = \nabla f \cdot \nabla g + g\Delta f.$$

By replacing $\nabla \cdot \boldsymbol{u}$ in equation (11.48) with $\nabla \cdot \boldsymbol{p}$ and by $\nabla \cdot \boldsymbol{q}$ separately, we have Green's first identity in the form

$$\iiint_V (\nabla f \cdot \nabla g + f\Delta g)\, dV = \iint_S f\frac{\partial g}{\partial n}\, dS \qquad \text{(Green's first identity)} \tag{11.50}$$

and

$$\iiint_V (\nabla f \cdot \nabla g + g\Delta f)\, dV = \iint_S g\frac{\partial f}{\partial n}\, dS. \tag{11.51}$$

Subtracting equation (11.51) from equation (11.50), we have

$$\iiint_V (f\Delta g - g\Delta f)\, dV = \iint_S \left(f\frac{\partial g}{\partial n} - g\frac{\partial f}{\partial n}\right) dS \qquad \text{(Green's second identity)}$$

$$\tag{11.52}$$

Equation (11.52) is known as Green's second identity. Moreover, if f and g satisfy Laplace's equation, i.e., $\Delta f = \Delta g = 0$, and therefore, f and g are harmonic functions by definition, then

$$\iint_S \left(f\frac{\partial g}{\partial n} - g\frac{\partial f}{\partial n}\right) dS = 0. \qquad \text{(Green's second identity for harmonic functions)}$$

$$\tag{11.53}$$

As a side note, if we set $f = g = \phi$, we obtain from equation (11.51) the following:

$$\iiint_V (\nabla\phi \cdot \nabla\phi + \phi\Delta\phi)\, dV = \iint_S \phi\frac{\partial\phi}{\partial n}\, dS. \tag{11.54}$$

We will now consider equation (11.53). Let us assume that f is a source potential (called the Rankine source) and g is a velocity potential in three dimensions, i.e.,

$$f = \frac{1}{r}, \quad g = \phi, \quad r = \left[(x_1 - \xi)^2 + (x_2 - \eta)^2 + (x_3 - \zeta)^2\right]^{1/2}, \tag{11.55}$$

and (x_1, x_2, x_3) are the coordinates of a point of interest (or the field point) and (ξ, η, ζ) are the coordinates of the source point where the singularity is located (e.g., $r = 0$). Note that both f and g satisfy Laplace's equation except at the singularity.

Since at $r = 0, f \to \infty$, in equation (11.55), there is a singularity. To remove this difficulty in the use of Green's second identity, equation (11.53) in our case, consider a very small sphere of radius ϵ about a singular point as shown in Figure 11.4, and substitute equation (11.55) in equation (11.53) as follows:

$$\iint_{S_r} \left[\frac{1}{r}\frac{\partial\phi}{\partial n} - \phi\frac{\partial}{\partial n}\left(\frac{1}{r}\right)\right] dS_r + \iint_{S_\epsilon} \left[\frac{1}{r}\frac{\partial\phi}{\partial n} - \phi\frac{\partial}{\partial n}\left(\frac{1}{r}\right)\right] dS_\epsilon = 0. \tag{11.56}$$

Note that $S = S_\epsilon \cup S_r$, and $dS_\epsilon = \epsilon^2 d\Omega$ and $d\Omega$ is the solid angle of the sphere whose radius is $r = \epsilon$. Then, equation (11.56) becomes (keeping in mind that $\partial/\partial n = -\partial/\partial r$ since the normal vector on the sphere is pointing out of the fluid or into the singularity)

$$\iint_{S_r} \left[\frac{1}{r}\frac{\partial\phi}{\partial n} - \phi\frac{\partial}{\partial n}\left(\frac{1}{r}\right)\right] dS_r - \epsilon\iint_{S_\epsilon} \frac{\partial\phi}{\partial r}\, d\Omega - \iint_{S_\epsilon} \phi\, d\Omega = 0. \tag{11.57}$$

As $\epsilon \to 0$, the second integral in equation (11.57) vanishes. Now consider the last integral in equation (11.57), and note that

$$-\iint_{S_\epsilon} \phi\, d\Omega = -\phi(P)\iint_{S_\epsilon} d\Omega = -4\pi\phi(P), \tag{11.58}$$

since $P(x_1, x_2, x_3) \in V$, and not of S_ϵ, and therefore the potential could be moved out of the integral sign. Therefore, equation (11.57) becomes, for a field point, P, inside the fluid,

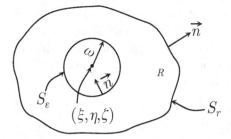

Figure 11.4 A sphere around a singular point located at (ξ, η, ζ).

$$\phi(x_1, x_2, x_3, t) = \phi(P) = -\frac{1}{4\pi} \int\int_S \left[\phi \frac{\partial}{\partial n}\left(\frac{1}{r}\right) - \frac{1}{r}\frac{\partial\phi}{\partial n} \right] dS, \quad P \in V. \quad (11.59)$$

If the point P is on the boundary S_r, everything will be the same except the constant in front of the integral sign in equation (11.59) will change, i.e.,

$$\phi(x_1, x_2, x_3, t) = \phi(P) = -\frac{1}{2\pi} \int\int_S \left[\phi \frac{\partial}{\partial n}\left(\frac{1}{r}\right) - \frac{1}{r}\frac{\partial\phi}{\partial n} \right] dS, \quad P \in S, \quad (11.60)$$

since the solid angle of a hemisphere is 2π. The function $\phi(P)$, in the integral over S_ϵ in equation (11.58), could still be moved out of the integral sign since in the limit $\epsilon \to 0$, $\phi(P)$ will be constant. Equation (11.60) is called the Fredholm integral equation of the second kind since the unknown ϕ appears both inside and outside the integral sign.

If the point P is outside the domain, then the left-hand sides of equations (11.59) and (11.60) vanish. Note that the point $P = (x_1, x_2, x_3)$ in equation (11.59) must be inside the domain, whereas in equation (11.60) it must be on the boundaries.

One has to solve equation (11.60) to determine $\phi(P)$ on the boundaries and then use equation (11.59) to determine $\phi(P)$ anywhere in the fluid domain. Equation (11.60) can be solved by discretizing the boundary by, for example, triangular or quadrilateral elements (the characteristic length of which is in the order of 1/20th or less of the wavelength) and solving for the unknown velocity potential from the integral equation which can be reduced to a set of simultaneous linear equations. This method is called, in general, the boundary-element method (see, e.g., Brebbia (1978)).

In an unbounded fluid, it suffices to solve the integral on the body surface only (why?). Then equation (11.60) could be approximated by

$$\phi(P_i) + \frac{1}{2\pi}\sum_{j=1}^{N} \phi(P_j)\frac{\partial}{\partial n}\left(\frac{1}{r_{ij}}\right)\Delta_j S = \frac{1}{2\pi}\sum_{j=1}^{N}\frac{1}{r_{ij}}\frac{\partial\phi(P_j)}{\partial n}\Delta_j S. \quad (11.61)$$

In equation (11.61), P_i and P_j are on the boundary and r_{ij} is the distance between P_i and P_j. Note that the right-hand side of equation (11.61) is known from the boundary condition on the body, i.e., from equation (11.26).

In general, there may be boundaries other than the body surface if the fluid is *not* unbounded. A free surface is a typical example to this. When there are many boundaries, it is clear that the solution of equation (11.61) will require a large computational (or CPU) time, although this is becoming of a less concern as the computers become faster with large memories and storage spaces. Therefore, if one can find a function that satisfies the conditions on certain boundaries, such as the one on the sea floor or the free surface, then all one has to do is to solve the integral equation (11.60) on the body surface only. For linear problems, there is such a function called the Green function which has the desired properties. This function is also called a source function.

To introduce such a Green function, let

$$\hat{G}(x_1, x_2, x_3; \xi, \eta, \zeta, t) = \Re\{G(x_1, x_2, x_3; \xi, \eta, \zeta)\}F(t), \quad (11.62)$$

where

$$G(x_1, x_2, x_3; \xi, \eta, \zeta) = \frac{1}{r} + H(x_1, x_2, x_3; \xi, \eta, \zeta) \qquad (11.63)$$

is a Green function (complex) with the property that $\Delta G = 0$ if $(x_1, x_2, x_3) \neq (\xi, \eta, \zeta)$, and $\Delta G = \infty$ if $(x_1, x_2, x_3) = (\xi, \eta, \zeta)$, and $\mathfrak{R}$ denotes the real part. Or we can write

$$\Delta G = \delta(x_1 - \xi, x_2 - \eta, x_3 - \zeta), \qquad (11.64)$$

where δ is the Dirac delta function which is by definition zero if $(x_1, x_2, x_3) \neq (\xi, \eta, \zeta)$ and infinity if $(x_1, x_2, x_3) = (\xi, \eta, \zeta)$. The Green function method is used in many applied mathematics problems, and it is a powerful method for certain boundary-value problems that can be encountered in hydrodynamics (see, e.g., Wehausen and Laitone (1960)), heat conduction, structural mechanics, chemical engineering, and so forth. More information on the mathematical aspects of the Green function method can be found, e.g., in Stakgold (1979), Roach (1982) and Mei (1989).

If the problem is two-dimensional and in an unbounded fluid region, then recalling that a velocity potential is proportional to $\log r$ (see equation (11.32)), we should be able to write

$$G(x_1, x_2; \xi, \eta) = -\log r. \qquad (11.65)$$

Note that the coefficient $-1/4\pi$ in equation (11.59) becomes $-1/2\pi$, and the coefficient $-1/2\pi$ in equation (11.60) becomes $-1/\pi$ because in two dimensions, we are dealing with a plane surface (circle) rather than a volume (sphere) for our domain.

In the three-dimensional case, after substitution of equation (11.63) into equations (11.59) and (11.60), we have

$$\iint\limits_{S} \left(\phi \frac{\partial G}{\partial n} - G \frac{\partial \phi}{\partial n} \right) dS = 0, \qquad P \notin V, S$$

$$= -2\pi \phi(x_1, x_2, x_3, t), \qquad P \in S \qquad (11.66)$$

$$= -4\pi \phi(x_1, x_2, x_3, t), \qquad P \in V,$$

depending on whether $P = (x_1, x_2, x_3)$ is outside, on or inside the closed boundary S, respectively.

One of the alternative methods for solving hydrodynamic problems is the finite-element method, (FEM). The disadvantage of this method compared with the Green function method, provided that a particular Green function exists and is available, is that ϕ has to be calculated everywhere inside the domain. On the other hand, (i) FEM results in a banded matrix that can be solved efficiently, (ii) the function calculated in FEM is simpler, and (iii) FEM requires less knowledge of mathematics and fluid mechanics than the Green function method does. Another numerical method that can be used is the finite-difference method (FDM) although it is not used frequently in fluid–structure interaction problems involving floating bodies.

Once the velocity potential is determined, one can use Euler's integral to find the pressure on a body. Therefore, the force and moment acting on a body can be obtained from

$$\boldsymbol{F} = \iint\limits_{S_B} p\, \boldsymbol{n}\, dS, \quad \boldsymbol{M} = \iint\limits_{S_B} p\, (\boldsymbol{r} \times \boldsymbol{n})\, dS, \qquad (11.67)$$

where the body-surface unit normal vector $\boldsymbol{n}$ is positive by convention if pointing out of the fluid volume or into the body, and $\boldsymbol{r}$ is the position vector from the origin of the coordinate system to a point on the surface, S_B, of the body.

Now recall Euler's integral in equation (11.18) with $C(t)$ absorbed in ϕ, and without the hydrostatic term (without loss in generality), to write the expression for the dynamic pressure:

$$p_d = -\rho \left(\frac{\partial \phi}{\partial t} + \frac{1}{2} |\nabla \phi|^2 \right). \tag{11.68}$$

Substituting equation (11.68) in equation (11.67), we obtain

$$\boldsymbol{F} = -\rho \iint_{S_B} \left[\frac{\partial \phi}{\partial t} + \frac{1}{2} |\nabla \phi|^2 \right] \boldsymbol{n} \, dS, \tag{11.69}$$

$$\boldsymbol{M} = -\rho \iint_{S_B} \left[\frac{\partial \phi}{\partial t} + \frac{1}{2} |\nabla \phi|^2 \right] (\boldsymbol{r} \times \boldsymbol{n}) \, dS. \tag{11.70}$$

If one applies Gauss' theorem and Reynolds' transport theorem, one can obtain various versions of these equations, see, e.g., Newman (1978). Note that, in a *linear problem*, the dynamic pressure is given by

$$p = -\rho \frac{\partial \phi}{\partial t}. \tag{11.71}$$

In Chapter 13, we will come back to the Green function method, and to the hydrodynamic forces and moments when there is free surface, and the body, whether fully submerged or on the free surface, is placed in the fluid domain.

11.5   Concluding Remarks

In this chapter, we have discussed briefly the ideal-fluid flow (inviscid and incompressible fluid) in an unbounded domain, i.e., no boundaries are present except the boundary of the body itself. The governing equations and boundary conditions are discussed first. Then, Euler's equations for the velocity components and the pressure are introduced. With the additional assumption that the vorticity vector vanishes in the fluid domain and on its boundaries, the concept of the velocity potential is introduced. And hence, the irrotational-flow equations are derived.

We have seen that the set of four Euler's equations to be solved for an ideal fluid reduced to the solution of a single equation for the unknown velocity potential, namely Laplace's equation for ϕ. This means that we can, in principle, solve for ϕ by use of Laplace's equation subject to appropriate boundary conditions, and then use the momentum equations (which now reduce to a single equation called Euler's integral) to determine the unknown pressure anywhere in the fluid domain.

The assumptions behind the derivation of Euler's integral and Bernoulli's equation are stated. We have seen that the assumptions necessary in the derivation of Euler's integral are that the fluid is incompressible and the flow is irrotational. It is shown that Euler's integral does not require that the fluid be inviscid or that the flow be steady. Whereas Bernoulli's equation for the velocity components and pressure requires that the flow be steady and the fluid be inviscid. It is also shown that Bernoulli's equation does not require that the flow be irrotational or even that the fluid be incompressible (although we have assumed that ρ is constant here, this was not at all necessary).

The no-slip boundary condition used in the case of a viscous fluid must be relaxed here because the momentum equations for the pressure are one order lower than the N-S equations due to the lack of viscous terms. As a result, in an irrotational flow of an ideal fluid, we require that the normal components of the fluid and body-surface velocities be the same, but the tangential components may be different.

Following the fundamental equations of an ideal-fluid flow, we have discussed some elementary singularities, such as a source, sink, and dipole, that can be used to construct a body in the flow field. In this powerful approach, the unknown strengths of the singularities are determined by imposing the body-boundary condition. The superposition of elementary singularities is possible because Laplace's equation that governs the velocity potential is linear.

Either in a 2-D flow or a 3-D axisymmetric flow, one can use the resources of complex-variable analysis to determine the velocity potential or its counterpart, the stream function, through the Cauchy–Riemann equations.

It is possible to distribute elementary singularities not only inside a body but also on its boundary. With this alternative approach, and the use of Green's identities, it can be shown that the solutions of Laplace's equation can be sought by solving an integral equation for the velocity potential. This could be done by paying special attention to the integrals that include the singularity itself. Upon the removal of the singularity, we have seen that the potential can be determined on the boundary by enforcing the boundary condition to determine the unknown strengths of the singularities.

In conjunction with the above approach, which can be termed the boundary-element method, we have discussed briefly the Green-function method. This method will be discussed further in Chapter 13. As an example, we have considered the fundamental source singularity in two dimensions to establish the integral equations for ϕ. Once the potential is solved on the boundary, the potential can easily be calculated anywhere inside the fluid domain, and thus the velocity vector and the pressure can be known. With the pressure known from Euler's integral, one can integrate the pressure in the body-normal direction and over the entire wetted-body surface to determine the force and moment vectors acting on a submerged body.

11.6   Self-Assessment

11.6.1

Consider uniform flow around a fixed sphere in three dimensions, see equation (11.31). The fluid is inviscid and the flow is irrotational. The unit normal vector, n, on the sphere surface is pointing out of the fluid.

1. Enforce the body-boundary condition (no-flux) in this ideal flow to determine the doublet strength, μ, for a sphere of radius a. Hint: The flow is axisymmetric, therefore it suffices to use the spherical polar coordinates in the x_1–x_2 plane only, i.e., $x_1 = r \cos \theta$, $x_2 = r \sin \theta$.

2. If the particle velocity components in spherical coordinates, for this 3-D axisymmetric flow, are given by

$$u_r = \frac{1}{r^2 \sin \theta} \frac{\partial \psi}{\partial \theta} = \frac{\partial \phi}{\partial r}, \qquad u_\theta = -\frac{1}{r \sin \theta} \frac{\partial \psi}{\partial r} = \frac{1}{r} \frac{\partial \phi}{\partial \theta},$$

   determine the stream function, ψ.

3. Prove that there are two stagnation points located at $r = a$, $\theta = 0$ and $r = a$, $\theta = \pi$.

11.6.2

A source of strength m is located at $z = a$, and another source of equal strength is located at $z = -a$, where a is a real number, see Figure 11.5. Set $z - a = r_1 e^{i\theta_1}$ and $z + a = r_2 e^{i\theta_2}$, where $r_1 = \sqrt{r^2 - 2ax + a^2}$ and $r_2 = \sqrt{r^2 + 2ax + a^2}$.

1. Show, in this two-dimensional, steady and ideal-flow problem, that the velocity components on the y ($x = 0$) axis are given by

$$u_r = \frac{\partial \phi}{\partial r} = \frac{m}{\pi} \frac{r}{r^2 + a^2}, \qquad u_\theta = \frac{1}{r} \frac{\partial \phi}{\partial \theta} = 0.$$

2. Comment on the meaning of $u_\theta = 0$ anywhere along the y axis.

11.6.3

A source is located at $x_1 = -1.0\,m$, $x_2 = x_3 = 0$, and a sink is located at $x_1 = 1.0\,m$, $x_2 = x_3 = 0$. The problem is three-dimensional. If the free-stream velocity is $U = 3.0\,m/s$, and the strength of each singularity is $m = 3.0\,m^3/s$, calculate the maximum body radius and the length of the Rankine ovoid in three dimensions. Hint: The total discharge rate of the fluid flowing from the source to the sink is

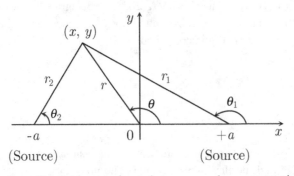

Figure 11.5 Two sources located at $z = -a$ and $z = a$, respectively.

$$m = 2\pi \int_0^{r_0} u(r)r \, dr,$$

where r_0 is the unknown body radius.

11.6.4

The complex potential for uniform flow past a circle with circulation is given by

$$F(z) = U\left(z + \frac{a^2}{z}\right) + i\frac{\Gamma}{2\pi}\log\left(\frac{z}{a}\right),$$

where a is the radius of the circle, U the uniform-flow velocity far from the circle and in the x direction, Γ the strength of the circulation, and $z = x + iy$, $i = \sqrt{-1}$.

1. Derive the expressions for the velocity potential, ϕ, and the stream function, ψ, as functions of the r and θ coordinates,
2. Derive the expressions for the fluid pressure on the circle. Note that Bernoulli's equation dictates that on $r = a$, $(u_r^2 + u_\theta^2)/2 + p/\rho$ is equal to $U^2/2 + p_\infty/\rho$, where p_∞ is the ambient pressure at ∞ (this is because the equation is equal to Bernoulli's constant on a streamline). In this case, the constant is equal to $U^2/2 + p_\infty/\rho$.
3. Show that the horizontal force $F_x = 0$ and the vertical force $F_y = \rho U\Gamma$ on the circle. You can then deduce that there is no force acting on the circle if the circulation $\Gamma = 0$ (potential flow predicts lift when there is circulation but predicts zero force (both the horizontal and vertical components) always in an unbounded fluid, and this is called D'Alembert's paradox).

Notes:

- $\int \cos\theta \sin\theta \, d\theta = -\frac{1}{4}\cos 2\theta$
- $\int \sin^3\theta \, d\theta = \frac{1}{3}\cos^3\theta - \cos\theta$
- $\int \cos\theta \sin^2\theta \, d\theta = \frac{\sin^3\theta}{3}$
- $\int \sin^2\theta \, d\theta = -\frac{1}{2}\sin\theta\cos\theta + \frac{1}{2}\theta$

11.6.5

The velocity potential for two-dimensional irrotational flow about a circle is given, in polar coordinates (r,θ), by

$$\phi = \left(r + \frac{a^2}{r}\right)U\cos\theta - \frac{\Gamma\theta}{2\pi},$$

where U is the uniform, free-stream velocity, a is the radius of the circle, and Γ is the circulation. If $a = 50$ cm and $U = 5$ m/s, determine the location of the stagnation points for $\Gamma = \pi U a$, i.e., determine θ_s and r_s when $u_r = u_\theta = 0$.

11.6.6

A *source* is located at $z = A = a + ib$, and a *sink* is located at $z = -A = -(a + ib)$. The source and sink strengths, m, are the same, and a and b are real constants. The flow is ideal and steady.

Obtain the velocity potential and stream function in terms of x, y, a, b, m, and sketch the equipotential lines and streamlines.

11.6.7

Starting with a source and a sink, infinitesimally close to each other, 2ϵ, on the x axis, obtain the velocity potential and the stream function for a dipole in two dimensions, i.e., the real and imaginary parts of equation (11.47). In your derivation, let $2m\epsilon = \mu$ as $\epsilon \to 0$ and $m \to \infty$, where μ is the dipole strength. You may want to start the derivation by use of the complex potential, $F(z)$, for a source and a sink, i.e., $F(z) = (m/2\pi)(\log(z + \epsilon) - \log(z - \epsilon))$.

11.6.8

Consider uniform flow around a sphere (in an unbounded fluid) of radius R; the flow is steady and irrotational, and the fluid is incompressible and inviscid.

1. Calculate the fluid velocity components in spherical polar coordinates.
2. By assuming that the velocity at $r = \infty$ is U_∞ and the pressure is p_∞, calculate the pressure distribution on the sphere.
3. Integrate the pressure on the sphere to determine the horizontal and vertical components of the force acting on the sphere (and hence, deduce D'Alembert's paradox).

11.6.9

Determine the velocity potential and stream function and velocity components for each of the following complex potentials (x, y, a, b, r, θ are all real quantities) and sketch the streamlines of the resulting two-dimensional flows:

1. $F = c\,z$, $c = a + ib$, $z = x + iy$, $a > 0$, $b > 0$
2. $F = z^2$, $z = x + iy$
3. $F = \frac{m}{2\pi} \log z$, $z = r\,e^{i\theta}$
4. $F = \frac{a}{z}$, $z = re^{i\theta}$, $a > 0$

11.6.10

There is a source of strength m located at $z = i\,h$ and a source of strength m located at $z = h$, where $i = \sqrt{-1}$ and h is an arbitrary (but real) constant (see Figure 11.6). Show that the $\theta = \pi/4$ axis is a streamline, or that $\partial\phi/\partial n = 0$ on this streamline where n

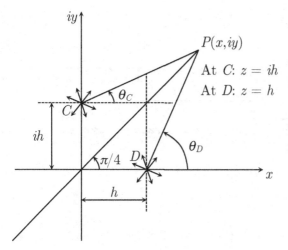

Figure 11.6 Two sources located at C and D.

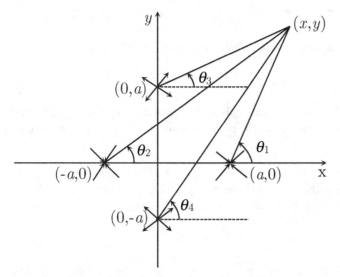

Figure 11.7 Superposition of two sources and two sinks.

is the normal to the streamline. To do this, write the complex potential to obtain the velocity potential ϕ and the stream function ψ first. Reminders: $\log(A) + \log(B) = \log(A\,B)$; $\log(A) - \log(B) = \log\left(\frac{A}{B}\right)$.

11.6.11

A source is located at $z = ia$ and another source is located at $z = -ia$ ($i = \sqrt{-1}$ and a is an arbitrary, but real, constant). Moreover, a sink is located at $z = a$ and another sink is located at $z = -a$ (see Figure 11.7). All source and sink strengths, m, are the same. The fluid is ideal.

Obtain the stream function ψ. What does $\psi = m/4$ represent?

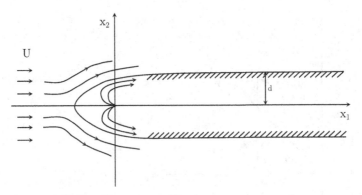

Figure 11.8 Uniform flow past a half body.

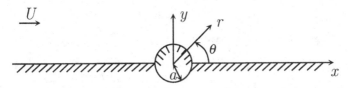

Figure 11.9 The cross section of a half-buried pipe on the sea floor.

11.6.12

Superimpose a source of strength m, located at the origin ($x_1 = x_2 = 0$), and uniform current U, see Figure 11.8. The problem is two-dimensional and the fluid is ideal. Show that

1. The stagnation point is located at $x_1 = -m/(2\pi U)$, $x_2 = 0$.
2. The half-width of the body on the x_2 axis ($x_1 = 0$) is $r = m/(4U)$.
3. The half-width of the body as $x_1 \to \infty$ is $d = m/(2U)$.

11.6.13

The fluid is inviscid and two-dimensional. The flow is irrotational. There are four source singularities that are located at $z_1 = a + ib$, $z_2 = a - ib$, $z_3 = -a + ib$, and $z_4 = -a - ib$, where a and b are real, positive constants and $i = \sqrt{-1}$. All sources have the same strength m.

Obtain the velocity potential, ϕ, and the stream function, ψ, in terms of

$$r_j = |z - z_j| = \sqrt{(x - a_j)^2 + (y - b_j)^2} \quad \text{and} \quad \theta_j = \text{Arctan}\left(\frac{y - b_j}{x - a_j}\right), \quad j = 1, 2, 3, 4,$$

where

$$z_j = a_j + ib_j, \quad \text{e.g.,} \quad a_1 = a, \ a_3 = -a, \ b_2 = -b, \ \text{etc., and} \ z - z_j = r_j\, e^{i\theta_j}$$
in complex polar form.

Sketch the equipotential lines and streamlines for $a = b = m = 1$.

11.6.14

A pipe of radius a is half buried on the ocean floor as shown in Figure 11.9. The water depth can be assumed infinite. The radius of the pipe is a. The fluid motion can be assumed two-dimensional, and the fluid is incompressible and inviscid, and the flow is irrotational and steady. Far from the pipe, the uniform flow velocity is U and the pressure is p_∞.

Determine

1. The complex potential, complex velocity, and the velocity components u_r and u_θ.
2. The pressure distribution on the pipe.
3. The force in the x direction (thrust) and in the y direction (lift).

11.6.15

The velocity components in a two-dimensional ($i = 1, 2$) flow are given by

$$u_1 = \frac{(x_1^2 - x_2^2)}{r^4}, \quad u_2 = \frac{(2x_1 x_2)}{r^4}, \quad \text{where } r = \sqrt{x_1^2 + x_2^2}.$$

1. Calculate the circulation

$$\Gamma = \oint_{ABCDA} (u_1\, dx_1 + u_2\, dx_2) = \oint_{ABCDA} V \cdot dx$$

around the contour of the rectangle given by the points A(1,2), B(3,2), C(3,6), and D(1,6).

2. Next, calculate the same circulation by evaluating

$$\Gamma = \int\int_S (\nabla \times V) \cdot dS = \int\int_S \omega \cdot dS = \int_2^6 \int_1^3 \left(\frac{\partial u_2}{\partial x_1} - \frac{\partial u_1}{\partial x_2} \right) dx_1 dx_2.$$

12 Water Waves

12.1 Introduction

In this chapter, we will review various concepts and equations of water-wave theory. This brief review is necessary since we will need later the tools of this chapter when we discuss the diffraction of waves by fixed obstacles and the motion problems of freely floating structures. The fundamental relations in water-wave theory will be reviewed first. In this section, the governing equation and the boundary conditions are presented for the nonlinear, progressive wave field. The perturbation expansion of the nonlinear velocity potential and other unknown functions, such as the free-surface elevation, are discussed in Section 12.3. The solution of the linearized problem is discussed in Section 12.4. A more detailed analysis of the subject can be found, e.g., in Wehausen, Webster, and Yeung (2016).

12.2 Fundamental Relations

We will assume that the fluid is incompressible and inviscid, and the flow is irrotational, so that the particle velocity is given by

$$\boldsymbol{u} = \boldsymbol{\nabla}\phi, \tag{12.1}$$

where ϕ is the velocity potential. Because of the incompressible fluid assumption, the continuity equation is given by

$$\boldsymbol{\nabla} \cdot \boldsymbol{u} = 0. \tag{12.2}$$

Recall that by substituting equation (12.1) in equation (12.2), we have Laplace's equation (11.14) as the governing equation of the problem:

$$\Delta\phi = 0. \tag{12.3}$$

We also have Euler's integral given by equation (11.18) to be used in the determination of the pressure. We may take the time-dependent function $C(t)$, in equation (11.18), which is the same anywhere in the fluid, as

$$C(t) = \frac{p_A}{\rho}, \tag{12.4}$$

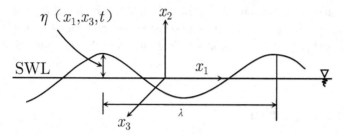

Figure 12.1 Free surface of a wave of length λ.

where p_A is the atmospheric pressure. This poses no problem since $C(t)$ can be absorbed in ϕ without altering the particle velocity field. We have to consider the boundary conditions next.

On any material surface, whether free or not, we have, the general kinematic boundary condition from equations (11.25) and (11.26),

$$u \cdot n = \frac{\partial \phi}{\partial n} = q \cdot n. \tag{12.5}$$

Clearly, equation (12.5) is also the body-boundary condition and the sea-floor boundary condition since it represents no-flux through the surface. By defining the boundary surface by the equation

$$F(x_1, x_2, x_3, t) = 0, \tag{12.6}$$

and requiring that the material derivative of F vanishes, we have

$$\frac{DF}{Dt} = \frac{\partial F}{\partial t} + u_1 \frac{\partial F}{\partial x_1} + u_2 \frac{\partial F}{\partial x_2} + u_3 \frac{\partial F}{\partial x_3} = F_t + \phi_{x_1} F_{x_1} + \phi_{x_2} F_{x_2} + \phi_{x_3} F_{x_3} = 0. \tag{12.7}$$

We can represent the function that describes the free surface by

$$F(x_1, x_2, x_3, t) = x_2 - \eta(x_1, x_3, t) = 0, \tag{12.8}$$

as shown in Figure 12.1. Substituting equation (12.8) in equation (12.7), we obtain the kinematic free-surface condition:

$$\frac{\partial \phi}{\partial x_2} - \frac{\partial \eta}{\partial t} - \frac{\partial \phi}{\partial x_1} \frac{\partial \eta}{\partial x_1} - \frac{\partial \phi}{\partial x_3} \frac{\partial \eta}{\partial x_3} = 0 \quad \text{on } x_2 = \eta. \tag{12.9}$$

Equation (12.9) is called the kinematic free-surface condition.

The dynamic condition on the free surface is that the pressure is continuous, i.e., $p = p_A \cong 0$ on $x_2 = \eta$, where we have taken the atmospheric pressure equal to zero without loss in generality. The surface tension is ignored, which means that the water waves we deal with here exclude the capillary waves (whose lengths are less than about 1.5 cm). Therefore, from Euler's integral equation (11.18), we obtain the dynamic free-surface condition:

$$\frac{\partial \phi}{\partial t} + \frac{1}{2} \|\nabla \phi\|^2 + g\eta = 0 \quad \text{on } x_2 = \eta, \tag{12.10}$$

where we have used equations (12.4) and (12.8). The two difficulties associated with these boundary conditions are: (i) they must be imposed on a boundary that is unknown in general, and (ii) they are nonlinear. Note, however, that the governing equation (12.3) is linear.

12.3   Perturbation Expansion

Because of the difficulties mentioned above, one usually assumes that the wave motion is "small" and therefore that the nonlinear terms can be discarded as a result of the argument that their magnitudes will be smaller than the magnitudes of the linear terms. In this respect, one can assume that the velocity potential, as well as the surface elevation, can be expanded in a perturbation series for which a perturbation parameter of ϵ is taken to be equal to, for example, Ak, where A is the wave amplitude and k is the wave number, i.e., $k = 2\pi/\lambda$, where λ is the wavelength. In shallow-water wave problems, one would rather deal with two small parameters: one representing the nonlinearity and the other representing the dispersion of long waves.

A perturbation series is a series expansion of an unknown function about a known function, provided that the deviation of the unknown function from the known function is "small" (say the known function is the potential that can, for example, be taken as constant – that is equal to zero everywhere; this corresponds to a quiescent fluid). Then, we can write

$$\phi = \epsilon\phi^{(1)} + \epsilon^2\phi^{(2)} + \epsilon^3\phi^{(3)} + \cdots, \quad \eta = \epsilon\eta^{(1)} + \epsilon^2\eta^{(2)} + \epsilon^3\eta^{(3)} + \cdots, \quad \epsilon = Ak,$$

or

$$\phi = \sum_{n=1}^{\infty} \epsilon^n \phi^{(n)}, \quad \eta = \sum_{n=1}^{\infty} \epsilon^n \eta^{(n)}, \tag{12.11}$$

where $\phi^{(1)}$ is called the first-order potential, $\phi^{(2)}$ the second-order potential, etc., and similarly for η, i.e., $\eta^{(1)}$ is called the first-order surface elevation, $\eta^{(2)}$ is called the second-order surface elevation, etc.

Now, if the wave motion is small, meaning $\epsilon \ll 1$, we may, without giving any necessary justification, discard all the higher-order terms after we substitute these expansions in the boundary conditions. Moreover, we can expand each of the terms of the boundary conditions in a Taylor series about the still-water surface, $x_2 = 0$. For example, the time derivative of the potential is written as

$$\frac{\partial\phi}{\partial t}(x_1, x_2 = \eta, x_3, t) = \frac{\partial\phi}{\partial t}(x_1, 0, x_3, t) + \eta\frac{\partial^2\phi(x_1, 0, x_3, t)}{\partial t\partial x_2} + \cdots. \tag{12.12}$$

Since η and ϕ are small, the higher-order terms can sometimes be ignored. This means that only the linear terms involving ϕ and η have to be evaluated on the still-water level (SWL) ($x_2 = 0$) instead of on the exact boundary surface, $x_2 = \eta(x_1, x_3, t)$. This is required to be consistent with the perturbation expansion. Therefore, we have the linearized versions of the boundary conditions, given by equations (12.9) and (12.10), as follows:

$$\frac{\partial \eta}{\partial t}(x_1,x_3,t) = \frac{\partial \phi}{\partial x_2}(x_1,0,x_3,t) \text{ (kinematic),} \qquad (12.13)$$

$$\eta(x_1,x_3,t) = -\frac{1}{g}\frac{\partial \phi}{\partial t}(x_1,0,x_3,t) \text{ (dynamic).} \qquad (12.14)$$

In equations (12.13) and (12.14), we have ignored the superscripts (1) which refer to the first-order potential and first-order wave elevation. The expansion equation (12.11) is such that when $\epsilon = 0$ there is no fluid motion, therefore ϕ and η vanish.

Similarly, the following expansions for the pressure and wave frequency can be written as

$$p = p^{(0)} + \epsilon p^{(1)} + \epsilon^2 p^{(2)} + \cdots, \qquad \omega = \omega^{(0)} + \epsilon\omega^{(1)} + \epsilon^2\omega^{(2)} + \cdots,$$

or

$$p = \sum_{n=0}^{\infty} \epsilon^n p^{(n)}, \qquad \omega = \sum_{n=0}^{\infty} \epsilon^n \omega^{(n)}, \qquad (12.15)$$

where $p^{(0)}$ and $\omega^{(0)}$ do not depend on ϵ or the wave amplitude A. In fact, $p^{(0)} = \rho g x_2$ in equation (12.15), and

$$\omega^{(0)} = \sqrt{gk\tanh(kh)} \qquad (12.16)$$

from the dispersion relation. In equation (12.16), h is the uniform water depth, k the wave number, and g the gravitational acceleration.

We must also consider the kinematic boundary condition on the sea floor. From equation (11.26), and assuming that the sea floor is not moving (this is not a necessary assumption), we have

$$\left.\frac{\partial \phi}{\partial n}\right|_{x_2=-h} = 0 \quad \text{or} \quad \left.\frac{\partial \phi}{\partial x_2}\right|_{x_2=-h} = 0, \qquad (12.17)$$

because we have assumed here that the sea floor is flat and parallel to the still-water surface.

We recall that the Taylor series expansion that, for instance, can lead to equation (12.12) is given by (for any function f which is continuous up to the mth derivative and when the expansion is about $x_2 = 0$)

$$f(x_1,x_2 = \eta,x_3,t) = f(x_1,0,x_3,t) + \frac{\eta}{1!}\frac{\partial f}{\partial x_2}(x_1,0,x_3,t) + \frac{\eta^2}{2!}\frac{\partial^2 f}{\partial x_2^2}(x_1,0,x_3,t) + \cdots$$

$$= \sum_{m=0}^{\infty} \frac{\eta^m}{m!}\frac{\partial^m f(x_1,0,x_3,t)}{\partial x_2^m}.$$

$$(12.18)$$

Because all functions given by equation (12.18) are represented by asymptotic expansions such as equation (12.11), we can write, e.g.,

$$(x_1, x_2 = \eta, x_3, t; \epsilon) = \sum_{n=1}^{\infty} \epsilon^n f^{(n)}(x_1, \eta, x_3, t),$$

or, by use of equation (12.18),

$$
\begin{aligned}
f &= \sum_{n=1}^{\infty} \epsilon^n \sum_{m=0}^{\infty} \frac{\eta^m}{m!} \frac{\partial^m f^{(n)}(x_1, 0, x_3, t)}{\partial x_2^m} \\
&= \sum_{n=1}^{\infty} \epsilon^n \left\{ \sum_{m=0}^{\infty} \left\{ \frac{1}{m!} \left[\sum_{k=1}^{\infty} \epsilon^k \eta^{(k)} \right]^m \left[\frac{\partial^m f^{(n)}(x_1, 0, x_3, t)}{\partial x_2^m} \right] \right\} \right\}.
\end{aligned}
\tag{12.19}
$$

Note that for pressure and frequency, the above series would start from $n = 0$.

If we collect the terms that belong to the same order, say up to $O(\epsilon^4)$, in equation (12.19), we have

$$
f(x_1, \eta, x_3, t; \epsilon) = \epsilon f^{(1)}(x_1, 0, x_3, t; \epsilon) + \epsilon^2 \left(f^{(2)} + f_{x_2}^{(1)} \eta^{(1)} \right)
$$
$$
+ \epsilon^3 \left(f^{(3)} + f_{x_2}^{(1)} \eta^{(2)} + \frac{1}{2} f_{x_2 x_2}^{(1)} \eta^{(1)2} + f_{x_2}^{(2)} \eta^{(1)} \right) + O(\epsilon^4),
\tag{12.20}
$$

where subscripts denote differentiation.

We now substitute the expansion (12.11) into equation (12.3) to obtain

$$
\Delta \phi = \Delta \left[\sum_{n=1}^{\infty} \epsilon^n \phi^{(n)} \right] = \epsilon \Delta \phi^{(1)} + \epsilon^2 \Delta \phi^{(2)} + \cdots = 0.
\tag{12.21}
$$

Since each of the terms in equation (12.21) is of different order, equation (12.21) is valid if and only if each term vanishes, i.e.,

$$
\epsilon \, \Delta \phi^{(1)} = 0 \Rightarrow \Delta \phi^{(1)} = 0, \qquad \epsilon^2 \Delta \phi^{(2)} = 0 \Rightarrow \Delta \phi^{(2)} = 0,
\tag{12.22}
$$

and so forth for higher orders. The first equation in (12.22) is called the governing equation of the first-order problem, the second equation is of the second-order problem, and so forth.

By use of the same procedure, the solid-boundary condition on the sea floor, equation (12.17), becomes

$$
\frac{\partial \phi}{\partial x_2}(x_1, -h, x_3, t; \epsilon) = \frac{\partial}{\partial x_2} \left(\sum_{n=1}^{\infty} \epsilon^n \phi^{(n)} \right) = 0
$$

or

$$
\sum_{n=1}^{\infty} \epsilon^n \frac{\partial \phi^{(n)}}{\partial x_2}(x_1, -h, x_3, t) = 0.
$$

Therefore,

$$
\frac{\partial \phi^{(1)}}{\partial x_2}(x_1, -h, x_3, t) = 0, \qquad \frac{\partial \phi^{(2)}}{\partial x_2}(x_1, -h, x_3, t) = 0,
\tag{12.23}
$$

and so forth for higher orders.

As a result, the first-order problem, $O(\epsilon)$ problem, becomes

$$\Delta\phi^{(1)}(x_1,x_2,x_3,t) = 0,$$
$$\phi_{x_2}^{(1)}(x_1,-h,x_3,t) = 0,$$
$$\phi_{x_2}^{(1)}(x_1,0,x_3,t) - \eta_t^{(1)}(x_1,x_3,t) = 0,$$
$$\phi_t^{(1)}(x_1,0,x_3,t) + g\eta^{(1)}(x_1,x_3,t) = 0.$$

(12.24)

The last two equations in (12.24) will formally be derived later (equations (12.33) and (12.35)).

For this $O(\epsilon)$ problem, the dynamic pressure is given by linearized Euler's integral. And, from equation (12.15), we have

$$p^{(1)}(x_1,x_2,x_3,t) = -\rho\phi_t^{(1)}(x_1,x_2,x_3,t), \quad x_2 < 0, \tag{12.25}$$

anywhere in the fluid. Note that in obtaining equation (12.24) or equation (12.25), we have collected the ϵ terms and then set them equal to zero, otherwise the terms with different orders of ϵ cannot balance each other.

Since from equation (12.11), we have, for the linear or first-order problem,

$$\phi \cong \epsilon\phi^{(1)}, \quad \eta \cong \epsilon\eta^{(1)}, \tag{12.26}$$

we can multiply equation (12.24) by ϵ and therefore drop the superscript (1). This was the justification in writing equation (12.13) or equation (12.14) before.

Let us now formalize the above procedure to systematically obtain the equations for the first- and second-order problems. But first, let us recall the order symbols (see also Section 10.8.1). If a function f which depends on a perturbation parameter ϵ tends to zero at the same rate that another function g tends to zero, we can write

$$f(\epsilon) = O[g(\epsilon)] \quad \text{as} \quad \epsilon \to 0 \quad \text{if} \quad \lim_{\epsilon \to 0}\frac{f(\epsilon)}{g(\epsilon)} = M, \quad 0 < |M| < \infty, \tag{12.27}$$

where M is a finite number. The function $g(\epsilon)$ is called the gauge function since it is used for comparison purposes. For example, the power sequence $\epsilon^0, \epsilon^1, \epsilon^2, \epsilon^3, \epsilon^4, \ldots$, is a collection of some gauge functions. Another example is the inverse powers of ϵ, i.e., $\epsilon^{-1}, \epsilon^{-2}, \epsilon^{-3}, \epsilon^{-4}, \ldots$. Note that if, as we have assumed, ϵ is small, i.e., $\epsilon \ll 1$, then $1 > \epsilon > \epsilon^2 > \epsilon^3 > \ldots$ A typical example to $f(\epsilon)$ equals to "big oh" of $g(\epsilon)$, as $\epsilon \to 0$, could be $\cos(\epsilon) = O(1)$, for we can expand $\cos(\epsilon)$ as

$$\cos(\epsilon) = 1 - \frac{\epsilon^2}{2!} + \frac{\epsilon^4}{4!} - \cdots,$$

and with $g(\epsilon) = \epsilon^0$,

$$\lim_{\epsilon \to 0}\frac{\cos(\epsilon)}{\epsilon^0} = \frac{1 - \frac{\epsilon^2}{2!} + \frac{\epsilon^4}{4!}}{1} = 1 = |M| < \infty.$$

Whereas the function $f(\epsilon) = 1 - \cos(\epsilon)$ is $O(\epsilon^2)$, since with $g(\epsilon) = \epsilon^2$,

$$\lim_{\epsilon \to 0}\frac{1 - \cos(\epsilon)}{\epsilon^2} = \frac{\frac{\epsilon^2}{2!} - \frac{\epsilon^4}{4!} + \cdots}{\epsilon^2} = \frac{1}{2!} = |M| < \infty.$$

In the examples given above, we have selected the gauge functions such that as $\epsilon \to 0$ we obtain a finite number $M > 0$.

Sometimes it may not be possible to determine the same rate at which a function tends to its limit but it may be possible to determine whether the rate is faster or slower than the rate of the gauge function approaching its limit. In such cases, the "little oh" symbol is used as follows:

$$f(\epsilon) = o[g(\epsilon)] \quad \text{as} \quad \epsilon \to 0 \quad \text{if} \quad \lim_{\epsilon \to 0} \frac{f(\epsilon)}{g(\epsilon)} = 0. \tag{12.28}$$

For example, $\cos(\epsilon) = o(\epsilon^{-1})$, since

$$\lim_{\epsilon \to 0} \frac{\cos(\epsilon)}{\epsilon^{-1}} = 0.$$

It is important to note that the mathematical orders discussed here are not necessarily the same as the physical order of magnitudes since M, which is the limit for "big oh," can be arbitrary. However one hopes that M is "close," to the physical order of magnitude.

We have already shown that each of the first-order, second-order, etc., potentials on the right-hand side of equation (12.11) satisfies Laplace's equation and the bottom-boundary condition individually (see equations (12.22) and (12.23)). Therefore, we need not consider them again.

Let us now consider the kinematic free-surface boundary condition equation (12.9) in two dimensions (for brevity), i.e.,

$$\frac{\partial \phi}{\partial x_2} - \frac{\partial \eta}{\partial t} - \frac{\partial \phi}{\partial x_1} \frac{\partial \eta}{\partial x_1} = 0 \quad \text{on} \quad x_2 = \eta. \tag{12.29}$$

Noting that the perturbation series for ϕ and η are given by equation (12.11), we now consider the term ϕ_{x_2}, and use equation (12.19) to write it as

$$\frac{\partial \phi}{\partial x_2}(x_1, \eta, t; \epsilon) = \epsilon \left\{ \phi_{x_2}^{(1)}(x_1, 0, t) + (\epsilon \eta^{(1)} + \epsilon^2 \eta^{(2)} + \cdots)^1 \times (\phi_{x_2 x_2}^{(1)}(x_1, 0, t)) \right.$$

$$\left. + \frac{1}{2!}(\epsilon \eta^{(1)} + \epsilon^2 \eta^{(2)} + \cdots)^2 \times (\phi_{x_2 x_2 x_2}^{(1)}(x_1, 0, t)) + \cdots \right\}$$

$$+ \epsilon^2 \{ \phi_{x_2}^{(2)}(x_1, 0, t)$$

$$+ (\epsilon \eta^{(1)} + \epsilon^2 \eta^{(2)} + \cdots)^1 \times (\phi_{x_2 x_2}^{(2)}(x_1, 0, t)) + \frac{1}{2!}(\epsilon \eta^{(1)} + \epsilon^2 \eta^{(2)} + \cdots)^2$$

$$\times (\phi_{x_2 x_2 x_2}^{(2)}(x_1, 0, t)) + \cdots \} + \epsilon^3 \{ \phi_{x_2}^{(3)}(x_1, 0, t) + \cdots \} + \epsilon^4 \{ \ldots \} + \cdots \tag{12.30}$$

We now collect the likewise terms in ϵ, i.e.,

$$\phi_{x_2}(x_1, \eta, t; \epsilon) = \epsilon \left\{ \phi_{x_2}^{(1)}(x_1, 0, t) \right\} + \epsilon^2 \left\{ \phi_{x_2 x_2}^{(1)}(x_1, 0, t) \eta^{(1)} + \phi_{x_2}^{(2)}(x_1, 0, t) \right\}$$

$$+ \epsilon^3 \left\{ \phi_{x_2 x_2}^{(1)}(x_1, 0, t) \eta^{(2)} + \frac{1}{2} \phi_{x_2 x_2 x_2}^{(1)}(x_1, 0, t) \eta^{(1)^2} \right.$$

$$\left. + \phi_{x_2 x_2}^{(2)}(x_1, 0, t) \eta^{(1)} + \phi_{x_2}^{(3)}(x_1, 0, t) \right\} + O(\epsilon^4). \tag{12.31}$$

We can also write

$$\frac{\partial \eta}{\partial t} + \frac{\partial \phi}{\partial x_1}\frac{\partial \eta}{\partial x_1} = \epsilon \eta_t^{(1)} + \epsilon^2 \eta_t^{(2)} + \cdots$$

$$+ \left\{\epsilon\phi_{x_1}^{(1)} + \epsilon^2\phi_{x_2x_1}^{(1)}\eta^{(1)} + \epsilon^2\phi_{x_1}^{(2)} + \cdots\right\}\left\{\epsilon\eta_{x_1}^{(1)} + \epsilon^2\eta_{x_1}^{(2)} + \cdots\right\}$$

$$= \epsilon\eta_t^{(1)} + \epsilon^2\left(\eta_t^{(2)} + \phi_{x_1}^{(1)}\eta_{x_1}^{(1)}\right)$$

$$+ \epsilon^3\left(\eta_t^{(3)} + \phi_{x_1}^{(1)}\eta_{x_1}^{(2)} + \phi_{x_2x_1}^{(1)}\eta^{(1)}\eta_{x_1}^{(1)} + \phi_{x_1}^{(2)}\eta_{x_1}^{(1)}\right) + O(\epsilon^4).$$

$$(12.32)$$

Therefore, equation (12.29) becomes

$$O(\epsilon) : \quad \phi_{x_2}^{(1)} - \eta_t^{(1)} = 0,$$

$$O(\epsilon^2) : \quad \phi_{x_2}^{(2)} - \eta_t^{(2)} = \phi_{x_1}^{(1)}\eta_{x_1}^{(1)} - \phi_{x_2x_2}^{(1)}\eta^{(1)}.$$

$$(12.33)$$

Note that once the first-order problem is solved, the right-hand side of the second-order boundary condition in equation (12.33) is known, and thus it can be treated as an applied or external pressure on the free surface located on the SWL. To see this, recall that the right-hand side of the first equation in (12.33) is zero because we have set the pressure on the free surface equal to $\rho C(t)$ in Euler's integral given by equation (11.18). We could as well set it to $p_A = C(t) = 0$ without loss in generality.

Let us consider the dynamic free-surface boundary condition given by equation (12.10) in two dimensions. Following the same procedure as before, we obtain

$$\phi_t = \epsilon\phi_t^{(1)} + \epsilon^2\left(\phi_t^{(2)} + \phi_{x_2t}^{(1)}\eta^{(1)}\right) + \epsilon^3\left(\phi_t^{(3)} + \phi_{x_2t}^{(1)}\eta^{(2)} + \cdots\right) + O(\epsilon^4)$$

$$\frac{1}{2}\left\{\phi_{x_1}^2 + \phi_{x_2}^2\right\} = \frac{1}{2}\left\{\left[\epsilon\phi_{x_1}^{(1)} + \epsilon^2\left(\phi_{x_2x_1}^{(1)}\eta^{(1)} + \phi_{x_1}^{(2)}\right) + \cdots\right]^2\right.$$

$$+ \left.\left[\epsilon\phi_{x_2}^{(1)} + \epsilon^2\left(\phi_{x_2x_2}^{(1)}\eta^{(1)} + \phi_{x_2}^{(2)}\right) + \cdots\right]^2\right\} + O\left(\epsilon^5\right). \quad (12.34)$$

Then, we have

$$O(\epsilon) : \phi_t^{(1)} + g\eta^{(1)} = 0,$$

$$O(\epsilon^2) : \phi_t^{(2)} + g\eta^{(2)} = -\phi_{x_2t}^{(1)}\eta^{(1)} - \frac{1}{2}\left(\phi_{x_1}^{(1)^2} + \phi_{x_2}^{(1)^2}\right).$$

$$(12.35)$$

The dynamic and kinematic free-surface conditions can be combined into one equation for each $O(\epsilon)$ and $O(\epsilon^2)$ as follows:

$$O(\epsilon) : \phi_{tt}^{(1)}(x_1,0,t) + g\phi_{x_2}^{(1)}(x_1,0,t) = 0,$$

$$O(\epsilon^2) : \phi_{tt}^{(2)}(x_1,0,t) + g\phi_{x_2}^{(2)}(x_1,0,t) = -\eta^{(1)}\left[\phi_{ttx_2}^{(1)} + g\phi_{x_2x_2}^{(1)}\right] - 2\left(\phi_{x_1}^{(1)}\phi_{tx_1}^{(1)} + \phi_{x_2}^{(1)}\phi_{tx_2}^{(1)}\right).$$

$$(12.36)$$

The first equation in (12.36) is the combined form of the third and fourth equations, respectively, in equation (12.24).

It is important to note that the first-order potential, which will be given explicitly in the next section, also satisfies the second-order problem if the water depth is infinite. In other words,

$$\phi(x_1,x_2,t) = \frac{gA}{\omega}e^{kx_2}\sin(kx_1 - \omega t) + O(\epsilon^3). \tag{12.37}$$

However, the second-order surface elevation, i.e.,

$$\eta = \epsilon\eta^{(1)} + \epsilon^2\eta^{(2)} + O(\epsilon^3) \tag{12.38}$$

is not the same as the first-order surface elevation given by equation (12.39) (see, e.g., Wiegel (1964)).

The higher-order infinitesimal wave theory based on the systematic power series expansion in $\epsilon = Ak$ is due to Stokes (1849). The proof of convergence can be found in Levi-Civita (1925). Schwartz (1974) obtained the infinite-depth expansion up to $O(\epsilon^{117})$ by use of a computer algorithm. However, because of the complexity of algebra and the rapid convergence of the asymptotic series, it is mostly unnecessary to consider problems of $O(\epsilon^4)$ and higher, unless, perhaps, if the water depth is very shallow. But then, the Stokes expansion in shallow water gives inaccurate results in general and thus should not be used if the water depth is shallow. Instead, a cnoidal wave theory can be used in shallow waters, see, e.g., Wiegel (1964) and Sarpkaya and Isaacson (1981).

Since a more detailed analysis of perturbation expansion is beyond the scope of this book, we will not pursue the matter further.

12.4   Long-Crested, Linear Progressive Waves

We assume that we have two-dimensional or long-crested linear water waves so that the associated functions do not depend on the x_3 coordinate. Of course, in the case of short-crested waves, which represent the real situation in the oceans, we cannot rule out the x_3 dependence. The word "linear" refers to the assumption that the amplitude of the waves is "small."

Let us now assume that a monochromatic (single frequency) wave, propagating in the positive x_1 direction, is given by

$$\eta(x_1,t) = A\cos(kx_1 - \omega t). \tag{12.39}$$

Here A is the wave amplitude. Equation (12.39) does not depend on time in a moving coordinate system, whose constant speed is given by

$$V_p = \frac{\omega}{k}. \tag{12.40}$$

In other words, the motion is steady in the moving coordinates. In a fixed coordinate system, η is a time-harmonic function. Because η is periodic, ϕ must also be periodic (see equation (12.24)), so that we can write

$$\phi(x_1,x_2,t) = \text{Re}\left\{Y(x_2)e^{i[kx_1 - \omega t]}\right\}. \tag{12.41}$$

Equation (12.41) is a result of the separation-of-variables technique used in solving linear partial differential equations. In other words, we have assumed that

$$\phi(x_1, x_2, t) = X(x_1)Y(x_2)T(t). \tag{12.42}$$

If we substitute equation (12.41) in Laplace's equation, (12.24), we obtain

$$\frac{d^2Y}{dx_2^2} - k^2Y = 0. \tag{12.43}$$

Equation (12.43) is a homogeneous second-order ordinary differential equation whose general solution can be written as

$$Y = Ee^{rx_2}, \tag{12.44}$$

where E and r are some constants to be determined. If we substitute equation (12.44) in equation (12.43), we obtain

$$Er^2e^{rx_2} - k^2Ee^{rx_2} = 0 ,$$

or $r^2 = k^2 \Rightarrow r_{1,2} = \pm k$. Since $\cosh(\xi) = (e^\xi + e^{-\xi})/2$, we can write equation (12.44) in the following form:

$$Y(x_2) = G\cosh(kx_2 + \beta), \tag{12.45}$$

where G and β are constants to be determined.

We can now use the bottom-boundary condition given in equation (12.24) to obtain

$$\frac{\partial \phi}{\partial x_2}(x_1, -h, t) = e^{i(kx_1 - \omega t)}Gk\sinh(-kh + \beta) = 0.$$

The above equation will be satisfied if $\beta = kh$ so that equation (12.45) becomes

$$Y(x_2) = G\cosh k(x_2 + h), \tag{12.46}$$

and ϕ, in equation (12.42), becomes

$$\phi = Re\left\{G\cosh[k(x_2 + h)]e^{i(kx_1 - \omega t)}\right\}. \tag{12.47}$$

Let us now use the dynamic free-surface condition in equation (12.24), and also use equation (12.39), to write

$$\eta = A\cos(kx_1 - \omega t) = \frac{1}{g}G\cosh(kh)i\omega \cos(kx_1 - \omega t) \quad \text{on } x_2 = 0. \tag{12.48}$$

So that G must satisfy

$$G = \frac{-Agi}{\omega\cosh(kh)}. \tag{12.49}$$

Substituting equation (12.49) in equation (12.47), and taking the real part, we obtain

$$\begin{aligned}
\phi(x_1, x_2, t) &= Re\left\{\left(\frac{-Agi}{\omega\cosh(kh)}\right)\cosh[k(x_2 + h)]e^{i(kx_1 - \omega t)}\right\} \\
&= \frac{gA}{\omega}\frac{\cosh[k(x_2 + h)]}{\cosh(kh)}\sin(kx_1 - \omega t).
\end{aligned} \tag{12.50}$$

We have not yet used the kinematic free-surface condition given by the third equation in (12.24). Taking the x_2 derivative of equation (12.50), and the time derivative of equation (12.39), we have

$$\frac{\partial \phi}{\partial x_2}(x_1, 0, t) = \frac{g A k}{\omega} \frac{\sinh[k(x_2 + h)]}{\cosh(kh)} \sin(kx_1 - \omega t), \qquad (12.51)$$

and

$$\frac{\partial \eta}{\partial t} = A\omega \sin(kx_1 - \omega t). \qquad (12.52)$$

Therefore, the third equation in (12.24) becomes (on $x_2 = 0$)

$$\{(g A k / \omega) \tanh(kh) - A\omega\} \sin(kx_1 - \omega t) = 0,$$

or

$$\omega^2 = gk \tanh(kh). \qquad (12.53)$$

Equation (12.53), which is the same as equation (12.16), is called the dispersion relation and is a nonlinear function of k if the water depth is finite. Note that if the water depth is infinite, equation (12.53) becomes

$$\omega^2 = \lim_{h \to \infty} gk \tanh(kh) = gk. \qquad (12.54)$$

For this deep-water case, the real part of the velocity potential of the incoming wave (or incident wave potential) becomes

$$\phi(x_1, x_2, t) = \frac{g A}{\omega} e^{kx_2} \sin(kx_1 - \omega t). \qquad (12.55)$$

Because a more detailed analysis of water waves is beyond the scope of this book, we will not go into many related concepts such as wave energy, group velocity, particle paths, etc. However, it is useful to give particle velocity components (linear) for the finite water depth case by use of equation (12.50):

$$
\begin{aligned}
u_1 &= \frac{\partial \phi}{\partial x_1} = \frac{g A k}{\omega} \frac{\cosh[k(x_2 + h)]}{\cosh(kh)} \cos(kx_1 - \omega t), \\
u_2 &= \frac{\partial \phi}{\partial x_2} = \frac{g A k}{\omega} \frac{\sinh[k(x_2 + h)]}{\cosh(kh)} \sin(kx_1 - \omega t).
\end{aligned}
\qquad (12.56)
$$

The total pressure (linear) can then be obtained from Euler's integral as

$$
\begin{aligned}
p(x_1, x_2, t) &= -\rho \frac{\partial \phi}{\partial t} - \rho g x_2 \\
&= \rho g A \frac{\cosh k(x_2 + h)}{\cosh(kh)} \cos(kx_1 - \omega t) - \rho g x_2,
\end{aligned}
\qquad (12.57)
$$

where the first term on the right-hand side represents the dynamic and the second term represents the hydrostatic pressure.

Note that in very deep water, for which $\lambda/h < 1/2$ ($\lambda = 2\pi/k$), we can introduce the following approximation:

$$\frac{\cosh k(x_2 + h)}{\cosh(kh)} \to e^{kx_2} \text{ as } h \to \infty. \qquad (12.58)$$

And also, $\omega^2 = gk$ from equation (12.54). As a result, the finite-water depth equations given above will somewhat be simplified in infinite water depth. Of course, this is also true for water particle accelerations (linear):

$$\frac{Du_1}{Dt} \cong \frac{\partial u_1}{\partial t} = gAk\frac{\cosh[k(x_2 + h)]}{\cosh(kh)}\sin(kx_1 - \omega t), \qquad (12.59)$$

$$\frac{Du_2}{Dt} \cong \frac{\partial u_2}{\partial t} = -gAk\frac{\sinh[k(x_2 + h)]}{\cosh(kh)}\cos(kx_1 - \omega t). \qquad (12.60)$$

In the following chapters, we will use some of the expressions given here when studying the wave-diffraction and floating-body problems. For a detailed analysis of many physical quantities involved in linear water waves, see, e.g., Sarpkaya and Isaacson (1981) and Dean and Dalrymple (1991).

12.5   Concluding Remarks

The main purpose of this chapter is to review the basic equations of the water-wave theory to be used in subsequent chapters. The governing equation and the boundary conditions of irrotational water waves in an inviscid and incompressible fluid are presented first. These equations clearly show that the boundary-value problem under consideration is a nonlinear one.

Although the linearity assumption (that the wave slope is "small") will be used in subsequent chapters, we have briefly discussed the perturbation-expansion procedure used in water-wave theory to show, in a somewhat rigorous manner, how the linearized boundary-value problem can be established. Following that section, Laplace's equation is solved in two dimensions by use of the separation of variables technique and enforcing the kinematic and dynamic boundary conditions on the free surface and the sea floor. We have seen that the dispersion relation that relates the wave frequency to wave number (or wavelength) is a result of satisfying the kinematic free-surface boundary condition. As a result of this solution procedure, the velocity potential for long-crested, progressive water waves is obtained. The velocity potential, which is valid for constant water depth, is then used to determine the particle velocity vector and the pressure anywhere inside the fluid domain and on its boundaries.

The water waves that we have discussed in this chapter are freely propagating, meaning that there is no obstruction in the flow field that may cause diffraction or scattering. It is also important to emphasize that the boundary conditions on the free surface are applied at the still-water surface rather than the actual free surface; this is a result of the formal perturbation procedure which uses the Taylor series expansion of a function about a known function.

12.6   Self-Assessment

12.6.1

Starting with equation (12.9) in two dimensions (x_1, x_2 plane), derive the third-order kinematic free-surface boundary condition.

12.6.2

Starting with equation (12.10) in two dimensions (x_1, x_2 plane), derive the third-order dynamic free-surface boundary condition.

12.6.3

Combine the kinematic and dynamic free-surface boundary conditions derived in Probs. 12.6.1 and 12.6.2 (that are correct up to third order) to obtain the combined free-surface boundary condition at the third order.

13 Wave Diffraction and Wave Loads

13.1 Introduction

In this chapter, we will discuss the diffraction of water waves when an obstruction is present in the flow field. In the diffraction problem, the obstruction is fixed even though the derivations carried out are applicable to the case of a freely floating body (or a body that has prescribed motions) if the boundary-value problem is linearized. Typical examples of wave diffraction problems are shown in Figure 13.1.

The reason why the diffraction-potential solution discussed in this chapter can also be used in the case of bodies that are not fixed is because the (complicated) hydrodynamic problem of a floating body moving among waves can be decomposed into a number of smaller problems that are easier to solve. Such a decomposition can be made because the governing equation (Laplace's equation) and the boundary conditions of the mathematical problem are (or can be made) linear. We then take the complicated problem of a floating body on the free surface and decompose the total velocity potential into

1. Incoming-wave potential, Φ_I
2. Diffraction potential, Φ_D
3. Radiation potentials, Φ_{Ri}, $i = 1, \dots, 6$

The incoming wave potential was discussed in Chapter 12 and the diffraction potential will be discussed in this chapter. The radiation potential, on the other hand, is the solution to the radiation problem and should not be confused with the radiation condition that will be discussed soon. The radiation problem is the problem in which a body makes prescribed motions in an otherwise calm water, i.e., there are no incoming waves but the rigid body makes six different harmonic motions, one at a time. These six motions are shown in Figure 13.2. The radiation problem will be discussed in Chapter 14.

There are some additional potentials that can be added to the list of the potentials above. For example, if a ship is moving with forward velocity, one needs to calculate the steady forward potential that is created by the steady forward motion of a body in the absence of any other motions and incoming waves. This may be the case for a ship steadily moving (slowly) in calm water. The steady forward potential, Φ_F, problem, which is sometimes called the wave-resistance problem, will not be further discussed.

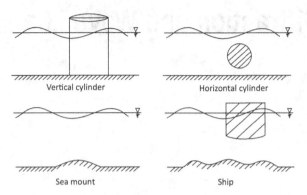

Vertical cylinder   Horizontal cylinder

Sea mount   Ship

Figure 13.1 Some wave diffraction problems in which the obstructions are fixed and waves diffract around them.

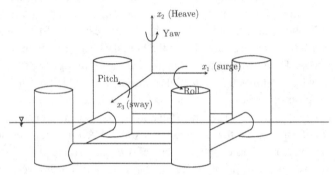

Figure 13.2 Six degrees of freedom motions of a rigid body. Rotations are positive by use of the right-hand rule.

We will first formulate the diffraction problem by assuming that the incoming waves are harmonic and satisfy the linearized boundary-value problem. This will be followed by a case study, where we present the problem of the diffraction of long-crested, linear waves by a vertical, circular cylinder in finite water depth. In that section, the horizontal force and overturning moment will be obtained in terms of some analytic functions.

We will revisit in Section 13.4 the Green-function method introduced in Chapter 11 with an application of the method to arbitrarily shaped (in the plan view), but vertical-sided obstructions, such as detached breakwaters in constant water depth.

Finally, we will discuss the exciting forces and moments due to waves when the obstruction is a "slender" body, such as a tubular member of an offshore structure. This will lead us to Morison's equation that includes two empirically determined coefficients: the inertia coefficient and the drag coefficient. In other words, Morison's equation incorporates viscous effects empirically, unlike the potential theory that is based on the assumption of an inviscid fluid. Potential theory will be discussed first. Current and wind loads that may be present in the area are not considered here. They can be reviewed through a number of works, including Ertekin and Rodenbusch (2016).

13.2   Formulation of the Mathematical Problem

We already have discussed most of the equations that are needed to formulate a wave-diffraction problem. Note that we deal with a boundary-value problem since various functions, such as the velocity potential or the wave elevation, depend on time only periodically, i.e., they are time-harmonic with an angular frequency $\omega = 2\pi/T$, where T is the period. In other words, the time dependence of the functions is known. If this was not the case, then we would have an initial-boundary-value problem which would require that the initial condition(s) must be specified in addition to the boundary conditions. A typical example to an initial-boundary-value problem can be the transient response of a freely floating body, e.g., a ship.

We will assume that the waves are "small" (or the fluid particle displacements are "small") such that linear water-wave theory, discussed in Chapter 12 briefly, is applicable. This allows us to use the linearized free-surface boundary conditions.

As determined before (see equation (12.24)), for an incompressible and inviscid fluid, and irrotational flow, the linear, total velocity potential has to satisfy Laplace's equation and the linear boundary conditions:

$$\Delta\Phi(x_1,x_2,x_3,t) = 0 \quad \text{in the fluid domain,}$$

$$\Phi_{tt}(x_1,0,x_3,t) + g\Phi_{x_2}(x_1,0,x_3,t) = 0 \quad \text{on the still-water surface,} \qquad (13.1)$$

$$\frac{\partial\Phi(x_1,x_2,x_3,t)}{\partial n} = 0 \quad \text{on the fixed boundaries.}$$

Note that the combined free-surface condition in equation (13.1) is obtained by combining the last two equations in (12.24). In addition, there is a far-field condition, called the radiation condition, that part of Φ has to satisfy. This condition will be discussed later in this section.

We already have obtained the solution of equation (13.1) for the incoming wave velocity potential, Φ_I, in Chapter 12, for a sinusoidal wave propagating in finite water depth, i.e.,

$$\eta_I(x_1,x_3,t) = A\cos(kx_1 - \omega t),$$

$$\Phi_I(x_1,x_2,x_3,t) = \frac{Ag}{\omega}\frac{\cosh[k(x_2+h)]}{\cosh(kh)}\sin(kx_1 - \omega t). \qquad (13.2)$$

Clearly, we have written in equation (13.2) the real part of the complex potential Φ_I, see equations (12.47) and (12.49). The function η_I represents an incoming wave that propagates toward the positive x_1 direction. Due to the fact that the waves are modified because of the presence of an obstruction, Φ in equation (13.1) is clearly not the same as Φ_I in equation (13.2). Recall from Chapter 10 that if we have a linear system, we can decompose a function into several parts, each of which satisfies the governing equation and the boundary conditions given by equation (13.1). In other words, we can decompose the total velocity potential as

$$\Phi = \Phi_I + \Phi_D \equiv \Phi_T, \qquad (13.3)$$

where Φ_D is called the diffraction potential or scattering potential resulting from the disturbance of the incoming or incident wave by the presence of the *fixed* obstruction, and Φ_T is the total potential. If we were dealing with moving obstructions or bodies, we could add other potentials to the right-hand side of equation (13.3), such as the radiation potential[1], Φ_R, and the steady forward potential, Φ_F, as was briefly discussed in Section 13.1.

We remark that the only interaction between the incoming wave potential and the diffraction potential (or the scattering potential) comes from the fixed-obstruction boundaries, i.e., from equation (13.1):

$$\frac{\partial \Phi_T}{\partial n} = \frac{\partial}{\partial n}(\Phi_I + \Phi_D) = 0 \quad \Rightarrow \quad \frac{\partial \Phi_I}{\partial n} = -\frac{\partial \Phi_D}{\partial n}. \tag{13.4}$$

Is there a condition other than the body-boundary and free-surface conditions that Φ_D has to satisfy in order that it can uniquely be determined? The answer to this question is "yes"; Φ_D must also satisfy a condition called the radiation condition or the Sommerfeld condition. The radiation condition ensures that the scattered waves are outgoing with a frequency ω and wave number $k = 2\pi/\lambda$, where λ is the wavelength, such that they die out infinitely far away from the obstruction. This condition can also be obtained by physical (energy) arguments as will be described in the following text. It can be shown that the solution to Φ_D is not unique without the radiation condition, and therefore, the boundary-value problem would have been ill posed without it (see, e.g., Sommerfeld (1949) and Sokolnikoff (1966)). without it. Let us derive the radiation condition next.

Consider a cylinder in the flow field as shown in Figure 13.3. The dotted lines represent the outgoing waves generated by the presence of a fixed obstruction or by a floating obstruction's motion itself. From energy conservation (see equation (13.11)), the energy of each ring-shaped wave should be constant, i.e.,

$$2\pi r \, A(r)^2 = \text{constant} \equiv c, \tag{13.5}$$

where $A(r)$ is the amplitude of the wave at a radial distance r from the center of the cylinder. Therefore,

$$A(r) \propto \frac{c}{\sqrt{r}}. \tag{13.6}$$

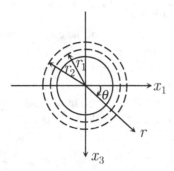

Figure 13.3 Radial outgoing waves.

---

[1] The radiation potential will be discussed in Chapter 14.

If the linear diffraction potential, Φ_D, is periodic with a frequency ω, and hence, is represented by

$$\Phi_D = Re\left\{\phi_D(x_1, x_2, x_3)e^{-i\omega t}\right\}, \tag{13.7}$$

where ϕ_D is the spatial part of the complex diffraction potential, then we can write the following for this spatial part of the potential (that represents a progressive wave), and also for the total diffraction potential:

$$\phi_D = \frac{c}{\sqrt{r}}e^{ikr} \propto A(r)e^{ikr}, \quad \Phi_D = Re\left\{\frac{c}{\sqrt{r}}e^{i(kr-\omega t)}\right\}. \tag{13.8}$$

Moreover, since

$$\frac{\partial \phi_D}{\partial r} = -\frac{1}{2}\frac{c}{r^{3/2}}e^{ikr} + \frac{c}{\sqrt{r}}ike^{ikr} \quad \text{or} \quad \sqrt{r}\left(\frac{\partial \phi_D}{\partial r}\right) - c\,i\,k\,e^{ikr} + \frac{1}{2}c\frac{e^{ikr}}{r} = 0,$$

and also $\sqrt{r}\phi_D = ce^{ikr}$ from equation (13.8), we must have, as $r \to \infty$,

$$\lim_{r \to \infty}\left\{\sqrt{r}\left[\frac{\partial}{\partial r} - ik\right]\phi_D\right\} = 0. \tag{13.9}$$

This equation basically states that $u_r \to 0$ as $r \to \infty$ and that the waves are outgoing. Equation (13.9) is called the radiation condition (or the Sommerfeld condition) for a three-dimensional body of bounded extent in a fluid (with a free surface) of unbounded extent on the horizontal plane. Note that $r = \sqrt{x_1^2 + x_3^2}$ on the horizontal plane.

In two dimensions, it can be shown that $\phi_D(x_1, x_2)$ must satisfy

$$\lim_{x_1 \to \pm\infty}\left\{\frac{\partial \phi_D}{\partial x_1} \mp ik\phi_D\right\} = 0. \tag{13.10}$$

Note the reversal of the sign of the second term in equation (13.10) as $x_1 \to \pm\infty$. Equation (13.10) can be obtained by assuming that

$$\Phi_D = Re\left\{\phi_D e^{-i\omega t}\right\}, \quad \phi_D(x_1, x_2) = \mp Ae^{ikx_1} \quad x_1 \to \pm\infty.$$

Therefore,

$$\frac{\partial \phi_D}{\partial x_1} + Aike^{ikx_1} = \frac{\partial \phi_D}{\partial x_1} \mp ik\phi_D = 0.$$

We *emphasize* that the incident wave potential, Φ_I, is excluded from satisfying the radiation condition since incoming waves are assumed to originate at $r = \infty$ in three dimensions and at $x_1 = \pm\infty$ in two dimensions, depending on whether the incoming waves propagate toward the negative x_1 axis or positive x_1 axis, respectively. This is the reason why we did not consider the radiation condition in Chapter 12. It is also important to note here that the radiation condition, equation (13.9) or equation (13.10), must also be satisfied by the radiation potentials that will be discussed in Chapter 14. The necessity for a radiation condition arises from the fact that we are dealing with a true-harmonic problem. In this connection, it should be mentioned that one way of obtaining a radiation condition is to formulate an initial-boundary-value problem

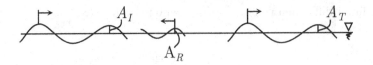

Figure 13.4 The incoming, reflected, and transmitted waves.

and then take the limit as $t \to \infty$ to predict the behavior of waves at $r = \infty$ in a time-harmonic problem. For an in-depth analysis of the subject, see, e.g., Mei (1989).

Next, let us consider equation (13.10), the two-dimensional radiation condition that shows that there are two plane waves moving in opposite directions far from the fixed obstruction. The scattered wave that translates toward the incoming wave is called the reflected wave. The wave that translates in the direction of the incoming wave is called the transmitted wave. The total mean (time-averaged) energy density of linear waves is made up of the potential and kinetic energies and is given by

$$\bar{E} = \frac{1}{2}\rho g A^2. \tag{13.11}$$

Therefore, to have the energy be conserved (no energy loss due to bottom friction is considered here since we have an inviscid fluid), we must write

$$A_I^2 = A_R^2 + A_T^2, \tag{13.12}$$

where A_R and A_T are the amplitudes of the reflected and transmitted waves, and A_I is the amplitude of the incoming wave, see Figure 13.4.

Let us now summarize the three-dimensional boundary-value problem for the time-independent part of the diffraction potential, ϕ_D:

$$\Delta\phi_D(x_1, x_2, x_3) = 0 \quad \text{in the fluid domain,}$$

$$-\omega^2\phi_D(x_1, 0, x_3) + g\frac{\partial\phi_D}{\partial x_2}(x_1, 0, x_3) = 0 \quad \text{on the still-water surface,} \tag{13.13}$$

$$\frac{\partial\phi_D(x_1, x_2, x_3)}{\partial n} = 0 \quad \text{on the fixed sea floor.}$$

$$\frac{\partial\phi_T}{\partial n} = \frac{\partial}{\partial n}(\phi_I + \phi_D) = 0 \quad\Rightarrow\quad \frac{\partial\phi_D}{\partial n} = -\frac{\partial\phi_I}{\partial n} \quad \text{on the fixed body/obstruction surface.} \tag{13.14}$$

$$\lim_{r\to\infty}\left\{\sqrt{r}\left[\frac{\partial}{\partial r} - ik\right]\phi_D\right\} = 0, \quad r = \sqrt{x_1^2 + x_3^2}, \ r \to \infty. \tag{13.15}$$

13.3   A Diffraction Example: Vertical Circular Cylinder

There are a few problems in wave hydrodynamics for which analytical solutions exist ("analytical solution" refers to a solution obtained in terms of tabulated or known functions, e.g., the Bessel function). These solutions are mostly for the linear wave-diffraction problems. In this section, we will consider a vertical cylinder which extends to the sea floor as shown in Figure 13.5 and obtain the solution of the wave-diffraction problem. This solution is called the MacCamy and Fuchs' (1954) solution.

The surface elevation and the velocity potential of the incoming waves far from the cylinder are given by equation (13.2). Note that Φ_I in equation (13.2) is the real part of the complex potential given by

$$\Phi_I = -\frac{igA}{\omega}\frac{\cosh[k(x_2+h)]}{\cosh(kh)}e^{i(kx_1-\omega t)}, \tag{13.16}$$

see equation (12.50).

It is convenient in this problem to solve for Φ_D, the diffraction potential, and hence the total potential Φ_T, in cylindrical coordinates because of the geometry of the cylinder (clearly, it is easier to satisfy the body boundary condition in cylindrical coordinates). Thus, we introduce the (r, θ, x_2) coordinates such that $x_1 = r\cos\theta$, $x_2 = x_2$, and $x_3 = r\sin\theta$. We can then write the exponential term, $\exp\{ikx_1\}$ in equation (13.16) as

$$e^{i(kx_1-\omega t)} = e^{i(kr\cos\theta-\omega t)} = \cos(kr\cos\theta-\omega t)+i\sin(kr\cos\theta-\omega t) = e^{-i\omega t}\cdot e^{ikr\cos\theta}$$

$$= e^{-i\omega t}\left\{J_0(kr)+2\sum_{m=1}^{\infty}i^m J_m(kr)\cos(m\theta)\right\},$$

by use of the identity given, e.g., in Abramowitz and Stegun (1972, p. 361). Here, $i = \sqrt{-1}$, and $J_m(kr)$ is the Bessel function of the first kind (of order m) with its argument kr, see also Section 6.2. Substituting this expansion in equation (13.16), we obtain

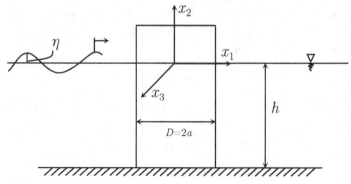

Figure 13.5 A vertical pile of circular cross section extending to the sea floor in water depth of h.

$$\Phi_I = -\frac{igA}{\omega}\frac{\cosh[k(x_2 + h)]}{\cosh(kh)}\left\{J_0(kr) + 2\sum_{m=1}^{\infty} i^m J_m(kr)\cos(m\theta)\right\}e^{-i\omega t}. \quad (13.17)$$

Let us now turn our attention to Φ_D. One can assume that the scattering (or diffraction) potential, Φ_D, that represents the disturbance due to the presence of the cylinder, admits an expansion similar to equation (13.17). Furthermore, Φ_D must be symmetric with respect to $\theta = 180°$ or the x_1 axis, and represent a periodic disturbance moving *outward* with frequency ω and wave number k. The scattering potential Φ_D must also vanish at $r = \infty$. Then, it was shown by MacCamy and Fuchs (1954) that the functional form of Φ_D must contain the Hankel function of the first kind, i.e.,

$$H_m^{(1)}(kr) = J_m(kr) + iY_m(kr), \quad (13.18)$$

where $J_m(kr)$ is the Bessel function of the first kind and $Y_m(kr)$ is the Bessel function of the second kind (of order m). Equation (13.18) is appropriate since

$$H_m^{(1)}(kr) \to \sqrt{\frac{2}{\pi kr}}\exp\left\{i(kr - (\frac{2m+1}{4})\pi)\right\} \quad \text{for} \quad kr \gg 1, \text{ and}$$

$$H_m^{(1)}(kr) \to 0 \quad \text{as} \quad kr \to \infty, \quad (13.19)$$

so that the radiation condition, equation (13.9), also is satisfied.

The form of Φ_D that satisfies the radiation condition equation (13.9), the flat seafloor condition in equation (13.1) (note that $\partial\Phi_I/\partial n = \partial\Phi_I/\partial x_2 = 0$ on the sea floor, and therefore, $\partial\phi_D/\partial x_2 = 0$ as well), and the free-surface condition in equation (13.1) can then be written as

$$\Phi_D = -\frac{igA}{\omega}\frac{\cosh[k(x_2 + h)]}{\cosh(kh)}e^{-i\omega t}\sum_{m=0}^{\infty}(\Phi_D)_m, \quad (13.20)$$

where

$$(\Phi_D)_m = B_m\cos(m\theta)H_m^{(1)}(kr), \quad (13.21)$$

where B_m are unknown.

The unknown complex coefficients B_m in equation (13.21) must now be determined by use of the last unused boundary condition which is the solid-boundary or non-permeability or no-flux condition on the cylinder, i.e., that the normal velocity of the fluid on the cylinder should be zero since the cylinder is rigid and fixed.[2] We have already given this condition by equation (13.4) on $r = a$, where a is the radius of the cylinder. Since $\partial/\partial n = -\partial/\partial r$, we have

$$\frac{\partial\Phi_I}{\partial r} = -\frac{\partial\Phi_D}{\partial r} \quad \text{on } r = a. \quad (13.22)$$

If we take the derivative of equations (13.17) and (13.20) with respect to r, and use equation (13.22), we can obtain

$$m = 0, \quad B_o = -J_0'(ka)/H_0^{(1)'}(ka),$$
$$m \geq 1, \quad B_m = -2i^m J_m'(ka)/H_m^{(1)'}(ka), \quad (13.23)$$

---

[2] If it is not fixed, the normal velocity of the fluid should be the same as the normal velocity of the cylinder.

where primes denote differentiation with respect to kr.[3] Substituting equations (13.21) and (13.23) in equation (13.20), one can show that

$$
\Phi_D = \frac{ig A}{\omega} \frac{\cosh[k(x_2 + h)]}{\cosh(kh)} \left\{ \frac{J_0'(ka)}{H_0^{(1)'}(ka)} H_0^{(1)}(kr) \right.
$$
$$
\left. +2 \sum_{m=1}^{\infty} i^m \frac{J_m'(ka)}{H_m^{(1)'}(ka)} H_m^{(1)}(kr) \cos(m\theta) \right\} e^{-i\omega t}.
$$

(13.24)

It can be shown of course that equation (13.24) satisfies all the boundary conditions, as well as Laplace's equation (see Prob. 13.7.4).

As a result of the foregoing derivations, the total potential Φ_T is given by the sum of equations (13.17) and (13.24), or

$$
\Phi_T = \frac{-ig A}{\omega} \frac{\cosh[k(x_2 + h)]}{\cosh(kh)} \left\{ \sum_{m=0}^{\infty} \beta_m [J_m(kr) \right.
$$
$$
\left. - \frac{J_m'(ka)}{H_m^{(1)'}(ka)} H_m^{(1)}(kr)] \cos(m\theta) \right\} e^{-i\omega t},
$$

(13.25)

where, for convenience, the coefficients β_m are introduced:

$$
\beta_0 = 1, \quad \beta_m = 2i^m \quad \text{if } m \geq 1.
$$

(13.26)

This completes the solution of the total velocity potential for the diffraction problem of a vertical circular cylinder in monochromatic waves (long-crested waves with a single frequency) and one that extends to the sea floor of finite depth. The solution is due to MacCamy and Fuchs (1954). The solution of this problem when the water is very deep, $h \to \infty$, was first given by Havelock (1940), and by Omer and Hall (1949) when the water is very shallow.

13.3.1   Wave Force and Moment

We are now ready to determine the force and moment acting on the cylinder. To do this, we consider Euler's integral for the linearized problem (see equation (12.57)), and write

$$
p(x_1, x_2, x_3, t) = -\rho \frac{\partial \Phi_T}{\partial t} - \rho g x_2, \quad x_2 \leq 0, \quad r \geq a.
$$

(13.27)

The second term on the right-hand side of equation (13.27) is the hydrostatic term, and it does not contribute to the horizontal wave force or overturning moment when integrated around the circular cylinder in linear theory, where the integration is done up to the SWL; therefore we ignore it. The sectional (per unit length along the axis of the cylinder, i.e., along the x_2 direction) force in the x_1 direction is then given by

---

[3] Note that k is constant since we solve the problem for each k, and there is only one k for each ω and h from the dispersion relation, equation (12.16) or equation (12.53).

$$\delta F(x_2,t) = \oint_C p\, n_{x_1}\, dC \quad (dC = r\, d\theta,\ r = a,\ n_{x_1} = -\cos\theta)$$

$$= -2a \int_0^\pi p(r = a, \theta, x_2, t) \cos\theta\, d\theta. \tag{13.28}$$

The integration is from 0 to π since the wave force is symmetric with respect to the x_1–x_2 plane, and we have a minus sign on the right-hand side of equation (13.28) since n is pointing out of the fluid (in the opposite direction of the r coordinate) or into the cylinder.

The pressure that is needed in equation (13.28) is given by equation (13.27), and Φ_T is given by equation (13.25). However, it is instructive to consider the individual forces separately due to Φ_I and Φ_D (in light of equation (13.3)). To do this, we write equation (13.28) as

$$\delta F(x_2,t) = 2a\rho \int_0^\pi \left(\frac{\partial \Phi_I}{\partial t} + \frac{\partial \Phi_D}{\partial t} \right)\bigg|_{r=a} \cos\theta\, d\theta. \tag{13.29}$$

We also have, from equation (13.17),

$$\frac{\partial \Phi_I}{\partial t}\bigg|_{r=a} = -i\omega Q(x_2,t) q_m(a,\theta), \tag{13.30}$$

where we have defined

$$Q(x_2,t) = -\frac{igA}{\omega} \frac{\cosh[k(x_2+h)]}{\cosh(kh)} e^{-i\omega t},$$

$$q_m(a,\theta) = J_0(ka) + 2\sum_{m=1}^\infty i^m J_m(ka) \cos(m\theta). \tag{13.31}$$

Similarly, from equation (13.24)

$$\frac{\partial \Phi_D}{\partial t}\bigg|_{r=a} = -i\omega Q(x_2,t) s_m(a,\theta), \tag{13.32}$$

where we have defined

$$s_m(a,\theta) = -\left\{ \frac{J_0'(ka)}{H_m^{(1)'}(ka)} H_0^{(1)}(ka) + 2\sum_{m=1}^\infty i^m \frac{J_m'(ka)}{H_m^{(1)'}(ka)} H_m^{(1)}(ka) \cos(m\theta) \right\}. \tag{13.33}$$

Substituting equations (13.30) and (13.32) in equation (13.29), we obtain

$$\delta F(x_2,t) = -2a\rho i\omega Q \left\{ \int_0^\pi q_m \cos\theta\, d\theta + \int_0^\pi s_m \cos\theta\, d\theta \right\}. \tag{13.34}$$

Performing the integrals in equation (13.34) (see, e.g., Gradshteyn and Ryzhik (1980)), and by use of the orthogonality condition (see, e.g., Abramowitz and Stegun (1972)) given by

$$\int_0^\pi \cos(\theta)\cos(m\theta)d\theta = \frac{\pi}{2} \quad \text{if } m = 1,$$

(13.35)

$$= 0 \quad \text{if } m > 1,$$

one obtains the total sectional horizontal force:

$$\delta F(x_2, t) = 2a\rho\omega Q\pi \left\{ J_1(ka) - \frac{J_1'(ka)H_1^{(1)}(ka)}{H_1^{(1)'}(ka)} \right\} \equiv \delta F_{F-K} + \delta F_S, \quad (13.36)$$

where δF_{F-K} is the sectional (not the total) horizontal force due to the incoming waves alone (ϕ_I) and δF_S is the sectional horizontal force due to the scattered (or diffracted) waves alone (ϕ_D) (see also equation (13.40)). δF_{F-K} is called the Froude–Krylov force. It is important to note that this force is obtained as if the obstruction (cylinder) is transparent (does not exist). The sum of these two forces is called the wave-exciting force. Note that equation (13.36) is a complex equation.

By noting that (see, e.g., Abramowitz and Stegun (1972, p. 361))

$$J_1(ka)Y_1'(ka) - J_1'(ka)Y_1(ka) = \frac{2}{\pi ka}, \quad (13.37)$$

we can write equation (13.36) as

$$\delta F(x_2, t) = \frac{4i\rho\omega Q}{kH_1^{(1)'}(ka)}. \quad (13.38)$$

Or by use of equation (13.31), and taking the real part, we obtain

$$\delta F(x_2, t) = \frac{4\rho g A}{k} \frac{\cosh[k(x_2 + h)]}{\cosh(kh)} \left\{ \frac{\cos(\omega t)J_1' - \sin(\omega t)Y_1'}{J_1'^2 + Y_1'^2} \right\}, \quad (13.39)$$

which can be written as

$$\delta F(x_2, t) = \frac{4\rho g A}{k} \frac{\cosh[k(x_2 + h)]}{\cosh(kh)} R(ka)\cos(\omega t + \alpha), \quad (13.40)$$

where we have defined

$$R(ka) = \left\{ J_1'^2(ka) + Y_1'^2(ka) \right\}^{-1/2}, \quad \alpha(ka) = \tan^{-1}\left\{ \frac{Y_1'}{J_1'} \right\}. \quad (13.41)$$

Note that the derivatives of J_1 and Y_1 can be obtained from the recurrence relations given, e.g., by Abramowitz and Stegun (1972):

$$J_1'(ka) = \frac{1}{2}\{J_0(ka) - J_2(ka)\}, \quad J_2(ka) = \frac{2}{ka}J_1(ka) - J_0(ka)$$

or

$$J_1'(ka) = J_0(ka) - \frac{1}{ka}J_1(ka).$$

Also note that $Y_1'(ka) = \frac{1}{2}\{Y_0(ka) - Y_2(ka)\}$. The numerical values of the Bessel functions are readily available (in table form, see, e.g., Abramowitz and Stegun (1972)), but they can also be calculated by, e.g., "Numerical Recipes" algorithms

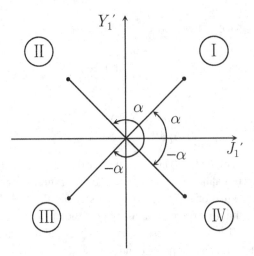

Figure 13.6 The signs of the Bessel functions.

(see, e.g., Press et al. (1992)), or by a symbolic mathematics program, such as MATLAB, MAPLE or MATHEMATICA.

Recalling that the incoming surface elevation, η_I, is given by equation (13.2), we see that the exciting force given by equation (13.40) lags the wave crest by a phase angle of α. Note that one should consider the signs of J_1' and Y_1' when calculating α (see Figure 13.6). In other words,

$$\alpha = \tan^{-1}\left\{\frac{Y_1'}{J_1'}\right\} + \alpha_1, \quad \text{where}$$

$$\alpha_1 = 0 \text{ in the 1st and 4th quadrants,}$$

$$\alpha_1 = \pi \text{ in the 2nd quadrant, and}$$

$$\alpha_1 = -\pi \text{ in the 3rd quadrant.}$$

From equation (13.36) (and after taking the real part), we also have

$$\delta F_{F-K} = -2a\rho g\pi A \frac{\cosh[k(x_2 + h)]}{\cosh(kh)} J_1(ka)\cos(\omega t + \delta), \tag{13.42}$$

where

$$\delta = -\frac{\pi}{2} \tag{13.43}$$

is the constant phase angle of the Froude–Krylov force. This phase shift is with respect to the wave crest at the origin (recall that the incoming wave is given by equation (12.39)). Similarly, after taking the real part of equation (13.36)

$$\delta F_S = 2a\rho g\pi A \frac{\cosh[k(x_2 + h)]}{\cosh(kh)} J_1'(ka)\left(\frac{J_1^2 + Y_1^2}{J_1'^2 + Y_1'^2}\right)^{1/2}\cos(\omega t + \epsilon), \tag{13.44}$$

where

$$\epsilon = \tan^{-1}\left\{-\frac{\pi ka}{2}(J_1 J_1' + Y_1 Y_1')\right\} + \epsilon_1,$$

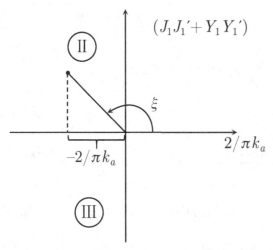

Figure 13.7 The signs of the combination of the Bessel functions.

$$\epsilon_1 = 0 \text{ in the 2nd quadrant,} \tag{13.45}$$

$$\epsilon_1 = -\pi \text{ in the 3rd quadrant,}$$

and where ϵ is the phase angle of the diffraction force (see Figure 13.7). Note that the sum of equations (13.42) and (13.44) is equal to equation (13.40).

So far, the force expressions that we have given are per unit length (differential slice) of the cylinder in the vertical direction. Next, we shall integrate δF_{F-K} and δF_S along the cylinder axis to obtain their contributions to the total horizontal force. From equation (13.42), we have

$$F_{F-K} = \int_{-h}^{0} \delta F_{F-K}(x_2,t)dx_2 = -\frac{2a\rho g \pi A}{k} \tanh(kh)J_1(ka)\cos(\omega t + \delta), \tag{13.46}$$

where the phase angle δ is given by equation (13.43). The upper limit of integration is up to the SWL, $x_2 = 0$, since we are dealing with the theory of infinitesimal amplitude, or linear waves, in which the free-surface conditions are enforced on the SWL. Equation (13.46) can be written in a dimensionless form as

$$C_{F-K} = \frac{|F_{F-K}|}{2\rho gahA} = \left| -\pi J_1(ka)\frac{\tanh(kh)}{kh} \right|, \tag{13.47}$$

where C_{F-K} is called the dimensionless Froude–Krylov force coefficient, and vertical bars around the Froude–Krylov force denote the amplitude of the force. The product $2\rho gahC_{F-K} = |F_{F-K}|/A$ is the Froude–Krylov force per unit wave amplitude. Such unit-amplitude functions are often called transfer functions (which may be dimensional, as here, or dimensionless in other cases), and they can be used to obtain (linear) forces in irregular (or random) seas, as we will see in Chapter 15.

Similarly, from equation (13.44), we write

$$F_S = \int_{-h}^{0} \delta F_S dx_2 = 2a\rho g \frac{\pi A}{k} \tanh(kh)J_1'(ka) \left(\frac{J_1^2 + Y_1^2}{J_1'^2 + Y_1'^2} \right)^{1/2} \cos(\omega t + \epsilon), \tag{13.48}$$

where the phase angle ϵ is given by equation (13.45). We can now introduce a dimensionless force coefficient:

$$C_S = \frac{|F_S|}{2\rho g a h A} = \left| \pi J_1'(ka) \frac{\tanh(kh)}{kh} \left(\frac{J_1^2 + Y_1^2}{J_1'^2 + Y_1'^2} \right)^{1/2} \right|, \qquad (13.49)$$

where C_S is called the dimensionless diffraction or scattering force coefficient

Therefore, summing up equations (13.46) and (13.48), we have

$$F = 4\rho g A h \frac{R(ka)\tanh(kh)}{k^2 h} \cos(\omega t + \alpha), \quad C_T = \frac{|F|}{2\rho g A h a} = \left| \frac{2R(ka)}{ka} \frac{\tanh(kh)}{kh} \right|, \qquad (13.50)$$

where C_T is called the dimensionless total-force coefficient, and $R(ka)$ and α were defined by equation (13.41).

The total overturning moment (with respect to the sea floor) is then given by

$$M = \int_{-h}^{0} \delta F(x_2, t)(x_2 + h)\, dx_2, \qquad (13.51)$$

where δF was given by equation (13.40), or

$$M = 4\rho g A \frac{R(ka)}{k^3} \left\{ \frac{kh \sinh(kh) + 1 - \cosh(kh)}{\cosh(kh)} \right\} \cos(\omega t + \alpha). \qquad (13.52)$$

In terms of the overturning moment coefficient, we have

$$C_O = \frac{|M|}{2\rho g A a h^2} = \frac{2R(ka)}{ka} \left\{ \frac{kh \sinh(kh) + 1 - \cosh(kh)}{k^2 h^2 \cosh(kh)} \right\}. \qquad (13.53)$$

C_O is also known as the toppling-moment coefficient. Note that the origin of the coordinate system is at the SWL.

Equations (13.50) and (13.52), respectively, are the solutions obtained by Mac-Camy and Fuchs (1954) for the horizontal force and overlapping moment on a vertical, circular cylinder. The comparison of the experiments with theory is discussed in the next section.

13.3.2   Surface Elevation and Run-up

The surface elevation can be obtained from the dynamic free-surface boundary condition given by equation (12.24), i.e.,

$$\eta_T(x_1, x_3, t) = -\frac{1}{g} \frac{\partial \Phi_T}{\partial t}(x_1, 0, x_3, t), \qquad (13.54)$$

since the atmospheric pressure is assumed to be negligible on the SWL. By use of equations (13.18), (13.25), and (13.37), equation (13.54) becomes (on the cylinder, $r = a$)

$$\left. \frac{\eta_T}{A} \right|_{r=a, x_2=0} = \frac{2i}{\pi ka} \left\{ \frac{1}{H_0^{(1)'}(ka)} + 2\sum_{m=1}^{\infty} i^m \frac{\cos(m\theta)}{H_m^{(1)'}(ka)} \right\} e^{-i\omega t}. \qquad (13.55)$$

Note that η_T in equation (13.55) is complex. It is possible to put this equation in a trigonometric form by redefining it (after taking the real part of equation (13.55)) as

$$\frac{\eta_T}{A} \equiv \frac{A_T}{A}\cos(\omega t + \psi),\qquad(13.56)$$

where

$$\frac{A_T}{A} = \frac{4}{\pi ka}(D^2 + E^2)^{1/2},\qquad(13.57)$$

$$D = \frac{J_0'}{2L_0} + \frac{\cos(\theta)Y_1'}{L_1} - \frac{\cos(2\theta)J_2'}{L_2} - \frac{\cos(3\theta)Y_3'}{L_3} + \frac{\cos(4\theta)J_4'}{L_4}$$
$$+ \frac{\cos(5\theta)Y_5'}{L_5} - \cdots,\qquad(13.58)$$

$$E = \frac{Y_0'}{2L_0} - \frac{\cos(\theta)J_1'}{L_1} - \frac{\cos(2\theta)Y_2'}{L_2} + \frac{\cos(3\theta)J_3'}{L_3} + \frac{\cos(4\theta)Y_4'}{L_4}$$
$$- \frac{\cos(5\theta)J_5'}{L_5} - \frac{\cos(6\theta)Y_6'}{L_6} + \cdots,\qquad(13.59)$$

$$L_i = (J_i'^2 + Y_i'^2), \quad i = 0, 1, 2, \ldots, \quad \psi = \tan^{-1}\left\{\frac{D}{E}\right\}.\qquad(13.60)$$

In equation (13.60), ψ is the phase angle measured relative to the incoming wave, and as before, the location of D and E on the Argand diagram should be accounted for in calculating ψ. Equation (13.55) is also known as the wave-run-up, i.e.,

$$R(\omega, \theta) = \left.\frac{(\eta_T)_{max}}{A}\right|_{r=a} = \left.\frac{A_T}{A}\right|_{r=a}.\qquad(13.61)$$

Some sample run-up values for a vertical cylinder are shown in Figure 13.8, as functions of wave period and cylinder polar angle θ. Since we can write the following in this linear problem,

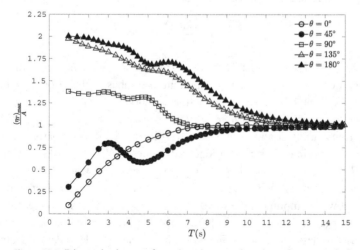

Figure 13.8 Dimensionless surface elevation as a function of wave period and cylinder angle.

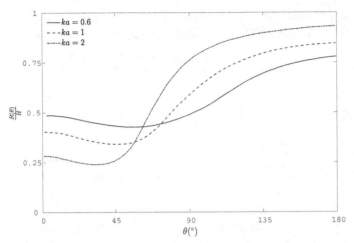

Figure 13.9 Run-up profiles around a circular cylinder as functions of the incoming wave angle, θ, and the dimensionless wave number (adapted from Sarpkaya and Isaacson (1981)).

$$\eta_T = \eta_I + \eta_S,$$

the scattered wave elevation on the vertical cylinder becomes

$$\eta_S|_{r=a} = A_T \cos(\omega t + \psi) - A\cos(ka\cos\theta - \omega t). \qquad (13.62)$$

However, η_S is not needed in general.

The run-up profiles around a circular cylinder for various values of ka are shown in Figure 13.9 (from Sarpkaya and Isaacson (1981)) as a function of the cylinder polar angle θ. Note that 180° corresponds to the front (wave side) of the cylinder and 0° to the rear (lee side) of the cylinder. The maximum values of $R(\omega,\theta)$ correspond to $\theta = 180°$, and these values are plotted as a function of ka in Figure 13.10 (from Sarpkaya and Isaacson (1981)). Note that

$$ka = \frac{2\pi}{\lambda}\frac{D}{2} = \pi\frac{D}{\lambda}, \qquad (13.63)$$

so that the ratio of the cylinder diameter to wavelength is an important parameter called the diffraction parameter. It is also important to emphasize that λ or k is a function of water depth through the dispersion relation equation (12.53).

These solutions discussed here are due to MacCamy and Fuchs (1954). Figure 13.11 (from Sarpkaya and Isaacson (1981)) illustrates the comparison between the theory and experiments (see Mogridge and Jamieson (1975) and Hogben and Standing (1974)). The agreement is generally good especially when $ka < 0.7$. Keeping the cylinder diameter the same, if we increase ka, λ becomes smaller, and therefore, for a fixed wave amplitude, the wave slope increases and thus nonlinearities become more important, i.e., the linear solution starts to break down as one would expect. The dimensionless moment coefficient for a vertical cylinder, adapted from Rahman (1988), is shown in Figure 13.12.

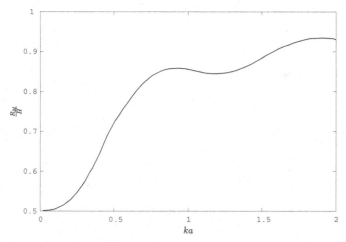

Figure 13.10 Maximum dimensionless run-up per unit wave height, H, as a function of the diffraction parameter (adapted from Sarpkaya and Isaacson (1981)).

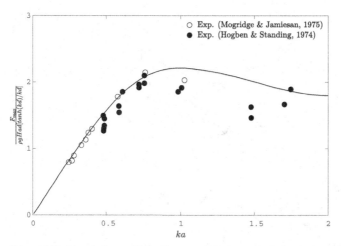

Figure 13.11 Comparison of MacCamy and Fuchs analytical solution with experimental data (adapted from Sarpkaya and Isaacson (1981)).

13.4   Green Function Method Revisited: The Island Problem

In Section 11.4, we have discussed, in rather general terms, the Green function method to solve the fluid–body interaction problems in an inviscid fluid of infinite extent. As an application of a diffraction problem, we consider next an island whose geometry in the x_1–x_3 plane is arbitrary (see Figure 13.13).

    The island (say, a detached breakwater) is assumed to have vertical sides and the incoming waves make an angle of $-\gamma$ with the x_1 axis. If the island is circular, then the solution given here should reduce to MacCamy and Fuchs' solution given in

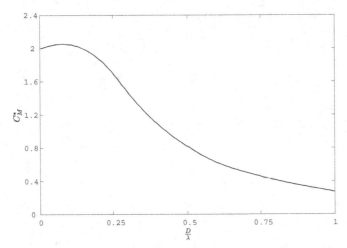

Figure 13.12 Dimensionless moment coefficient as a function of the cylinder diameter over wavelength ratio (adapted from Rahman (1988)).

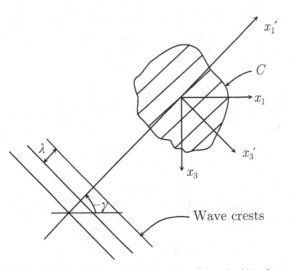

Figure 13.13 Waves around an island (or a detached breakwater, C).

Section 13.3 provided, of course, that the water depth is constant. The incoming wave potential is given by

$$\Phi_I(P,t) = \Re\left\{\phi_I(P)e^{-i\omega t}\right\}$$
$$= \frac{gA}{\omega}\frac{\cosh[k(x_2 + h)]}{\cosh(kh)} \sin\left\{k(x_1 \cos\gamma + x_3 \sin\gamma) - \omega t\right\}, \quad (13.64)$$

where P represents the coordinates of a point in the fluid, i.e., $P \equiv (x_1, x_2, x_3)$, and we have used equation (12.50). The surface profile is given by equation (12.39) (for waves propagating in the x_1' direction):

$$\eta_I = A\cos\{k(x_1\cos\gamma + x_3\sin\gamma) - \omega t\}. \tag{13.65}$$

Recall that $\Phi_T = \Phi_I + \Phi_D$ must satisfy the combined free-surface condition, the bottom condition, body condition, and the radiation condition (clearly, the last two conditions are satisfied by ϕ_D only). Because the depth dependence of Φ_D is the same as of Φ_I (see the sea-floor condition in equation (13.1)), we write Φ_D as

$$\Phi_D = Re\{\phi_D e^{-i\omega t}\},$$

where

$$\phi_D(P) = \hat{\phi}_D(x_1, x_3)\frac{\cosh[k(x_2 + h)]}{\cosh(kh)}. \tag{13.66}$$

Note that $\hat{\phi}_D$ is not a function of x_2.

It is easy to verify that ϕ_D in equation (13.66) satisfies the free-surface condition and the sea-floor (or bottom) condition given in equation (13.1). However, Laplace's equation in equation (13.1) now becomes

$$\Delta_3\Phi_D = \left[\frac{\partial^2\hat{\phi}_D}{\partial x_1^2} + k^2\hat{\phi}_D + \frac{\partial^2\hat{\phi}_D}{\partial x_3^2}\right]\frac{\cosh k(x_2 + h)}{\cosh(kh)} = 0$$

or

$$\hat{\phi}_{Dx_1x_1} + \hat{\phi}_{Dx_3x_3} = -k^2\hat{\phi}_D, \tag{13.67}$$

where Δ_3 is the three-dimensional Laplace operator. Equation (13.67) is called the two-dimensional Helmholtz equation. Therefore, the problem to be solved becomes

$$\Delta_2\hat{\phi}_D + k^2\hat{\phi}_D = 0 \quad \text{(inside the fluid domain)}$$

$$\lim_{r\to\infty} r^{1/2}\left(\frac{\partial\hat{\phi}_D}{\partial r} - ik\hat{\phi}_D\right) = 0 \quad \text{(at infinity and on the } x_2 = 0 \text{ plane)} \tag{13.68}$$

$$\frac{\partial\hat{\phi}_D}{\partial n} = -\frac{\partial\phi_I}{\partial n} \quad \text{(on the island surface below the SWL)},$$

where Δ_2 is the two-dimensional Laplace operator on the horizontal plane (x_1, x_3). All other boundary conditions (on the free surface and sea floor) have already been satisfied.

Now consider a Green function in the form of

$$G(P, Q) = -\log r + H(x_1, x_3; \xi, \zeta), \tag{13.69}$$

such that it satisfies the radiation condition, see equation (13.9):

$$\lim_{r\to\infty}\sqrt{r}\left\{\frac{\partial G}{\partial r} - ikG\right\} = 0, \tag{13.70}$$

and the Helmholtz equation, (13.67). If we consider the integral equation (11.66) on the rigid boundary, where $\partial\phi/\partial n = 0$ and $P \in C$, we have, on the horizontal plane (two-dimensional),

$$\pi\hat{\phi}_D(P) = -\int_C \left(\hat{\phi}_D(Q)\frac{\partial G}{\partial n}(P,Q) - G(P,Q)\frac{\partial\hat{\phi}_D(Q)}{\partial n}\right) dC$$

$$= -\int_C \left(\hat{\phi}_D(Q)\frac{\partial G}{\partial n}(P,Q) + G(P,Q)\frac{\partial\phi_I(Q)}{\partial n}\right) dC, \tag{13.71}$$

where the body-boundary condition in equation (13.68) is used in writing the second equality. As a result,

$$\hat{\phi}_D(P) + \frac{1}{\pi}\int_C \hat{\phi}_D(Q)\frac{\partial G(P,Q)}{\partial n} dC = -\frac{1}{\pi}\int_C \frac{\partial\phi_I(Q)}{\partial n}G(P,Q)\, dC. \tag{13.72}$$

$Q \equiv (\xi,\zeta)$ is a point of singularity on the boundary C, $P \equiv (x_1,x_3)$ is any point on the boundary C, and the integrals are now line integrals along the body contour in the plan view, rather than the surface integrals we had before in three dimensions.

The Green function, equation (13.69), that satisfies the Helmholtz equation is known as Weber's solution and is given by

$$G(P,Q) = i\frac{\pi}{2}H_0^{(1)}(kr), \tag{13.73}$$

where $H_0^{(1)}$ is the Hankel function of the first kind and of order zero (see, e.g., Harms (1979) for references). Therefore, equation (13.72) becomes

$$\hat{\phi}_D(P) + \frac{i}{2}\int_C \hat{\phi}_D(Q)\frac{\partial H_0^{(1)}}{\partial n} dC = -\frac{i}{2}\int_C \frac{\partial\phi_I}{\partial n}H_0^{(1)}(kr)\, dC. \tag{13.74}$$

This equation gives $\hat{\phi}_D$ on C. Once $\phi_D(P)$, $P \in C$, is calculated from equation (13.74), the diffraction potential can be obtained from

$$\hat{\phi}_D(P) = -\frac{i}{4}\int_C \left[\hat{\phi}_D(Q)\frac{\partial H_0^{(1)}}{\partial n}(kr) + \frac{\partial\phi_I}{\partial n}(Q)H_0^{(1)}(kr)\right] dC, \tag{13.75}$$

which gives $\hat{\phi}_D$ anywhere outside C, $P \notin C$. For the application of this method to detached breakwaters and harbor oscillations, such as the one shown in Figure 13.14, see, e.g., Lee (1971) and also Harms (1979). Once $\hat{\phi}_D$ is determined, then Φ_D is known from equation (13.66), and therefore, the total velocity potential can be obtained, and thus, one can easily calculate the pressure, surface elevation, velocities, etc.

The surface elevation is obtained from the dynamic free-surface boundary condition as we have done before (see equation (13.54)). This would allow one to determine the diffraction coefficient given by

$$K'(\omega) = \frac{|\eta_T(x_1,x_3,t)|_{max}}{|\eta_I(x_1,x_3,t)|_{max}} = \text{maximum}\left(\frac{A_T}{A}\right), \tag{13.76}$$

where $K'(\omega)$ is also called the transfer function or response function of the maximum wave elevation amplitude, and it is a function of the wave frequency, ω. In equation (13.76), $\eta_T = \eta_I + \eta_D$ is the total wave elevation.

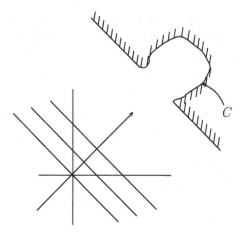

Figure 13.14 Waves propagating toward a harbor.

13.5 Wave Forces on Slender Cylinders: Morison's Equation

In this section, we will discuss wave forces on cylindrical structural members, such as the pontoons or columns of a platform or a pile that extends to the sea floor, when both the inertia and viscous forces are important. Because of the nonlinear nature of the N-S equations and the boundary conditions, many researchers have attempted to simplify the computation of wave forces on cylindrical piles. Most of the studies have been based on the experimental determination of the inertia and drag coefficients that appear in Morison's equation (see Morison et al. (1950)). Morison's equation was developed as an ad hoc approach to a limited set of experimental data. However, because of the importance of cylindrical piles both in offshore engineering (e.g., jack-up platforms) and coastal engineering (e.g., pier piles, bridge columns, etc.), there have been many investigations on the proper coefficients to be used in almost every different case that one can imagine. A few of the different cases include inclined cylinders, group of cylinders, roughened cylinders, and horizontal cylinders (see, e.g., Sarpkaya and Isaacson (1981)).

Let us first state the original Morison's equation given by

$$\delta F = (1 + C_a)\rho\pi \left(\frac{D}{2}\right)^2 \dot{u}_1 + \frac{1}{2}\rho D C_d u_1 |u_1|, \qquad (13.77)$$

where δF is the sectional (in the x_2 (vertical) direction) horizontal force, D is the diameter of the vertical pile, u_1 is the wave horizontal velocity (see Chapter 12), $\dot{u}_1$ is the wave horizontal acceleration, C_a is called the added-mass or virtual-mass coefficient or and C_d the form-drag coefficient. C_a and C_d are dimensionless. $C_m = (1 + C_a)$ is called the inertia coefficient. Note that u_1 is the particle velocity only due to the incoming wave, i.e., there is no diffraction effects accounted for, unlike in the theory of MacCamy and Fuchs discussed in Section 13.3. The justification for this is that the diffraction effects are small because the cylinder is "slender," therefore u_1 and $\dot{u}_1$ are evaluated along the cylinder axis since the errors made would be "small."

Let us now look at how the force depends on various physical quantities. Consider an oscillatory flow, in an *unbounded fluid (no free surface or sea floor)*, with a period $T = 2\pi/\omega$, and velocity $u_1(t) = U\cos(\omega t)$, where U is the amplitude of velocity. The functional dependence of the force (per unit length) acting on the body, whose characteristic length is D, can be written as

$$\delta F = f(U, D, \rho, \nu, T). \tag{13.78}$$

By applying dimensional analysis to equation (13.78), in which the set (ρ, U, D) is chosen as dimensionally independent, one can obtain

$$\frac{\delta F}{\frac{1}{2}\rho D U^2} = C_D\left(\frac{UD}{\nu}, \frac{UT}{D}\right), \tag{13.79}$$

or for the total force,

$$\frac{F}{\frac{1}{2}\rho D^2 U^2} = C_D(Re, KC), \tag{13.80}$$

where $Re = UD/\nu$ is the usual diameter-based Reynolds number, and $KC = UT/D$ is called the Keulegan–Carpenter number (see Keulegan and Carpenter (1958)). For large values of the KC number, $KC \gg 1$, we expect the force coefficient, C_D, in equation (13.80) to approach the value of the steady (form-)drag coefficient which was discussed earlier (see Chapter 10) since large KC numbers correspond to long periods. In other words,

$$C_D(Re, KC) \to C_d(Re) \quad \text{if} \quad KC \gg 1. \tag{13.81}$$

On the other hand, when $KC \ll 1$, the inertia effects will dominate viscous effects since the end of the time duration necessary to develop the boundary layer and flow separation could not have been reached. Therefore, we have basically an inviscid fluid since $Re \to \infty$, or

$$C_D(Re, KC) \to C_d(KC), \quad Re \gg 1, \quad KC \ll 1. \tag{13.82}$$

If, on the other hand, we now consider the presence of a *periodic free surface*, we can then anticipate that the force depends on

$$F = f(h, H, \lambda, D, \rho, g, \nu), \tag{13.83}$$

where h is the water depth, λ wavelength, and H wave height. In fact, it should not be difficult to obtain

$$\frac{F}{\rho g H D^2} = f\left(\frac{h}{\lambda}, \frac{H}{\lambda}, \frac{D}{\lambda}, Re\right) \tag{13.84}$$

by use of dimensional analysis. If the problem is linear and the fluid is inviscid, we can write equation (13.84) as

$$\frac{F}{\rho g H D^2} = f_1\left(\frac{D}{\lambda}, \frac{D}{h}\right). \tag{13.85}$$

Therefore, if the fluid is inviscid, the force coefficient given in equation (13.84) does depend only on the ratio D/λ for a fixed-body geometry and water depth, i.e., D/h is

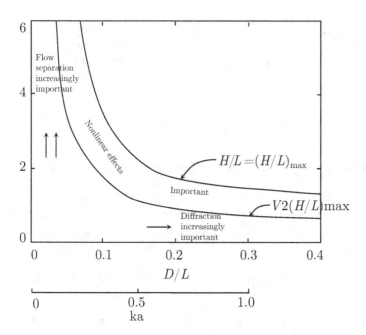

Figure 13.15 Different wave force regimes (adapted from Sarpkaya and Isaacson (1981)).

constant. So that $F = F(\omega)$ only, where ω is the angular wave frequency related to the wavelength through the dispersion relation, equation (12.53). If we had explicitly included the frequency (or period) in the equations above, the KC number would also appear. One should try to show this by having the dispersion relation in mind.

To see the effect of body dimension, wave height, and wavelength on the forces on a vertical, circular cylinder, consider Figure 13.15 (from Sarpkaya and Isaacson (1981)). In this figure, $L \equiv \lambda$, $K \equiv KC$, and H is the wave height, and we see that as D/λ becomes small, the viscous effects (i.e., separation) become important. For a circular cylinder, and if $D/\lambda > 0.15$, the diffraction effects are more important. The KC number given in Figure 13.15 is

$$K \equiv KC = \frac{\pi H/\lambda}{\frac{D}{\lambda}\tanh(kh)}, \qquad (13.86)$$

and it is evaluated at the SWL and is for finite water depth. In infinite water, equation (13.86) becomes $KC = \pi H/D$ since $\tanh kh \to 1.0$. Also, one can deduce from Figure 13.15 that if $H/D > 1$, the viscous effects become important for a fixed ratio of D/λ.

Therefore, in engineering calculations, it is recommended that, for a vertical circular cylinder that extends to the sea floor,

$$\text{if } \frac{H}{D} < 1.0 \text{ and } \frac{D}{\lambda} \geq 0.15 \quad \text{diffraction theory be used,}$$

$$\text{if } \frac{H}{D} \geq 1.0 \text{ and } \frac{D}{\lambda} < 0.15 \quad \text{Morison's equation be used,}$$

$$\text{if } \frac{H}{D} < 1.0 \text{ and } \frac{D}{\lambda} < 0.15 \quad \text{both methods be used.}$$

It is noted that these should be cautiously used for other body geometries. In Figure 13.15, $(H/\lambda)_{max}$ is the maximum wave steepness (before wave breaking occurs) given by (for finite but large – meaning not shallow – water depth, h)

$$\frac{H}{\lambda} = 0.14 \tanh(kh). \tag{13.87}$$

In deep water, this clearly becomes $H/\lambda = 0.14$.

13.5.1   Inertia Terms: Apparent Mass and the Froude–Krylov Force

In the case of the inviscid limit given by equation (13.82), and whether there is free surface or not, the force acting on a body is only proportional to the mass of the body times the fluid acceleration. To show this, and therefore obtain the inertial part of the force in equation (13.77), consider the kinetic energy of a moving fluid:

$$T = \frac{1}{2}\rho \int_V \boldsymbol{u} \cdot \boldsymbol{u} \, dV = \frac{1}{2}\rho \int_V \boldsymbol{\nabla}\phi \cdot \boldsymbol{\nabla}\phi \, dV. \tag{13.88}$$

If we now use Green's first identity given by equation (11.54), equation (13.88) becomes

$$T = \frac{1}{2}\rho \int_S \phi \frac{\partial \phi}{\partial n} \, dS. \tag{13.89}$$

However, it can be shown (see, e.g., Currie (1974)) that S ($S = S_b \cup S_\infty$, where S represents the body surface and S_∞ represents the control surface enclosing the body) can be replaced by S_b alone, and that, for $\boldsymbol{n}$ pointing out of the fluid volume V, we have

$$T = \frac{1}{2}\rho \int_{S_b} \phi \frac{\partial \phi}{\partial n} dS. \tag{13.90}$$

For convenience, we now define the apparent mass (or added mass) of the fluid as the mass of the fluid that, if it were moving with the same velocity as the body, would have the same kinetic energy of the fluid given by equation (13.90).[4] We then set

$$\frac{1}{2}C_a \rho \forall u_1^2 = \frac{1}{2}\rho \int_{S_b} \phi \frac{\partial \phi}{\partial n} dS,$$

to obtain the added-mass coefficient:

$$C_a = \frac{1}{\forall u_1^2} \int_{S_b} \phi \frac{\partial \phi}{\partial n} dS, \tag{13.91}$$

---

[4] Physically speaking, an accelerating body must displace the fluid around it and this would require additional energy to be transferred to the fluid. This additional energy can be seen as equivalent to the energy input by a "larger" body such that the difference between the mass of the larger body and the original body may be called the "added mass."

where V denotes the underwater volume of the *body*. To obtain equation (13.91), the kinetic energy was written in terms of the added-mass coefficient C_a, i.e.,

$$T = \frac{1}{2}\rho C_a V u_1^2. \tag{13.92}$$

The work done in accelerating the body that moves through a distance dx_1 must be equal to the change in the fluid energy, i.e.,

$$dT = \delta F dx_1 = \delta F u_1 dt. \tag{13.93}$$

On the other hand, we have, from equation (13.92),

$$\frac{dT}{dt} = \rho V C_a u_1 \frac{du_1}{dt}. \tag{13.94}$$

Hence, from equations (13.93) and (13.94), we obtain the inertial part of the force in Morison's equation, i.e.,

$$\delta F = \rho V C_a \frac{du_1}{dt}, \tag{13.95}$$

which shows that the force is proportional to added mass times the acceleration of fluid particles. The added mass can be calculated analytically in some cases or it can be calculated numerically. It can also be determined experimentally.

Equation (13.95) should be valid whether there is a free surface or not. There is an additional component of the force if a free surface is present. This force is proportional to the volume of the fluid displaced, i.e., the Froude–Krylov force, which is associated with the pressure change on the body surface due to the passage of a wave by assuming that the presence of the body does not alter the flow field (this excludes the hydrostatic pressure). This force has already been calculated (see equation (13.42)). Let us show next the relation between equation (13.42) and the Froude–Krylov part of Morison's equation.

First note that if $ka \ll 1$ (slender body) or $a/\lambda \ll 1$, then since (see Abramowitz and Stegun (1972, p. 360)),

$$J_1(X) = \frac{X}{2} - C_1\frac{X^3}{3^2} + C_2\frac{X^5}{3^4} + 0(X^7), \tag{13.96}$$

we must have

$$J_1(ka) = \frac{ak}{2} + 0((ak)^3), \tag{13.97}$$

where a is the radius and k is the wave number (actually if $a/\lambda < 1/20$ the error will be less than 1% in these calculations). Even if k is not very small, we assume that a is small and therefore ka is small. The smallness of a leads to u_1 or du_1/dt being evaluated at $r = 0$ or $x = 0$ (along the cylinder axis).

Substituting equation (13.97) in equation (13.42), and recalling that $\cos(\omega t - \pi/2) = \sin(\omega t)$, we obtain

$$\delta F_{F-K} = -a^2\rho g\pi Ak\frac{\cosh[k(x_2 + h)]}{\cosh(kh)}\sin(\omega t). \tag{13.98}$$

But the horizontal velocity of a fluid particle was given by equation (12.56):

$$u_1 = \frac{\partial \phi_I}{\partial x_1} = \Re \left\{ \frac{g A k}{\omega} \frac{\cosh[k(x_2 + h)]}{\cosh(kh)} e^{i(kx_1 - \omega t)} \right\}$$

$$= \frac{g A k}{\omega} \frac{\cosh[k(x_2 + h)]}{\cosh(kh)} \cos(kx_1 - \omega t),$$

(13.99)

and the horizontal acceleration in linear theory, from equation (12.59), is therefore, equal to

$$\left. \frac{Du_1}{Dt} \right|_{x_1=0} \cong \left. \frac{\partial u_1}{\partial t} \right|_{x_1=0} = -g A k \frac{\cosh[k(x_2 + h)]}{\cosh(kh)} \sin(\omega t) \qquad (13.100)$$

along the cylinder axis located at $x_1 = 0$ (because the cylinder is "slender") of the circular cylinder when we use the linear theory. Then the differential mass in the x_2 direction (vertical) times the acceleration gives equation (13.98) exactly. In other words,

$$\delta F_{F-K} = \rho \frac{\forall}{L} \frac{\partial u_1}{\partial t}, \qquad (13.101)$$

where $\forall/L = \pi a^2$. We can now sum equation (13.95) (per unit depth) and equation (13.101) to get

$$\delta F_I = (1 + C_a) \rho \pi \left(\frac{D}{2} \right)^2 \dot{u}_1, \qquad (13.102)$$

where the subscript I refers to the inertia force (per unit depth), and the superimposed dot denotes the local time derivative in linear theory. Equation (13.102) is nothing but the inertia term on the right-hand side of equation (13.77). Note again that the first term on the right-hand side of equation (13.102) is due to the presence of the free surface (Froude–Krylov force) and the second term is due to the acceleration of fluid particles (added-mass force), and that the local time derivatives (accelerations) are used instead of the full material derivative (that also includes the spatial acceleration terms) because of the linearity of the problem.

13.5.2   Drag Term

In the viscous limit, given by equation (13.81), the sectional drag force can be represented by

$$\delta F_d = \frac{1}{2} \rho D C_d u_1 |u_1|, \qquad (13.103)$$

where u_1 is given by equation (13.99). The magnitude of u_1 used in equation (13.103) is to ensure that the flow direction is properly taken into account. Recall that in non-oscillatory or steady problems discussed in Chapter 10, it was not necessary to use $u_1|u_1|$; we simply used u_1^2 since the velocity was not harmonic, i.e., $u_1 \neq f(t)$. Note that C_d is generally taken as equal to 1.0 for a circular cylinder. For some other shapes, see Figure 13.19.

Morison's equation, (13.77), is now obtained by summing equations (13.102) and (13.103). There can be no mathematical basis for this summation, especially if we recall that the coefficients C_a and C_d are intended to be constants. In reality, they also are functions of wave frequency. However, this formula is widely used in offshore and coastal engineering to determine wave and current forces on slender cylinders, such as drilling risers or jetty piles.

Note that the drag term is nonlinear. This causes a slight problem in the use of Morison's equation especially when the body is freely floating since the location where u_1 is calculated is unknown (as the body is moving), and, therefore, one must use the total *relative velocity* rather than u_1 alone if the body is freely floating (see equation (13.114) later). To overcome this problem, and also to be consistent with the assumption of linearity, the drag term is linearized by defining a linear drag coefficient as will be discussed subsequently. There is one more reason for linearizing the drag force, and it is related to the use of spectral analysis in irregular waves which require that the system be linear (see Chapter 15).

13.5.3 Total Force before Drag Linearization

It is possible to determine the total force on a vertical cylinder that extends to the sea floor at $x_2 = -h$ by

$$F|_{r=0} = \int_{-h}^{0} \delta F_I \, dx_2 + \int_{-h}^{0} \delta F_d \, dx_2 = -\frac{1}{4}\rho g \pi D^2 A (1 + C_a) \tanh(kh) \sin(\omega t)$$

$$+\frac{1}{4}\rho C_d D g A^2 \left\{ 1 + \frac{2kh}{\sinh(2kh)} \right\} \cos(\omega t)|\cos(\omega t)|.$$

$$(13.104)$$

Similarly, the overturning moment with respect to the sea floor can be determined. This is left as an exercise (see Prob. 13.7.6).

13.5.4 Linearization of the Drag Term

Let us turn to the question of the linear drag coefficient and write the horizontal component of the particle velocity as

$$u_1 = u_0 \cos(\omega t) \quad \text{at } x_1 = 0, \tag{13.105}$$

where u_0 is the amplitude of the velocity. The drag force in the direction of wave motion becomes

$$\delta F_d = \frac{1}{2}\rho D C_d u_0^2 \cos(\omega t)|\cos(\omega t)|. \tag{13.106}$$

The linear drag force can now be defined by equating the net work done (on a differential cylinder element) by the drag force δF_d, given by equation (13.106), and the linearized drag force, given by

$$\delta F_{dL} = \frac{1}{2}\rho D C_{dL} u_o \cos(\omega t). \qquad (13.107)$$

This means that the viscous energy dissipated (per wave cycle) is the same whether we use equation (13.106) or equation (13.107). Therefore, by requiring equal energy dissipation per wave cycle, we must have $E_d = E_{dL}$, i.e.,

$$E_d = \int_0^\lambda \frac{1}{2}\rho C_d D u_0^2 \cos(\omega t)|\cos(\omega t)| dx_1 = E_{dL} = \int_0^\lambda \frac{1}{2}\rho C_{dL} D u_0 \cos(\omega t) dx_1.$$

$$(13.108)$$

Recalling also that $u_1 = dx_1/dt$ or $dx_1 = u_0 \cos(\omega t) dt$, and $|\cos(\omega t)| \cos(\omega t) = \cos^2(\omega t)$ if $0 \le t \le T/4$, and that the total energy is equal to four times the energy per quarter of a wavelength for constant C_d and C_{dL}, we obtain, from equation (13.108),

$$C_{dL} = C_d \frac{\int_0^{T/4} u_0^3 \cos^3(\omega t) dt}{\int_0^{T/4} u_0^2 \cos^2(\omega t) dt},$$

or

$$C_{dL} = \frac{8}{3\pi} C_d u_0. \qquad (13.109)$$

Note that C_{dL} has the dimension of velocity, unlike C_d (which is dimensionless).
The amplitude of the horizontal velocity, u_0, is given by (from equation (13.99))

$$u_0 = \frac{g A k}{\omega} \frac{\cosh[k(x_2 + h)]}{\cosh(kh)}. \qquad (13.110)$$

Substituting equation (13.109) in equation (13.107), we obtain

$$\delta F_{dL} = \frac{4}{3\pi}\rho D C_d u_0^2 \cos(\omega t). \qquad (13.111)$$

We need to note that there are other methods of linearization of the drag term, different from the method discussed here. Also, see, e.g., Paulling (1979a, 1979b) and Krolikowski and Gay (1980) on the linearization of the drag force when a current is present, and when the waves are random. See also Ertekin and Rodenbusch (2016).

13.5.5   Total Force after Drag Linearization

Then the total horizontal force (that uses the linearized drag) on a vertical cylinder is given by

$$F|_{r=0} = \int_{-h}^{0} \delta F_{dL} dx_2 + \int_{-h}^{0} \delta F_I dx_2 = 4\rho g AD \tanh^2(kh) \times$$

$$\times \left\{ \frac{g Ak^2 C_d}{3\pi\omega^2 \sinh^2(kh)} \left[\frac{h}{2} + \frac{1}{4k} \sinh(2kh) \right] \cos(\omega t) - \frac{(1 + C_a)\pi D}{16 \tanh(kh)} \sin(\omega t) \right\}.$$

$$(13.112)$$

It is seen from equation (13.112) that the drag force and the inertia force are 90° out of phase.

The toppling (or overturning) moment with respect to the sea floor can be calculated from an integral like equation (13.51). This is discussed in Prob. 13.7.6. Also, it is noted that the force equation above is written for finite water depth. It can be specialized for the deep water ($kh \to \infty$) case, see Prob. 13.7.1.

There may also be current in the vicinity of the cylinder. Let us denote the steady shear-current velocity by

$$u_c = u_c(x_2). \qquad (13.113)$$

Also, it is possible that the cylinder is moving. In the case of a cylinder that extends to the sea floor, the only possible mode of motion is in the horizontal plane, $x_1 - x_3$, and this is called surging. However, Morison's equation is used not only to determine the forces on fixed platforms, such as jack-up rigs, but also on floating platforms, such as semi-submersibles which may have tubular structural members such as columns, pontoons, or braces. These members may also be inclined, rather than just be vertical or horizontal. Therefore, in general, one can have three translational and three rotational (or angular) velocity components for the structure motion as was shown in Figure 13.2. In such cases, the velocity calculated in equations (13.103) and (13.109) must be the relative velocity, i.e.,

$$u_r = u_p + u_c - u_b, \qquad (13.114)$$

where u_r is the relative velocity, u_p is the particle velocity, u_c is the current velocity, and u_b is the body velocity.

13.5.6   Equivalent Forces

Let us now consider the force expression given by Morison's equation and the force expression given by MacCamy and Fuchs. Recall first, for the sole purpose of obtaining equal wave forces from the two different methods, the force given by the diffraction theory, i.e., equation (13.44):

$$\delta F_{MF}(x_2, t) = \frac{4\rho g A}{k} \frac{\cosh[k(x_2 + h)]}{\cosh(kh)} R(ka) \cos(\omega t + \alpha),$$

where

$$R(ka) = \left(J_1'^2(ka) + Y_1'^2(ka) \right)^{-1/2}, \qquad \alpha(ka) = \tan^{-1}\left(\frac{Y_1'}{J_1'} \right),$$

and Morison's equation with the linearized drag coefficient (sum of equations (13.102) and (13.111)):

$$\delta F_M(x_2,t) = (1 + C_a)\rho\pi(\frac{D}{2})^2\dot{u}_1 + \frac{1}{2}\rho DC_{dL}u_0\cos(\omega t)$$

or

$$\delta F_M(x_2,t) = -\rho\pi\left(\frac{D}{2}\right)^2 (1 + C_a)g Ak\frac{\cosh[k(x_2 + h)]}{\cosh(kh)}\sin(\omega t)$$

$$+ \frac{1}{2}\rho DC_{dL}\frac{g Ak}{\omega}\frac{\cosh[k(x_2 + h)]}{\cosh(kh)}\cos(\omega t). \qquad (13.115)$$

The differential (or sectional) MacCamy–Fuchs force δF_{MF} may be written as

$$\delta F_{MF} = \frac{4\rho Ag}{k}\frac{\cosh[k(x_2 + h)]}{\cosh(kh)}R(ka)\{\cos(\omega t)\cos(\alpha) - \sin(\omega t)\sin(\alpha)\}. \quad (13.116)$$

If we now set equation (13.115) equal to equation (13.116), with the sole purpose of obtaining the same force, we obtain

$$\bar{C}_{dL} = \frac{8\omega R(ka)}{k^2 D}\cos(\alpha), \qquad \bar{C}_a = \frac{16R(ka)}{\pi k^2 D^2}\sin(\alpha) - 1, \qquad (13.117)$$

where $\bar{C}_{dL}$ and $\bar{C}_a$ are the equivalent linear drag and added mass coefficients, respectively. One can now use equation (13.117) in place of C_{dL} and C_a in equation (13.115). Note that as in the case of Morison's equation, which lacks a mathematical basis, the above approach can also be rejected easily since one cannot have both the inviscid and viscous fluid assumptions at the same time. However, $\bar{C}_{dL}$ may be thought of representing the damping of scattered waves due to the transport of energy toward infinity (scattering problem). Note that $(1 + \bar{C}_a)$ represents the inertia-force coefficient as before.

13.5.7   Long-Period Limit

Physically, one would expect that as the waves become longer and longer, i.e., as $T \to \infty$ or $\omega \to 0$, the free surface would virtually become a rigid lid. To see this, consider the combined free-surface boundary condition given by equation (13.1) in two dimensions:

$$\frac{\partial^2\Phi}{\partial t^2}(x_1,0,t) + g\frac{\partial\Phi}{\partial x_2}(x_1,0,t) = 0.$$

If we write Φ as

$$\Phi(x_1,x_2,t) = \Re\left\{\phi(x_1,x_2)e^{-i\omega t}\right\}, \qquad (13.118)$$

and substitute it in the free-surface condition, we obtain

$$-\omega^2\phi(x_1,0) + g\frac{\partial\phi}{\partial x_2}(x_1,0) = 0. \qquad (13.119)$$

And as $\omega \to 0$, equation (13.119) becomes

$$\left.\frac{\partial\phi}{\partial x_2}\right|_{x_2=0} = 0. \qquad (13.120)$$

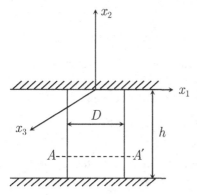

Figure 13.16 A circular cylinder between two parallel plates.

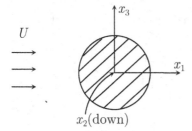

Figure 13.17 Cross section of the cylinder in Figure 13.16.

This means that the vertical component of the particle velocity is zero or the free surface becomes a rigid lid (or plate), as $\omega \to 0$. Then, the vertical cylinder is bounded by two rigid surfaces as shown in Figure 13.16. Since $-\infty \le x_3 \le +\infty$, the flow will be two-dimensional in the x_1-x_3 plane as shown in Figure 13.17 ($A - A'$ cut).

To find the added mass, C_a, as the wave period $T \to \infty$, we will first consider the velocity potential for a circular cylinder (fixed) in the presence of uniform flow, U (recall that $u_1(t) = U \cong$ constant when $T \to \infty$). We can do this since, in an ideal fluid, the velocities around a fixed circular cylinder placed in uniform oncoming flow and the velocities around a moving circular cylinder in an otherwise calm fluid are the same (this is clearly true for absolute velocities).

The velocity potential, ϕ, was given as the sum of a dipole equation (11.33) and uniform flow, i.e.,

$$\phi = \frac{\mu}{2\pi} \frac{x_1}{r^2} + U x_1 \tag{13.121}$$

(see, e.g., Lamb 1932, Art. 68), or in polar coordinates

$$\phi = \frac{\mu}{2\pi} \frac{\cos\theta}{r} + U r \cos\theta. \tag{13.122}$$

Since the rigid body condition must be satisfied on the cylinder, we have

$$\left.\frac{\partial\phi}{\partial r}\right|_{r=a} = -\frac{\mu}{2\pi} \frac{\cos\theta}{r^2} + U\cos\theta = 0 \tag{13.123}$$

or, as determined before in Chapter 11,

$$\mu = 2\pi U a^2. \tag{13.124}$$

If we now impress both on the fluid and the cylinder a velocity $-U$, we can obtain the potential for a cylinder moving with velocity U in the opposite direction of the x_1 coordinate (to the left), and in an otherwise calm fluid, i.e.,

$$\phi = \frac{Ua^2 x_1}{x_1^2 + x_2^2} = \frac{Ua^2 \cos\theta}{r}. \tag{13.125}$$

We do this to calculate the added-mass coefficient. Note that $\partial\phi/\partial r = -U\cos\theta$ on $r = a$. However,

$$\frac{\partial\phi}{\partial n}(r,\theta) = -\frac{\partial\phi}{\partial r}(r,\theta) = \frac{Ua^2 \cos\theta}{r^2}, \quad \phi\left.\frac{\partial\phi}{\partial n}\right|_{r=a} = U^2 a\cos^2\theta. \tag{13.126}$$

We can now use equation (13.126) in the expression for the kinetic energy of a fluid given by equation (13.90), i.e.,

$$T = \frac{1}{2}\rho \int_{S_b} \phi\frac{\partial\phi}{\partial n}\, dS = \frac{1}{2}\rho \int_0^{2\pi} (U^2 a\cos^2\theta)\, a\, d\theta$$

or

$$T = \frac{1}{2}\rho U^2 a^2 \int_0^{2\pi} \cos^2\theta\, d\theta = \frac{1}{2}\rho U^2 a^2\pi. \tag{13.127}$$

Therefore, from equations (13.92) and (13.127), we obtain the dimensionless added-mass coefficient:

$$C_a = \frac{2T}{\rho \forall U^2} = \frac{a^2\pi}{\forall} = 1, \tag{13.128}$$

since the volume, $\forall$, here is per unit length of the cylinder, and therefore, $\forall = \pi a^2$ (cross-sectional area), and $u_1 = U$. Therefore, the dimensional added mass is

$$C_a' = \rho\forall C_a = \rho\pi a^2, \tag{13.129}$$

for a circular cylinder moving perpendicular to its axis (also see, e.g., Newman (1978, Sec. 4.15)). The dimensional coefficient C_a' then represents the amount of fluid displaced by the motion of the circle. It is important to emphasize that added mass depends only on the body shape in an unbounded fluid.

Going back to Morison's equation (13.104) (or equation (13.112)) for the force, we see that if the waves are very long compared with the diameter, i.e., $D/\lambda \ll 1$, one can take $\bar{C}_a \cong 1$. Then, it is logical to expect that C_a, which was obtained by MacCamy and Fuchs' theory, should give the same result. Let us show this next.

If $ka \ll 1$ $(T \to \infty)$, then one obtains from equation (13.41) (see Abramowitz and Stegun 1972, Chap. 9)

$$R(ka) = \frac{1}{\sqrt{J_1'^2(ka) + Y_1'^2(ka)}} \cong \frac{\pi}{2}(ka)^2 \tag{13.130}$$

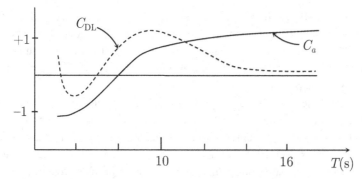

Figure 13.18 The variations of the drag and inertia coefficients with the wave period.

and

$$\sin \alpha = \frac{Y_1'}{\sqrt{J_1'^2 + Y_1'^2}} \cong \frac{\pi}{2}(ka)^2 \frac{2}{\pi(ka)^2} = 1. \qquad (13.131)$$

Therefore, $\bar{C}_a$ becomes (from equation (13.117))

$$\bar{C}_a = \frac{16R(ka)}{\pi k^2 D^2} \sin(\alpha) - 1 \cong 1, \quad ka \ll 1. \qquad (13.132)$$

Similarly, for $ka \ll 1$

$$\cos(\alpha) = \frac{J_1'}{\sqrt{J_1'^2 + Y_1'^2}} \cong \frac{\pi}{4}k^2 a^2, \qquad (13.133)$$

and

$$\bar{C}_{dL} \cong \frac{8\omega}{k^2 D}\frac{\pi}{2}(ka)^2\frac{\pi}{4}k^2 a^2 = \frac{\omega\pi^2 k^2 D^3}{16}, \quad ka \ll 1. \qquad (13.134)$$

Note that the damping is $O(k^2)$, and therefore, it is a small quantity if waves are relatively long, i.e., wave damping goes to zero as the frequency goes to zero.

Typical C_a and C_{dL} curves look something like in Figure 13.18. However, the inertia coefficient will be positive always (see Chakrabarti and Tam (1975)). See Koterayama (1984) and Sarpkaya (1977) for the variation of drag and inertia coefficients as functions of the Keulegan–Carpenter number, KC.

13.6   Concluding Remarks

The wave diffraction problem in the presence of fixed obstructions is formulated within the assumptions of the linear potential theory. In this formulation, we have seen that the total velocity potential is decomposed into the incoming-wave potential and the scattering- (or diffraction-)wave potential.

The incoming-wave potential has already been derived in Chapter 12 for the finite, constant water depth. It was shown that Φ_I satisfies Laplace's equation, the free-surface and the sea-floor boundary conditions. The diffraction potential, on the other hand, must satisfy not only Laplace's equation, the free-surface and sea-floor conditions, but also a far-field condition called the radiation condition. This condition ensures that the scattered waves generated by the obstruction are outgoing toward the infinity. This condition is derived by means of energy arguments and the time periodicity of the potential.

As an example of a diffraction problem, linear harmonic waves are considered in the presence of a circular, vertical cylinder in constant water depth. The diffraction potential is determined by enforcing the boundary conditions and solving Laplace's equation. As a result, the total velocity potential is obtained in terms of Bessel functions. Following the solution of the boundary-value problem for Φ_T, we have used Euler's integral to determine the dynamic pressure on the cylinder, and then integrated the pressure on the surface of the cylinder to determine the horizontal component of the wave force acting on the vertical cylinder. This force is shown to have two components: (i) the Froude–Krylov part due to the incoming-wave potential only, and (ii) the scattering or diffraction part due to the diffraction potential only. The overturning (or toppling) moment with respect to the base of the cylinder is also determined. The surface elevation is obtained by use of the dynamic free-surface condition. Once the surface elevation is known, the run-up on the cylinder can be obtained. The closed-form solution discussed here is due to MacCamy and Fuchs (1954).

It is not always possible to find a closed-form (or analytic) solution, as we could to the vertical, circular cylinder diffraction problem, and therefore, one must resort to a numerical or semi-analytical method to solve potential flow problems that involve three-dimensional obstructions. In Section 13.4, the Green function method is used to formulate the diffraction problem for obstructions that have arbitrary shapes in the plan view (on the still-water plane), but whose sides are vertical so that the obstruction is basically a vertical cylinder whose cross section in the plan view is arbitrary. Once the integral equation for the unknown potential is solved on the obstruction, by distributing sources on the wetted surface of the obstruction (up to the SWL since we use the linear theory), the potential anywhere can be determined from equation (11.66) for $P \in V$, and therefore, the free-surface elevation can be obtained at any desired point around or at the location of the obstruction. Note that the Green function used in the integral equation (13.74) satisfies all the boundary conditions except the body-boundary condition.

When the obstruction in the flow field is "slender," compared with the wavelength, and it has a circular cross section, such as a tubular member of a fixed or floating platform, it is possible, in principle, to use a simple method to determine the wave forces. This simple method, called Morison's equation method, is discussed in Section 13.5. Morison's equation, which is derived in an ad hoc way, contains experimentally determined inertia and viscous drag coefficients. The viscous part of the force, which is absent in the potential theory approach due to the assumption of inviscid fluid, is due to the fact that, in a viscous fluid, flow separation occurs at high Reynolds numbers, and therefore, there is an additional pressure force (since the pressure in the front and rear of the object is no longer the same because of the presence of a wake) termed the form-drag force leading to the form-drag coefficient. Note that this drag coefficient should not be confused with the skin-friction drag coefficient which is expected to be small compared with the form-drag coefficient when the flow is separated. Even though the inertia and drag coefficients depend on the Reynolds and KC numbers, the flow due to a wave field in water is generally a high-Reynolds-number flow. And the dependency of these coefficients on the KC number is rather weak for smooth cylinders when the Reynolds number is high. As a result, the coefficients are taken as constants in most cases.

Following the introduction of Morison's equation, the relation between the kinetic energy of the fluid and the added-mass coefficient is established in the absence of a free surface. The free-surface effect is contained in the Froude–Krylov force in the case of slender bodies. Finally, the sum of the inertia force, which includes the added-mass force due to particle accelerations, Froude–Krylov force, and drag force is formed to obtain the total force acting on a vertical, tubular cylinder.

Because the drag force is quadratic in particle velocity, this force is equivalently linearized. This procedure, which is based on equal-energy dissipation, resulted in a linearized drag coefficient which has the dimension of velocity (recall that the original drag coefficient is dimensionless).

It is further shown that by properly choosing the inertia and drag coefficients, one can obtain MacCamy–Fuchs' solution. However, it must be emphasized that this is again an ad hoc approach to obtaining the same force by two totally different methods, and that one cannot mathematically, and even physically, justify this equal-force approach since potential theory is based on the inviscid fluid assumption, whereas Morison's equation contains a viscous term. Nevertheless, one, perhaps, can convince him/herself that the scattering of a wave, which causes additional forces in the case of MacCamy–Fuchs' solution, may be considered equivalent to the energy- dissipation mechanism present in Morison's equation (the viscous-drag term) as far as obtaining the same force by the two different methods is concerned.

And finally, we have shown that the added-mass coefficient approaches 1.0 as waves becomes longer and longer, and therefore, that MacCamy–Fuchs' solution for the force approaches the solution obtained by Morison's equation. It is also shown that the damping force (due to wave scattering) is of $O(k^2)$ if $k \ll 1$, where k is the wave number, i.e., long waves.

Figure 13.19 (Delaney and Soreasen (1953) and Saunders (1957)) shows some additional drag coefficients and inertia coefficients for various cross sections that may be encountered in ocean engineering.

13.7   Self-Assessment

13.7.1

Obtain the deep-water version ($kh \rightarrow \infty$) of the force equation given by equation (13.112).

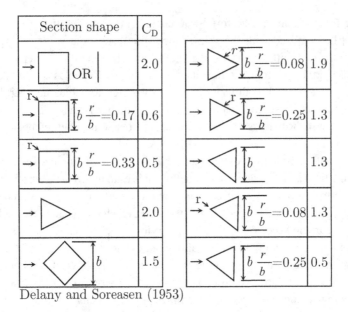

Delany and Soreasen (1953)

Inertia coefficients for common structural forms

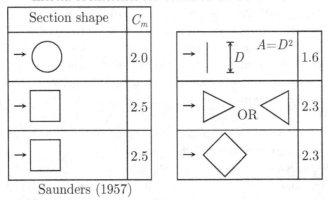

Saunders (1957)

Figure 13.19 Various drag, C_d, and inertia, $C_m = (1 + C_a)$, coefficients (from Delaney and Soreasen (1953) and Saunders (1957)).

13.7.2

Consider a completely submerged, circular cylinder in two dimensions (x_1–x_2 plane), i.e., the horizontal cylinder can be considered infinitely long in the x_3 direction. The diameter of the cylinder is $D = 4\,\text{m}$. The wave period is 10 s and the axis of the cylinder is 10 m below the SWL. The wave amplitude is 1.0 m, and the water can practically be assumed infinitely deep, and the waves are linear.

Use Morison's equation to find the horizontal force and the moment (per unit length of the cylinder in the x_3 direction) with respect to the SWL. What is the amplitude of the maximum force and maximum moment? Given: $C_a = 1.2$ and $C_d = 0.9$, in salt water of density $\rho = 1025\,\text{kg/m}^3$.

13.7.3

Use cylindrical polar coordinates to prove that Φ_I, given by equation (13.16), satisfies the boundary conditions, and that it is the solution of Laplace's equation.

13.7.4

Use cylindrical polar coordinates to prove that Φ_D, given by equation (13.24), satisfies the boundary conditions, and that it is the solution of Laplace's equation. (Of course, Φ_D satisfies a set of boundary conditions different from what Φ_I satisfies in Prob. 13.7.3.)

13.7.5

Consider a surface-piercing, vertical circular cylinder that extends to the sea floor. The water depth is $h = 100$ m, the cylinder diameter is $D = 40$ m, the wave height is $H = 20$ m, and the wave period is $T = 17$ s. Also, $\rho = 1025$ kg/m^3 and $g = 9.8$ m/s^2.

Calculate the amplitudes of the wave-exciting force and overturning moment by use of MacCamy and Fuch's theory. Compare the solutions you obtained with the published results in Sarpkaya and Isaacson (1981), as well as in Rahman (1988).

NOTES: The total force is in Newton and the moment is in Newton*m. Also, you can use the tabulated values for Bessel functions (e.g., in Abramowitz and Stegun's (1972) book) in a spreadsheet, or use subroutines (such as IMSL or the ones given in Press et al. (1992)) in your own computer program, or use a symbolic-math computer program such as MATLAB).

13.7.6

Show that the equation for the total overturning moment with respect to the sea floor acting on a vertical cylinder that extends to the sea floor, in water depth h, by use of Morison's equation (use the linearized drag coefficient) is given by

$$M = M_I + M_D = -(1 + C_a)\frac{\pi \rho g \, AD^2 h}{4}\left(\tanh kh - \frac{1}{kh} + \frac{1}{kh \cosh kh}\right)\sin \omega t$$

$$+ \frac{1}{3}\frac{\rho D}{\pi}C_d\frac{g^2 A^2}{\omega^2}\left(\frac{kh \sinh 2kh + \frac{1}{2} - \frac{1}{2}\cosh 2kh + (kh)^2}{\cosh^2 kh}\right)\cos \omega t.$$

13.7.7

Consider a surface-piercing, vertical circular cylinder (a pile) that extends to the sea floor. The water depth is $h = 20$ m, the cylinder diameter is $D = 3$ m, and the wave height is $H = 2$ m and wave period is $T = 10$ s. Also, $\rho = 1000$ kg/m^3 and $g = 9.8$ m/s^2. The inertia and drag coefficients are $C_m = 1.0 + C_a = 1.9$ and $C_d = 0.8$, respectively.

1. Use Morison's equation to determine the total horizontal force and overturning moment (with respect to the sea floor, see Prob. 13.7.6) acting on the pile. Use the linearized drag coefficient.
2. Use MacCamy and Fuchs' theory to determine the horizontal force and overturning moment (with respect to the sea floor) acting on the pile. You can use the tabulated values for Bessel functions (e.g., in Abramowitz and Stegun's (1972)'s book), or use subroutines (such as IMSL) in your own computer program, or use a symbolic-math computer program such as MATLAB.
3. Compare the solutions you obtained above (by use of both methods) with the published results in Sarpkaya and Isaacson's (1981) book, as well as in Rahman's (1988) book. Comment on the importance of the drag force on this pile compared with the inertia force. If the diameter of the pile is much more, e.g., such as an oil-storage tank whose diameter is $D = 30$ m, which result do you think will be more accurate (Morison's vs. MacCamy and Fuchs') and why?
4. Compare only the drag components of the forces given by equations (13.104) and (13.112) by plotting them from $\omega t = 0$ to $\omega t = 2\pi$. Comment on the mean and maximum values of the drag forces obtained by the nonlinear and linear forces.

13.7.8

Solve Prob. 13.7.5 by use of Morison's equation, and for $C_a = 1.0$ and $C_d = 1.0$. See Prob. 13.7.6 for the moment equation. Compare your solution with MacCamy and Fuch's solution and the experiments. Use the linearized drag coefficient, C_{dL}, see equation (13.109).

13.7.9

The following two parts of the problem are independent of each other.

1. The diameter of a circular, vertical cylinder that extends to the sea floor is 10m. The water depth is 20 m. The wave height is 2 m. Assume that the drag force is negligible, calculate the amplitude of the total horizontal force and total overturning moment (with respect to the sea floor) on the cylinder by use of Morison's equation for $C_a = 1.2$, $\rho = 1000$ kg/m^3 and the wave period of $T = 10$ s. Note that the water depth can be assumed "shallow" so that $kh \to 0$.
2. A two-dimensional (x_1–x_2 plane) circular cylinder (which is infinitely long in the x_3 direction) is situated 15 m below the SWL. The diameter of the cylinder is 2 m. The wave period is 12 s. The wave amplitude is 1.0 m. The water depth is infinite. The drag force can be neglected. Use Morison's equation to determine the amplitude of horizontal force per unit length of the cylinder for $C_a = 1.3$ and $\rho = 1025$ kg/m^3.

14 Prescribed Body Motions and Floating Bodies

14.1 Introduction

In the beginning of Chapter 13, we have stated that the total potential can be decomposed in such a way that the motion problem of a body which is acted upon by waves can be solved within the context of linearity by separating the associated problems of the incoming, diffraction and radiation potentials. If the body also has forward motion, an additional potential, namely the steady forward-speed potential, must be determined and added to the total potential. This forward-speed potential, which is related to the wave-resistance problem, will not be discussed here but see Wehausen, Webster, and Yeung (2016). We will deal here with floating bodies that have no forward speed. However, because of the linearity of the general problem, most of the concepts discussed here are also applicable to the case of a floating body that has forward speed.

In this chapter, we will first formulate the potential problem of a body making prescribed motions in an unbounded fluid. This formulation will lead us to the concept of added mass which was briefly discussed in Chapter 13. Subsequently, we will introduce the problem of small motions when a body moves in the presence of a free surface. This problem will surface the hydrodynamic damping coefficients. The remarkable relation between the hydrodynamic coefficients of added mass and damping, and the exciting forces, namely the Haskind–Hanaoka relationship will be derived. This will be followed by a discussion on the equations of motion for a rigid body in six degrees of freedom.

14.1.1 Kinematics of Rigid-Body Motion

As it was depicted in Figure 13.2, a rigid body has six degrees of freedom. To analyze the forces due to the motion of a body, it is essential that we review some rigid-body kinematics here.

Let us introduce two right-handed coordinate systems as shown in Figure 14.1. The coordinate system $Ox_1x_2x_3$ is fixed in the fluid (earth-bound or inertial coordinates.) The other coordinate system $\bar{O}\bar{x}_1\bar{x}_2\bar{x}_3$ is fixed in the body and, therefore, moves with the body. It is called the body coordinate system. The unit base vectors in each coordinate system are denoted by e_i and $\bar{e}_i$. Consider a point P, somewhere in the fluid or on or inside the body, that moves in time. We can describe the position of P either in the

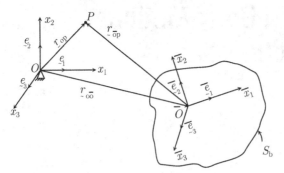

Figure 14.1 Position vectors from the origins of the earth-bound (global) and body-fixed coordinates to a point, P, in space.

fixed or the moving coordinate system. Since any vector r can be expressed in terms of either set of base vectors, one can write

$$r_{OP} = r_{O\bar{O}} + r_{\bar{O}P} = x_i e_i = \bar{x}_i \bar{e}_i. \tag{14.1}$$

If we choose to write $r_{O\bar{O}}$ in the fixed system and $r_{\bar{O}P}$ in the moving system, i.e.,

$$r_{O\bar{O}} = x_{O\bar{O}i} e_i, \qquad r_{\bar{O}P} = \bar{x}_{\bar{O}Pi} \bar{e}_i, \tag{14.2}$$

then equations (14.1) and (14.2) give

$$r_{OP} = x_{O\bar{O}i} e_i + \bar{x}_{\bar{O}Pi} \bar{e}_i. \tag{14.3}$$

The main reason for choosing to write $r_{\bar{O}P}$ in the moving coordinate system is because $\bar{x}_{\bar{O}Pi}$ represents the location of P when P is on the body surface in the moving coordinate system, i.e., $\bar{x}_{\bar{O}Pi}$ represents the geometry of the body which is known. Thus, the use of r_{OP} in equation (14.3) is for convenience.

The velocity of the point P can be written in either coordinate system. However, because we will mainly be interested in the case where P is attached to the body, it is more convenient to find the velocity of P with respect to the fixed system, $Ox_1x_2x_3$. Note that $\bar{e}_i$ are time independent in the moving system $\bar{O}\bar{x}_1\bar{x}_2\bar{x}_3$, but they are dependent on time when observed by an observer attached to the fixed system, $Ox_1x_2x_3$. Therefore, the time derivative of equation (14.3) gives us the velocity of the point P as follows:

$$q = \frac{dr_{OP}}{dt} = \frac{dx_{O\bar{O}i}}{dt} e_i + \frac{d\bar{x}_{\bar{O}Pi}}{dt} \bar{e}_i + \bar{x}_{\bar{O}Pi} \frac{d\bar{e}_i}{dt}. \tag{14.4}$$

Let us assume for the moment that equation (14.4) can be written as

$$q = q_{O\bar{O}} + q_{\bar{O}P} + \Omega \times r_{\bar{O}P}, \tag{14.5}$$

where

$$q_{O\bar{O}} = \frac{dx_{O\bar{O}i}}{dt} e_i, \qquad q_{\bar{O}P} = \frac{d\bar{x}_{\bar{O}Pi}}{dt} \bar{e}_i, \qquad i = 1,2,3,$$

$$\Omega = \frac{d\bar{x}_{\bar{O}Pj}}{dt} \bar{e}_{j-3}, \quad j = 4,5,6. \tag{14.6}$$

The three components of Ω represent the angular velocities with respect to $\bar{x}_i$, $i = 1,2,3$, axes. Note that if the point P is attached to the body, then $q_{\bar{O}P} = 0$ since

$\bar{x}_{\bar{O}Pi} \neq f(t)$ in the moving system; and in this case, $d\bar{x}_{\bar{O}P4}/dt$ represents the rolling velocity, $d\bar{x}_{\bar{O}P5}/dt$ represents the yawing velocity and $d\bar{x}_{\bar{O}P6}/dt$ represents the pitching velocity of the body (see Figure 13.2). The comparison of equations (14.4) through (14.6) shows that we must have

$$\bar{x}_{\bar{O}Pi}\frac{d\bar{e}_i}{dt} = \boldsymbol{\Omega} \times \boldsymbol{r}_{\bar{O}P} = \left(\frac{d\bar{x}_{\bar{O}Pj}}{dt}\bar{e}_{j-3}\right) \times (\bar{x}_{\bar{O}Pi}\bar{e}_i), \quad i = 1,2,3, \quad j = 4,5,6, \quad (14.7)$$

Let us next show that equation (14.7) is true.

The time derivative of $\bar{e}_i$, being also a vector, can be described in terms of the base vectors $\bar{e}_i$, i.e.,

$$\frac{d\bar{e}_i}{dt} = \left(\frac{d\bar{e}_i}{dt} \cdot \bar{e}_j\right)\bar{e}_j \equiv \bar{\gamma}_{ij}\bar{e}_j, \quad (14.8)$$

where, the second-order tensor $\bar{\gamma}_{ij}$, just defined, is a skew symmetric tensor since

$$\frac{d(\bar{e}_i \cdot \bar{e}_j)}{dt} = \frac{d\bar{e}_i}{dt} \cdot \bar{e}_j + \bar{e}_i \cdot \frac{d\bar{e}_j}{dt} = \frac{d(\delta_{ij})}{dt} = 0 \quad (14.9)$$

or

$$\bar{\gamma}_{ij} + \bar{\gamma}_{ji} = 0 \quad \Rightarrow \quad \bar{\gamma}_{ij} = 0 \quad \text{if } i = j, \qquad \bar{\gamma}_{ij} = -\bar{\gamma}_{ji} \quad \text{if } i \neq j.$$

As a result,

$$\bar{x}_{\bar{O}Pi}\frac{d\bar{e}_i}{dt} = \bar{x}_{\bar{O}Pi}\bar{\gamma}_{ij}\bar{e}_j = (\bar{x}_{\bar{O}P2}\bar{\gamma}_{21} + \bar{x}_{\bar{O}P3}\bar{\gamma}_{31})e_1 + (\bar{x}_{\bar{O}P1}\bar{\gamma}_{12} + \bar{x}_{\bar{O}P3}\bar{\gamma}_{32})e_2$$
$$+ (\bar{x}_{\bar{O}P1}\bar{\gamma}_{13} + \bar{x}_{\bar{O}P2}\bar{\gamma}_{23})e_3.$$
$$(14.10)$$

On the other hand, the last equality in equation (14.7) is

$$\left(\frac{d\bar{x}_{\bar{O}Pj}}{dt}\bar{e}_{j-3}\right) \times (\bar{x}_{\bar{O}Pi}\bar{e}_i) = \left(\frac{d\bar{x}_{\bar{O}P5}}{dt}\bar{x}_{\bar{O}P3} - \bar{x}_{\bar{O}P2}\frac{d\bar{x}_{\bar{O}P6}}{dt}\right)\bar{e}_1$$
$$- \left(\frac{d\bar{x}_{\bar{O}P4}}{dt}\bar{x}_{\bar{O}P3} - \bar{x}_{\bar{O}P1}\frac{d\bar{x}_{\bar{O}P6}}{dt}\right)\bar{e}_2 \quad (14.11)$$
$$+ \left(\frac{d\bar{x}_{\bar{O}P4}}{dt}\bar{x}_{\bar{O}P2} - \bar{x}_{\bar{O}P1}\frac{d\bar{x}_{\bar{O}P5}}{dt}\right)\bar{e}_3.$$

The comparison of equations (14.10) and (14.11) shows that we can now define

$$\frac{d\bar{x}_{\bar{O}P4}}{dt} = \bar{\gamma}_{23} = -\bar{\gamma}_{32}, \quad \frac{d\bar{x}_{\bar{O}P5}}{dt} = \bar{\gamma}_{31} = -\bar{\gamma}_{13}, \quad \frac{d\bar{x}_{\bar{O}P6}}{dt} = \bar{\gamma}_{12} = -\bar{\gamma}_{21}, \quad (14.12)$$

to show that equation (14.7) holds. Therefore, if the point P is attached to the body, one obtains from equation (14.5)

$$\boldsymbol{q} = \boldsymbol{q}_{o\bar{O}} + \boldsymbol{\Omega} \times \boldsymbol{r}_{\bar{O}P}, \qquad P \in S_b, \quad (14.13)$$

where $\boldsymbol{q}_{o\bar{O}}$ is the velocity of $\bar{O}$ with respect to $Ox_1x_2x_3$, and $\boldsymbol{\Omega}$ and $\boldsymbol{r}_{\bar{O}P}$ are given by equations (14.6) and (14.2), respectively.

In indicial notation, and by use of Einstein's summation convention, we can write equation (14.13) as

$$q_i = q_{O\bar{O}i} + \epsilon_{ijk}\frac{d\bar{x}_{\bar{O}P(j+3)}}{dt}\bar{x}_{\bar{O}Pk},$$

(14.14)

where $i, j, k = 1, 2, 3$, and ϵ_{ijk} is the permutation symbol defined by equation (11.10).

14.2   Unbounded Fluid and Added Mass Coefficients

Let us consider a body moving in an unbounded fluid of constant mass density, ρ, and zero viscosity, i.e., $\mu = 0$. The resulting flow is assumed to be irrotational, i.e., the vorticity vector, ω, vanishes everywhere in the fluid. The force and the moment acting on the body are given by

$$F = \int_{S_b} p\,n\,dS, \qquad M = \int_{S_b} p\,(r_{OP} \times n)\,dS,$$

(14.15)

where p is the pressure, and n is the unit normal vector on the body surface, pointing out of the fluid (or into the body). The force and moment in equation (14.15) above are measured in the earth-bound (fixed) system. However, they can also be resolved in the moving system as we will see later on.

The pressure that appears in equation (14.15) can be obtained from the linear form of Euler's integral (see the note below equation (14.17)), i.e.,

$$p = -\rho g x_2 - \rho\frac{\partial\phi}{\partial t},$$

(14.16)

where the first term, which represents the hydrostatic pressure, will be omitted here without loss in generality (but should be included in the final analysis). The force and moment then become

$$F = -\rho\frac{\partial}{\partial t}\int_{S_b} \phi\,n\,dS, \qquad M = -\rho\frac{\partial}{\partial t}\int_{S_b} \phi\,(r_{OP} \times n)\,dS.$$

(14.17)

It is important to note here that the expressions given by equation (14.17) are exact for the force and moment on a body in an unbounded fluid even though we used the approximate (linear) pressure given by equation (14.16). In other words, even if we had considered the exact (rather than the linear) pressure in Euler's integral, we would have obtained equation (14.17) (see, e.g., Newman (1978, Sections 4.12 and 4.13)).

To describe the above quantities in terms of the moving system, consider equation (14.1) and write it as

$$r_{OP} = x_{OPi}e_i = (x_{O\bar{O}i} + x_{\bar{O}Pi})e_i = (\bar{x}_{O\bar{O}i} + \bar{x}_{\bar{O}Pi})\bar{e}_i.$$

(14.18)

Note that the base vectors in the moving system can be described in terms of the base vectors in the fixed system, and vice versa, i.e.,

$$\bar{e}_i = (\bar{e}_i \cdot e_j)e_j \equiv \beta_{ij}e_j, \qquad e_i = (e_i \cdot \bar{e}_j)\bar{e}_j \equiv \beta_{ji}\bar{e}_j,$$

(14.19)

where β_{ij}s are called the direction cosines. Substituting equation (14.19) in equation (14.18), and interchanging the dummy variables i and j, one can obtain

$$x_{OPi} = x_{O\bar{O}i} + x_{\bar{O}Pi} = (\bar{x}_{O\bar{O}j} + \bar{x}_{\bar{O}Pj})\beta_{ji}. \tag{14.20}$$

It is noted for future use that we have $\partial x_{OPi}/\partial \bar{x}_{\bar{O}Pj} = \beta_{ji}$ from equation (14.20).

The absolute velocity, u, of the fluid particles may be referred to either one of the coordinate systems, i.e., we can write

$$u = \frac{\partial \phi}{\partial x_1} e_1 + \frac{\partial \phi}{\partial x_2} e_2 + \frac{\partial \phi}{\partial x_3} e_3 = \bar{u} = \frac{\partial \bar{\phi}}{\partial \bar{x}_1} \bar{e}_1 + \frac{\partial \bar{\phi}}{\partial \bar{x}_2} \bar{e}_2 + \frac{\partial \bar{\phi}}{\partial \bar{x}_3} \bar{e}_3, \tag{14.21}$$

where $\phi(x_1, x_2, x_3, t) = \bar{\phi}(\bar{x}_1, \bar{x}_2, \bar{x}_3, t)$. In other words, we can express the absolute velocity (measured with respect to an observer in the fixed coordinate system) either in terms of the fixed coordinates, x_i, or the moving coordinates, $\bar{x}_i$, by use of equation (14.20):

$$\phi(x_1, x_2, x_3, t) = \phi(\bar{x}_1 \beta_{11} + \bar{x}_2 \beta_{21} + \bar{x}_3 \beta_{31}, \bar{x}_1 \beta_{12} + \bar{x}_2 \beta_{22} + \bar{x}_3 \beta_{32},$$

$$\bar{x}_1 \beta_{13} + \bar{x}_2 \beta_{23} + \bar{x}_3 \beta_{33}, t) \tag{14.22}$$

$$\equiv \bar{\phi}(\bar{x}_1, \bar{x}_2, \bar{x}_3, t).$$

Note that $\bar{x}_i$ and x_i in equation (14.21) are the same as $\bar{x}_{OPi}$ and x_{OPi}, respectively, i.e., $x_i \equiv x_{OPi} = x_{O\bar{O}i} + x_{\bar{O}Pi}$, $\bar{x}_i \equiv \bar{x}_{OPi} = \bar{x}_{O\bar{O}i} + \bar{x}_{\bar{O}Pi}$.

Equation (14.21) can be proven as follows. Since,

$$\frac{\partial \bar{\phi}}{\partial \bar{x}_i} = \frac{\partial \phi}{\partial x_j} \frac{\partial x_j}{\partial \bar{x}_i} = \frac{\partial \phi}{\partial x_j} \beta_{ij},$$

we have

$$\frac{\partial \bar{\phi}}{\partial \bar{x}_i} \bar{e}_i = \frac{\partial \phi}{\partial x_j} \beta_{ij} \bar{e}_i = \frac{\partial \phi}{\partial x_j} e_j,$$

because $e_j = \beta_{ij} \bar{e}_i$ from equation (14.19). Recall that β_{ij} are the direction cosines, i.e.,

$$\beta_{ij} = \cos(\bar{x}_i, x_j). \tag{14.23}$$

And because $\beta_{ij} \beta_{ik} = \delta_{jk}$, the transformation is orthogonal or orthonormal This completes the proof of equation (14.21). Note that the direction cosines can also be written as $\beta_{ij} = \cos(n, x_j)$ since the unit normal vector on the body is attached to the body-fixed coordinates.

The velocity potential ϕ or $\bar{\phi}$ must also satisfy the kinematic boundary condition on S_b which represents the surface of the body:

$$q \cdot n = \frac{\partial \phi(x_j, t)}{\partial x_i} n_i(x_j, t) = \frac{\partial \bar{\phi}(\bar{x}_j, t)}{\partial \bar{x}_i} \bar{n}_i(\bar{x}_j), \quad x_j \equiv (x_1, x_2, x_3), \tag{14.24}$$

where $\bar{n}_i$ is independent of time in the body-fixed coordinate system. If we now substitute q (for P on the body surface) from equation (14.13) on the left-hand side of the above equation, we have

$$(q_{O\bar{O}} + \Omega \times r_{\bar{O}P}) \cdot n = q_{O\bar{O}} \cdot n + \Omega \cdot (r_{\bar{O}P} \times n), \tag{14.25}$$

since $(\Omega \times r_{\bar{O}P}) \cdot n = \Omega \cdot (r_{\bar{O}P} \times n)$ by vector calculus.

For convenience, let us write the following:

$$q_{O\bar{O}} = \frac{dx_{O\bar{O}i}}{dt}e_i = \frac{d\bar{x}_{O\bar{O}i}}{dt}\bar{e}_i \equiv \bar{q}_i\bar{e}_i, \quad i = 1,2,3,$$

$$\Omega = \frac{dx_{\bar{O}Pj}}{dt}e_{j-3} = \frac{d\bar{x}_{\bar{O}Pj}}{dt}\bar{e}_{j-3} \equiv \bar{q}_j\bar{e}_{j-3}, \quad j = 4,5,6,$$

(14.26)

where we have resolved the above two vectors in either coordinate system and also defined

$$\bar{q}_i = \frac{d\bar{x}_{O\bar{O}i}}{dt}, \quad i = 1,2,3; \qquad \bar{q}_j = \frac{d\bar{x}_{\bar{O}Pj}}{dt}, \quad j = 4,5,6.$$

(14.27)

Substituting equations (14.26) and (14.27) in equation (14.25), and by use of equation (14.24), one can obtain

$$\frac{\partial\bar{\phi}}{\partial\bar{x}_i}\bar{n}_i = \bar{q}_i\bar{n}_i + \bar{q}_j\bar{e}_{j-3}\cdot(\bar{x}_i\bar{e}_i \times \bar{n}_k\bar{e}_k)$$

or

$$\frac{\partial\bar{\phi}}{\partial\bar{x}_1}\bar{n}_1 + \frac{\partial\bar{\phi}}{\partial\bar{x}_2}\bar{n}_2 + \frac{\partial\bar{\phi}}{\partial\bar{x}_3}\bar{n}_3 = \bar{q}_1\bar{n}_1 + \bar{q}_2\bar{n}_2 + \bar{q}_3\bar{n}_3 + \bar{q}_4(\bar{x}_2\bar{n}_3 - \bar{x}_3\bar{n}_2)$$

$$+ \bar{q}_5(\bar{x}_3\bar{n}_1 - \bar{x}_1\bar{n}_3) + \bar{q}_6(\bar{x}_1\bar{n}_2 - \bar{x}_2\bar{n}_1).$$

(14.28)

For brevity, one can define

$$\bar{n}_4 = \bar{x}_2\bar{n}_3 - \bar{x}_3\bar{n}_2, \quad \bar{n}_5 = \bar{x}_3\bar{n}_1 - \bar{x}_1\bar{n}_3, \quad \bar{n}_6 = \bar{x}_1\bar{n}_2 - \bar{x}_2\bar{n}_1,$$

(14.29)

to obtain the following from equation (14.28):

$$\frac{\partial\bar{\phi}}{\partial\bar{n}} = \bar{q}_j\bar{n}_j, \quad j = 1,2,\ldots,6.$$

(14.30)

The boundary condition, equation (14.30), suggests (since $\bar{n}_j$ is a function of $\bar{x}_i$ only and $\bar{q}_j$ is a function of t only) that one can decompose the potential as follows:

$$\bar{\phi}(\bar{x}_i,t) = \phi_j(\bar{x}_i)\bar{q}_j(t), \quad j = 1,2,\ldots,6, \quad x_i \equiv (x_1,x_2,x_3).$$

(14.31)

This decomposition, which is called the Kirchhoff decomposition, is based on the fact that $\bar{n}_j$ is independent of time. Substituting equation (14.31) in equation (14.30), we have

$$\frac{\partial\phi_i(\bar{x}_j)}{\partial n} = \bar{n}_i, \quad \bar{x}_j \in S_b.$$

(14.32)

Note that this body-boundary condition is written for the prescribed *unit velocity* of the body in its ith mode of motion. Some authors prefer to use the body-boundary condition written in terms of the prescribed *unit displacement* of the body rather than the *unit velocity* as used here, and therefore, one should be careful about this.

A further condition on the form of ϕ_i at infinite distances away from the body is necessary. Considering a moving dipole in a 3-D irrotational flow of unbounded extent, one can show that (see equation (11.31))

$$\phi_i = O\left(\frac{1}{\bar{r}_{\bar{O}P}^2}\right) \quad \text{as } \bar{r}_{\bar{O}P} \to \infty, \tag{14.33}$$

which simply states that ϕ_i goes to zero, like $1/\bar{r}_{\bar{O}P}^2$ goes to zero, as $\bar{r}_{\bar{O}P} \to \infty$.

We are now ready to consider the force and moment equations given by equation (14.17) (note that we replace $\boldsymbol{n}$ by $\bar{\boldsymbol{n}} = \bar{n}_i \bar{\boldsymbol{e}}_i$), and use the Kirchoff decomposition, equation (14.31):

$$\boldsymbol{F} = -\rho \frac{\partial}{\partial t} \left\{ \int_{S_b} \bar{\phi}(\bar{x}_j, t) \bar{n}_i(\bar{x}_j) \bar{\boldsymbol{e}}_i(t) dS \right\} = -\rho \frac{\partial}{\partial t} \left\{ \int_{S_b} \phi_k(\bar{x}_j) \bar{q}_k(t) \bar{n}_i(\bar{x}_j) \bar{\boldsymbol{e}}_i ds \right\}$$

$$= -\rho \frac{\partial}{\partial t} \left\{ \bar{q}_k(t) \int_{S_b} \phi_k \bar{n}_i \bar{\boldsymbol{e}}_i dS \right\},$$

or from equation (14.32)

$$\boldsymbol{F} = -\rho \frac{\partial}{\partial t} \left\{ \bar{q}_k(t) \int_{S_b} \phi_k \frac{\partial \phi_i}{\partial n} \bar{\boldsymbol{e}}_i dS \right\}. \tag{14.34}$$

The moment, $\boldsymbol{M}$, can be obtained in a similar way:

$$\boldsymbol{M} = -\rho \frac{\partial}{\partial t} \left\{ \bar{q}_j \bar{x}_{O\bar{O}k} \int_{S_b} \phi_j \frac{\partial \phi_l}{\partial n} (\bar{\boldsymbol{e}}_k \times \bar{\boldsymbol{e}}_l) dS + \bar{q}_j \int_{S_b} \phi_j \frac{\partial \phi_{k+3}}{\partial n} \bar{\boldsymbol{e}}_k dS \right\}. \tag{14.35}$$

Let us now define a second-order tensor, called the added-mass tensor, by

$$\mu_{ij} \equiv \rho \int_{S_b} \phi_j \frac{\partial \phi_i}{\partial n} dS, \quad i, j = 1, 2, \dots, 6, \tag{14.36}$$

and also introduce

$$B_i \equiv \mu_{ij} \bar{q}_j, \quad i, j = 1, 2, \dots, 6, \qquad \boldsymbol{B} \equiv B_i \bar{\boldsymbol{e}}_i, \qquad \boldsymbol{I} \equiv B_{i+3} \bar{\boldsymbol{e}}_i, \quad i = 1, 2, 3. \tag{14.37}$$

Then, equations (14.34) and (14.35) become

$$\boldsymbol{F} = -\frac{\partial}{\partial t} \{ \bar{q}_k \mu_{ik} \bar{\boldsymbol{e}}_i \} = -\frac{d}{dt} \boldsymbol{B},$$
$$\boldsymbol{M} = -\frac{\partial}{\partial t} \left\{ (\bar{x}_{O\bar{O}k} \bar{\boldsymbol{e}}_k) \times (\bar{q}_j \mu_{lj} \bar{\boldsymbol{e}}_l) + \bar{q}_j \mu_{(k+3)j} \bar{\boldsymbol{e}}_k \right\} = -\frac{d}{dt} \{ \boldsymbol{r}_{O\bar{O}} \times \boldsymbol{B} + \boldsymbol{I} \}. \tag{14.38}$$

We have the time derivatives of the vectors $\boldsymbol{B}$ and $\boldsymbol{I}$ in equation (14.38). Let us evaluate these by considering the fact that any vector, say $\boldsymbol{A}$, can be expressed in terms of either set of base vectors, i.e., $\boldsymbol{A} = A_i \boldsymbol{e}_i = \bar{A}_i \bar{\boldsymbol{e}}_i$. The time derivative of the vector $\boldsymbol{A} = \boldsymbol{A}(t)$ is then

$$\frac{d\boldsymbol{A}}{dt} = \frac{d(A_i \boldsymbol{e}_i)}{dt} = \frac{d\bar{A}_i}{dt} \bar{\boldsymbol{e}}_i + \bar{A}_i \frac{d\bar{\boldsymbol{e}}_i}{dt}, \tag{14.39}$$

resolved in either coordinate system. This is true since $e_i \neq f(t)$ but $\bar{e}_i = f(t)$ in the earth-bound coordinate system. However, from equations (14.7) and (14.8), we have

$$\bar{A}_i \frac{d\bar{e}_i}{dt} = \Omega \times A, \quad \Omega = \bar{\gamma}_{23}\bar{e}_1 + \bar{\gamma}_{31}\bar{e}_2 + \bar{\gamma}_{12}\bar{e}_3 = \frac{1}{2}\bar{\gamma}_{ij}\epsilon_{ijk}\bar{e}_k. \tag{14.40}$$

Substituting equation (14.40) in equation (14.39), one obtains

$$\frac{dA}{dt} = \frac{d\bar{A}_i}{dt}\bar{e}_i + \Omega \times (\bar{A}_i\bar{e}_i). \tag{14.41}$$

The first term on the right-hand side of equation (14.41) represents the variation of A relative to the moving coordinate system, and the second term represents the variation of A due to the system (body coordinates) itself rotating (with respect to the inertial coordinate system).

The time derivatives in equation (14.38) can now be evaluated by use of equation (14.41). This results in

$$F = -\frac{dB}{dt} - \Omega \times B,$$

$$M = -\frac{dr_{O\bar{O}}}{dt} \times B - r_{O\bar{O}} \times \left(\frac{dB}{dt} + \Omega \times B\right) - \frac{dI}{dt} - \Omega \times I \tag{14.42}$$

$$= -q_{O\bar{O}} \times B + r_{O\bar{O}} \times F - \frac{dI}{dt} - \Omega \times I.$$

The last equality above is written because we want to take the moment with respect to the moving system, $\bar{O}\bar{x}_1\bar{x}_2\bar{x}_3$, so that both the force and moment in equation (14.42) refer to the moving system. Note also that all quantities in equation (14.42) are now in the body coordinate system.

The dimensions of each element μ_{ij} of the added-mass tensor are different since the dimensions of each $\bar{q}_i$ in equation (14.31) and, therefore, of each ϕ_i are different, i.e.,

$$[\bar{\phi}] = (2,0,-1); \quad [\phi_i] = (1,0,0), i = 1,2,3; \quad [\phi_i] = (2,0,0), i = 4,5,6;$$
$$[\mu_{ij}] = (0,1,0), i,j = 1,2,3 \quad \text{(added mass)};$$
$$[\mu_{ij}] = (1,1,0), i = 1,2,3, j = 4,5,6 \text{ or } i = 4,5,6, j = 1,2,3 \text{ (moment of added mass)};$$
$$[\mu_{ij}] = (2,1,0), i,j = 4,5,6 \text{ (moment of inertia of added mass)},$$
$$\tag{14.43}$$

where [ ] $\equiv (L, M, T)$ notation is used. Also, note that the first subscript i refers to the *direction of the force/moment* and the second subscript j refers to the *direction of the motion*.

In Chapter 13, we have defined briefly the "added mass" in terms of the kinetic energy of the fluid. Let us consider it again. The kinetic energy of the fluid is given by (see equation (13.90))

$$T = \frac{1}{2}\rho \int_V u \cdot u \, dV = \frac{1}{2}\rho \int_{S_b} \bar{\phi}\frac{\partial\bar{\phi}}{\partial\bar{n}} \, dS, \tag{14.44}$$

and by use of the Kirchhoff decomposition equation (14.31), we have

$$T = \frac{1}{2}\rho \int_{S_b} \phi_j \bar{q}_j \frac{\partial \phi_i}{\partial n} \bar{q}_i \, dS = \frac{1}{2}\bar{q}_i \bar{q}_j \rho \int_{S_b} \phi_j \frac{\partial \phi_i}{\partial n} \, dS = \frac{1}{2}\bar{q}_i \bar{q}_j \mu_{ij}. \tag{14.45}$$

Since i and j are dummy indices, we can write equation (14.45) as

$$T = \frac{1}{2}\bar{q}_j \bar{q}_i \mu_{ji} = \frac{1}{2}\bar{q}_i \bar{q}_j \mu_{ij}, \tag{14.46}$$

which shows that:

(i) μ_{ij} must be symmetric, i.e.,

$$\mu_{ij} = \mu_{ji}, \tag{14.47}$$

(ii) μ_{ij} must be the components of a positive definite matrix since $T \geq 0$. However, μ_{ij} itself may be negative in some cases.

The same result, i.e., that μ_{ij} is symmetric, can be shown by applying Green's second identity to ϕ_i, see equation (11.52):

$$\int_V \left(\phi_i \Delta \phi_j - \phi_j \Delta \phi_i \right) \, dV = \int_S \left(\phi_i \frac{\partial \phi_j}{\partial n} - \phi_j \frac{\partial \phi_i}{\partial n} \right) \, dS = 0,$$

since ϕ_i or ϕ_j is harmonic (they satisfy Laplace's equation). Considering μ_{ij}, defined by equation (14.36), equation (14.47) follows immediately. The volume integral given above is zero since ϕ_i satisfies Laplace's equation and the integral over all surfaces, including the control surface at infinity, and it reduces to an integral over the surface of the body only because of the far-field condition equation (14.33).

There are several methods of calculating the added-mass tensor (or matrix) of bodies. If the body has a simple geometry, such as a sphere, ellipsoid, or a circle, it is possible to obtain an analytic solution in an unbounded fluid. If the body geometry is not simple, one can use the Green function method as described in Chapter 11, the boundary-element method, the FEM, or the FDM to determine μ_{ij}. If the problem is two-dimensional, one can use a complex potential and the conformal-mapping method, to map the arbitrary geometry of the body to a simpler one whose added mass coefficients are known in an unbounded fluid. For example, a Joukowski aerofoil can be mapped onto a circle whose added mass coefficients are known. Then one can use the inverse transformation to obtain the added-mass coefficients of a Joukowski aerofoil. Kennard (1967), for example, gives added-mass (or inertia) coefficients for many 2-D and 3-D bodies. Some added-mass coefficients for two-dimensional shapes are (see, e.g., Newman (1978)):

$$m_{11} = m_{22} = \rho \pi a^2 \quad m_{66} = 0, \quad \text{circle of radius } a, \tag{14.48}$$

$$m_{11} = \rho \pi b^2 \quad m_{22} = \rho \pi a^2 \quad m_{66} = \frac{1}{8}\rho \pi (a^2 - b^2)^2,$$

$$\text{ellipse of major axis } a \quad \text{and minor axis } b, \tag{14.49}$$

$$m_{11} = m_{22} = 4.754 \rho \pi a^2 \quad m_{66} = 0.725 \rho a^4, \quad \text{square of each side } 2a, \tag{14.50}$$

14.3   Added Mass and Damping Coefficients in the Presence of a Free Surface

In this section, we will discuss the "small" motions of a body in the presence of a free surface. The body motions will be prescribed as in the unbounded-fluid case discussed in the last section, and we will assume that there is no forward motion. In other words, the body undergoes three translational and three rotational motions about the neighborhood of a fixed reference position. The position of the body when the body does not move is called the "mean" position, and the surface of the body at this mean position is denoted by S_m, as shown in Figure 14.2. We will see in this section that the presence of a free surface will produce the wave-damping coefficients (or simply "the damping coefficients"), in addition to the added-mass coefficients obtained in the last section. Since we are dealing with an inviscid fluid, the damping refers to the damping of the fluid motion due to the creation and subsequent propagation of waves outward.

Let us consider the two coordinate systems shown in Figure 14.2. One is fixed on the still-water surface, $Ox_1x_2x_3$, and the other is fixed somewhere on or inside the body, $\bar{O}\bar{x}_1\bar{x}_2\bar{x}_3$. The origin of the moving coordinate system can be taken conveniently at the center of gravity, G, of the body, i.e., $\bar{O} \equiv G$. When the body is not moving, we have the following relations between the two coordinate systems:

$$\bar{x}_1 = x_1, \qquad \bar{x}_2 = x_2 + d, \qquad \bar{x}_3 = x_3, \tag{14.51}$$

where d refers to the elevation of the center of gravity of the body, and $d > 0$ if O is above $\bar{O}$ and $d < 0$ if O is below $\bar{O}$ (see Figure 14.2). The coordinate system $Ox_1x_2x_3$ can describe conveniently the mean position of the body.

Let us denote the three translational displacements of the center of gravity of the body (with respect to the fixed coordinate system) by x_{G1}, x_{G2}, and x_{G3}, and the three

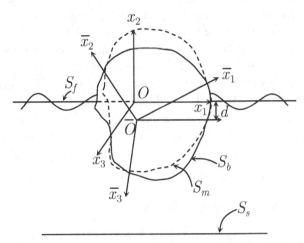

Figure 14.2 Mean and instantaneous positions of a floating body.

rotational displacements of the body (with respect to the moving coordinate system) by θ_1, θ_2, and θ_3. In other words, x_{G1} = surge, x_{G2} = heave, x_{G3} = sway, θ_1 = roll, θ_2 = yaw, and θ_3 = pitch. It can be shown that the relations between the fixed and moving coordinates is given by

$$x_1 = x_{G1} + \bar{x}_1 \cos\theta_2 \cos\theta_3 - \bar{x}_2 \cos\theta_2 \sin\theta_3 + \bar{x}_3 \sin\theta_2,$$
$$x_2 = x_{G2} + \bar{x}_1(\sin\theta_1 \sin\theta_2 \cos\theta_3 + \cos\theta_1 \sin\theta_3)$$
$$+ \bar{x}_2(-\sin\theta_1 \sin\theta_2 \sin\theta_3 + \cos\theta_1 \cos\theta_3) - \bar{x}_3 \sin\theta_1 \cos\theta_2,$$
$$x_3 = x_{G3} + \bar{x}_1(-\cos\theta_1 \sin\theta_2 \cos\theta_3 + \sin\theta_1 \sin\theta_3)$$
$$+ \bar{x}_2(\cos\theta_1 \sin\theta_2 \sin\theta_3 + \sin\theta_1 \cos\theta_3) + \bar{x}_3 \cos\theta_1 \cos\theta_2.$$

or

$$\{x_i\} = \{x_{Gi}\} + [Q]\{\bar{x}_i\}, \tag{14.52}$$

where the components of the 3×3 transformation matrix $[Q]$ can be seen above. The relations given by equation (14.52) can be obtained by first successively displacing the body parallel to the $\bar{x}_1, \bar{x}_2$, and $\bar{x}_3$ axes, and then rotating the body, again successively, about the $\bar{x}_1, \bar{x}_2$ and $\bar{x}_3$ axes. The angles, $\theta_1, \theta_2, \theta_3$, are called the Eulerian angles. The equation (14.52) is exact, i.e., no assumption on the magnitude of θ_i, $i = 1,2,3$, has been made.

Now, if we assume that x_{Gi} and θ_i are "small," so that any term which includes powers of these variables higher than the first one is discarded, equation (14.52) takes its linearized form:

$$(x_1,x_2,x_3) = (x_{G1},x_{G2},x_{G3}) + (\bar{x}_1,\bar{x}_2,\bar{x}_3) + (\theta_1,\theta_2,\theta_3) \times (\bar{x}_1,\bar{x}_2,\bar{x}_3)$$

or

$$x_i = x_{Gi} + \bar{x}_i + \epsilon_{ijk}\theta_j\bar{x}_k. \tag{14.53}$$

The inverse relationship can be obtained by noting that the transformation is orthogonal:

$$\bar{x}_i = -x_{Gi} + x_i - \epsilon_{ijk}\theta_j x_k. \tag{14.54}$$

The body-boundary condition then becomes

$$\frac{\partial\phi}{\partial n} = q \cdot n = \frac{dx_{Gi}}{dt}n_i + \epsilon_{ijk}\frac{d\theta_j}{dt}\bar{x}_k n_i, \quad \text{where } q_i = \frac{dx_{Gi}}{dt} + \epsilon_{ijk}\frac{d\theta_j}{dt}, \tag{14.55}$$

and since $d\bar{x}_i/dt = 0$, being measured in the moving-system attached to the body.

The boundary condition on the sea floor is

$$\frac{\partial\phi}{\partial n} = 0, \quad x_i \in S_s. \tag{14.56}$$

Since the motion of the body is small by assumption, we expect that the resulting waves generated by the body are also small. Hence, we can use the linearized (combined) free-surface condition given by (see equation (12.36) in two dimensions)

$$\phi_{tt}(x_1,0,x_3,t) + g\phi_{x_2} = 0, \quad x_i \in S_f, \tag{14.57}$$

again, the subscripts denote differentiation.

Recalling the Kirchhoff decomposition equations (14.31) and noting the body-boundary condition (14.55), we can write the Kirchhoff decomposition as

$$\bar{\phi}(\bar{x}_1, \bar{x}_2, \bar{x}_3, t) = q_i(t)\, \phi_i(\bar{x}_1, \bar{x}_2, \bar{x}_3) = \frac{dx_{Gi}}{dt}\phi_i + \frac{d\theta_i}{dt}\phi_{i+3}. \tag{14.58}$$

Therefore, from equation (14.55), equations (14.56) and (14.58) we must have

$$\frac{\partial \phi_i}{\partial n} = n_i \quad \text{if } x_i \in S_m; \qquad \frac{\partial \phi_i}{\partial n} = 0, \quad \text{if } x_i \in S_s. \tag{14.59}$$

Obviously, equation (14.58) must also satisfy the free-surface condition, equation (14.57). In other words, $\ddot{q}_i\phi_i + gq_i\partial\phi_i/\partial x_2 = 0$ for each $i = 1,2,\ldots,6$. But since $q_i = f(t)$ and $\phi_i \neq f(t)$, we must have

$$\frac{\ddot{q}_i(t)}{q_i(t)} = \frac{-g\phi_{ix_2}(x_j)}{\phi_i(x_j)} = \text{constant} = -\omega^2, \tag{14.60}$$

so that

$$\phi_i(x_1, 0, x_3) - \frac{g}{\omega^2}\frac{\partial \phi_i}{\partial x_2} = 0. \tag{14.61}$$

This means that we must have a simple harmonic motion whose frequency is ω in each mode. The angular frequency, ω, must therefore be the same for each mode of motion.

In addition to satisfying the boundary conditions, equations (14.59) and (14.61), ϕ_i must also satisfy the radiation condition discussed in Chapter 13. The Kirchoff decomposition equation (14.58), as it stands, is inadequate to satisfy the radiation condition. However, it can be modified in such a way that all the derivations carried out above still remain valid. To do this, assume that we can decompose the potential as follows:

$$\phi(x_1, x_2, x_3, t) = q_j(t)\phi_j^{(R)}(x_1, x_2, x_3) + \omega x_j(t)\phi_j^{(I)}(x_1, x_2, x_3), \tag{14.62}$$

where $q_j = dx_j/dt, j = 1,2,\ldots,6$ and $\phi_j^{(R)}$ and $\phi_j^{(I)}$ are the real and imaginary parts, respectively, of the velocity potential ϕ_j, i.e.,

$$\phi_j = \phi_j^{(R)} + i\phi_j^{(I)}, \quad i = \sqrt{-1}, \quad j=1,2,\ldots,6. \tag{14.63}$$

Clearly, we must have

$$\frac{\partial \phi_i^{(R)}}{\partial n} = n_i, \quad \frac{\partial \phi_i^{(I)}}{\partial n} = 0, \quad x_i \in S_m; \qquad \frac{\partial \phi_i^{(R)}}{\partial n} = \frac{\partial \phi_i^{(I)}}{\partial n} = 0, \quad x_i \in S_s. \tag{14.64}$$

Also, the radiation condition is given by equation (13.9) for the radiation problem we have here:

$$\lim_{r \to \infty} \left\{ \sqrt{r}\left[\frac{\partial}{\partial r} - ik\right]\phi \right\} = 0, \quad i = \sqrt{-1}, \tag{14.65}$$

where $r = \sqrt{x_1^2 + x_3^2}$. As a result, we must have

$$\lim_{r \to \infty} \left\{ \sqrt{r}\left[\frac{\partial \phi_j}{\partial r} - ik\phi_j\right] \right\} = 0, \tag{14.66}$$

which shows that $\phi_j = O(1/\sqrt{r})$ as $r \to \infty$.

Let us now go back to the decomposition equation (14.62) and note that we have a simple harmonic motion with frequency ω. This allows us to write the six displacements of the body in the following form:

$$x_j = |x_j|e^{-i\omega t}, \quad i = \sqrt{-1}, \quad j = 1, 2, \ldots, 6, \tag{14.67}$$

where $|x_j|$ is the magnitude (or amplitude) of the jth displacement. Therefore, one can show that

$$\phi(x_i) = q_j \phi_j = q_j(\phi_j^{(R)} + i\phi_j^{(I)})$$

$$= q_j \phi_j^{(R)} + i\frac{dx_j}{dt}\phi_j^{(I)} = q_j \phi_j^{(R)} + \omega x_j \phi_j^{(I)},$$

where $i = \sqrt{-1}$, $j = 1, 2, \ldots, 6$, which is nothing but equation (14.62). Therefore, equation (14.62) also satisfies the free-surface condition equation (14.61) and the radiation condition equation (14.66) provided that the dispersion relation given by equation (12.16) is also satisfied.

We are now ready to calculate the forces and moments acting on a body which makes the prescribed motions. We can write the forces and moments, by use of a compact notation, as

$$F_i = -\rho \int_{S_m} \frac{\partial \phi}{\partial t} n_i dS, \quad i = 1, 2, \ldots, 6, \tag{14.68}$$

where n_i were given by equation (14.29) for $i = 4, 5, 6$. Therefore, by use of equations (14.62) and (14.64), we have

$$F_i = -\frac{dq_j}{dt}\rho \int_{S_m} \phi_j^{(R)} \frac{\partial \phi_i^{(R)}}{\partial n} dS - \omega q_j \rho \int_{S_m} \phi_j^{(I)} \frac{\partial \phi_i^{(R)}}{\partial n} dS,$$

or, recalling μ_{ij} from equation (14.36), and introducing λ_{ij},

$$F_i = -\mu_{ij}\frac{dq_j}{dt} - \lambda_{ij}q_j, \quad i, j = 1, 2, \ldots, 6, \tag{14.69}$$

where

$$\mu_{ij} = \rho \int_{S_m} \phi_j^{(R)} \frac{\partial \phi_i^{(R)}}{\partial n} dS, \quad \lambda_{ij} = \rho\omega \int_{S_m} \phi_j^{(I)} \frac{\partial \phi_i^{(R)}}{\partial n} dS. \tag{14.70}$$

The components of the second-order tensor μ_{ij} are called the added-mass coefficients as before, and λ_{ij} are called the wave-damping coefficients or simply, damping coefficients. Note that, in the unbounded-fluid case, there is no free surface and, as a result, there is no wave generation. And therefore, the damping coefficients do not exist since no energy is carried by waves toward infinity. See, for example, Ertekin et al. (1993) for some added-mass and damping coefficients for a floating structure of a semisubmersible type.

By considering a control volume enclosed by the free surface, the body surface, sea floor, and an imaginary cylinder at $r = \infty$, one can show that $\mu_{ij} = \mu_{ji}$ and $\lambda_{ij} = \lambda_{ji}$

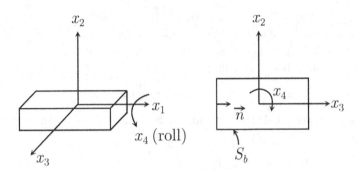

Figure 14.3 A 3-D pontoon which is symmetric with respect to the x_1–x_2 plane.

by use of Green's second identity as mentioned in Section 14.2. Note that these results are valid if there is no forward motion. It can also be shown that the average rate at which work is done by the body upon the fluid is directly proportional to λ_{ij}, and, therefore, that the matrix λ_{ij} must be positive definite for all ω. In particular, $\lambda_{ii} > 0$ for all $i = 1, 2, \ldots, 6$ and ω, but in some cases $\lambda_{ij}, i \neq j$ may be negative (see, e.g., Wehausen (1971)).

Some of the hydrodynamic coefficients may vanish if there are one or more symmetry planes in the body geometry. As an example, let us consider a 3-D pontoon which is symmetric with respect to the $x_1 - x_2$ plane as shown in Figure 14.3. We can write the components of the body normal vector $\boldsymbol{n}$ as

$$
\begin{aligned}
n_2(x_1, x_2, x_3) &= n_2(x_1, x_2, -x_3), \\
n_3(x_1, x_2, x_3) &= -n_3(x_1, x_2, -x_3), \\
n_4(x_1, x_2, x_3) &= x_2 n_3 - x_3 n_2 = -n_4(x_1, x_2, -x_3),
\end{aligned}
\tag{14.71}
$$

since $\boldsymbol{n}^{2D} = n_2 \boldsymbol{e}_2 + n_3 \boldsymbol{e}_3$ and $n_4 = (\boldsymbol{r} \times \boldsymbol{n}^{2D}) \cdot \boldsymbol{e}_1$, where $\boldsymbol{n}^{2D}$ is the 2-D normal vector in the $x_2 - x_3$ plane, see equation (14.29). In other words, n_2 is an even function of x_3, whereas n_3 and n_4 are odd functions of x_3. If we now consider the body-boundary condition, $\partial \phi_i / \partial n = n_i$, then we must have the following:

$$
\begin{aligned}
\frac{\partial \phi_2(x_1, x_2, x_3)}{\partial n} &= n_2(x_1, x_2, x_3) = n_2(x_1, x_2, -x_3) = \frac{\partial \phi_2}{\partial n}(x_1, x_2, -x_3), \\
\frac{\partial \phi_3(x_1, x_2, x_3)}{\partial n} &= n_3(x_1, x_2, x_3) = -n_3(x_1, x_2, -x_3) = -\frac{\partial \phi_3}{\partial n}(x_1, x_2, -x_3), \\
\frac{\partial \phi_4}{\partial n}(x_1, x_2, x_3) &= n_4(x_1, x_2, x_3) = -n_4(x_1, x_2, -x_3) = -\frac{\partial \phi_4}{\partial n}(x_1, x_2, -x_3).
\end{aligned}
\tag{14.72}
$$

Therefore, ϕ_2 is an even function of x_3, whereas ϕ_3 and ϕ_4 are odd functions of x_3.

Let us write the added-mass coefficients as

$$
\mu_{ij} = \rho \int_{S_b} \phi_j \frac{\partial \phi_i}{\partial n} dS = \rho \int_{S_b} \phi_j n_i \, dS,
\tag{14.73}
$$

and let $S_b = S_{b+} \cup S_{b-}$, $S_{b+} = S_{b+}(x_2, x_3)$, $S_{b-} = S_{b-}(x_2, -x_3)$, i.e., S_{b+} denotes the starboard-side surface of the body and S_{b-} denotes the port-side surface of the body. Then equation (14.73) can be written as

$$\mu_{ij} = \rho \left\{ \int_{S_{b+}} \phi_j n_i dS + \int_{S_{b-}} \phi_j n_i dS \right\}. \tag{14.74}$$

Recall that i refers to the direction of the force (or moment) and j refers to the direction of the motion (translational or rotational).

Let $i = 2$ refer to the heave force and $j = 3$ refer to the sway motion. Equation (14.74) becomes

$$\mu_{23} = \rho \left\{ \int_{S_{b+}} \phi_3(x_1, x_2, x_3) n_2(x_1, x_2, x_3) dS + \int_{S_{b-}} \phi_3(x_1, x_2, -x_3) n_2(x_1, x_2, -x_3) dS \right\}$$

$$= \rho \left\{ \int_{S_{b-}} -\phi_3(x_1, x_2, -x_3) n_2(x_1, x_2, -x_3) dS + \int_{S_{b-}} \phi_3(x_1, x_2, -x_3) n_2(x_1, x_2, -x_3) dS \right\}$$

$$= 0 = \mu_{32}.$$

Therefore, heave and sway are uncoupled, meaning the sway motion does not cause any heave force, and vice versa. Note that, in the second step above, the integral over S_{b+} is replaced by the integral over S_{b-}. Note that these results hold because of the symmetry of the body with respect to the x_1–x_2 plane.

Let us next consider $i = 2$ (heave) and $j = 4$ (roll) case. We have

$$\mu_{24} = \rho \left\{ \int_{S_{b-}} -\phi_4(x_1, x_2, -x_3) n_2(x_1, x_2, -x_3) dS \right.$$

$$\left. + \int_{S_{b-}} \phi_4(x_1, x_2, -x_3) n_2(x_1, x_2, -x_3) dS \right\}$$

$$= 0 = \mu_{42}.$$

Therefore, heave and roll are also uncoupled, meaning the roll motion does not cause heave force, and vice versa.

Finally, let us consider $i = 3$ (sway) and $j = 4$ (roll). We have

$$\mu_{34} = \rho \left\{ \int_{S_{b-}} -\phi_4(x_1, x_2, -x_3)[-n_3(x_1, x_2, -x_3)] dS \right.$$

$$\left. + \int_{S_{b-}} \phi_4(x_1, x_2, -x_3) n_3(x_1, x_2, -x_3) dS \right\}$$

$$= 2\rho \int_{S_{b-}} \phi_4(x_1, x_2, -x_3) n_3(x_1, x_2, -x_3) dS$$

$$\neq 0 \Rightarrow \mu_{34} = \mu_{43} \neq 0.$$

Therefore, sway and roll are coupled, meaning the roll motion causes sway force, and vice versa.

It is noted that in cases where there is one or more symmetry planes (e.g., there may also be symmetry with respect to the $x_2 - x_3$ plane, i.e., fore-aft symmetry), one can use the fact that certain added-mass coefficients must vanish to check the accuracy of computations in a computer program which is based on, for example, the Green function method. As an example, if the calculated μ_{24} is not very close to zero, when there is starboard-port symmetry, then either there is a programming error or the panel (or facet) size is too large, and thus, numerical inaccuracies are significant.

14.4   Haskind–Hanaoka Relationship

There is a remarkable relationship between the diffraction and radiation potentials such that the exciting force and moment on a body can be obtained without knowing the diffraction potential, ϕ_D, if the radiation potentials, ϕ_i, are known. Let us consider this relation by recalling the diffraction forces and moments from Chapter 13:

$$F = -\rho \int_{S_b} \left(\frac{\partial \Phi_I}{\partial t} + \frac{\partial \Phi_D}{\partial t} \right) n \, dS, \quad M = -\rho \int_{S_b} \left(\frac{\partial \Phi_I}{\partial t} + \frac{\partial \Phi_D}{\partial t} \right) (r_{\bar{O}P} \times n) \, dS.$$

$$(14.75)$$

Because the fluid-particle motion is harmonic (with an angular frequency ω) everywhere, we can write

$$\Phi_I = \mathcal{R}\{\phi_I(x_j)e^{-i\omega t}\}, \quad \Phi_D = \mathcal{R}\{\phi_D(x_j)e^{-i\omega t}\}, \quad i = \sqrt{-1}, \quad (14.76)$$

and substitute these in equation (14.75), to obtain

$$F = \mathcal{R}\left\{ i\rho\omega e^{-i\omega t} \int_{S_b} (\phi_I + \phi_D)n \, dS \right\},$$

$$(14.77)$$

$$M = \mathcal{R}\left\{ i\rho\omega e^{-i\omega t} \int_{S_b} (\phi_I + \phi_D)(r_{\bar{O}P} \times n) \, dS \right\},$$

where $\mathcal{R}$ indicates the real part.

By use of equation (14.29), the above two equations can conveniently be written as a single indicial equation:

$$F_j = \mathcal{R}\left\{ i\rho\omega e^{-i\omega t} \int_{S_b} (\phi_I + \phi_D)n_j \, dS \right\}, \quad i = \sqrt{-1}, \quad j = 1,2,\ldots,6, \quad (14.78)$$

where n_j, $j = 4,5,6$, are given by equation (14.29) and n_j, $j = 1,2,3$, are the usual components of the unit normal vector on the mean body surface. Note that by considering the integral in equation (14.78) that involves ϕ_D, and the body-boundary condition in equation (14.59), we can write

$$\int_{S_b} \phi_D n_j \, dS = \int_{S_b} \phi_D \frac{\partial \phi_j}{\partial n} \, dS. \quad (14.79)$$

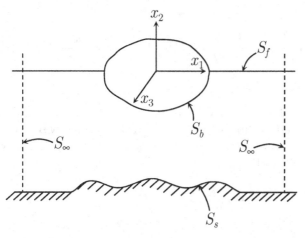

Figure 14.4 A control volume in the fluid, bounded by the body, still-water surface, sea floor, and a large control cylinder of radius R.

Next, let us consider Figure 14.4, which shows a control volume in the fluid, bounded by the body, still-water surface, sea floor, and a large control cylinder of radius R. Since both ϕ_D and ϕ_j are harmonic, and they satisfy Laplace's equation, Green's second identity, equation (11.52), gives

$$0 = \int_{S_b \cup S_f \cup S_s \cup S_\infty} \left(\phi_D \frac{\partial \phi_j}{\partial n} - \phi_j \frac{\partial \phi_D}{\partial n} \right) dS. \tag{14.80}$$

However,

$$\int_{S_f} \left(\phi_D \frac{\partial \phi_j}{\partial n} - \phi_j \frac{\partial \phi_D}{\partial n} \right) dS = \int_{S_f} \left(\phi_D \frac{\partial \phi_j}{\partial x_2} - \phi_j \frac{\partial \phi_D}{\partial x_2} \right) dS$$

$$= \int_{S_f} \left(\phi_D \frac{\omega^2}{g} \phi_j - \phi_j \frac{\omega^2}{g} \phi_D \right) dS = 0$$

because both the diffraction and radiation potentials satisfy the linearized free-surface condition equation (14.57) applied on the SWL.

As far as the radiation boundary, S_∞, goes, it is left as an exercise to show that

$$\int_{S_\infty} \left(\phi_D \frac{\partial \phi_j}{\partial n} - \phi_j \frac{\partial \phi_D}{\partial n} \right) dS = \lim_{R \to \infty} \int_{-h}^{0} dx_2 \int_{0}^{2\pi} R \left(\phi_D \frac{\partial \phi_j}{\partial r} - \phi_j \frac{\partial \phi_D}{\partial r} \right) d\theta = 0,$$

when one uses the fact that both ϕ_D and ϕ_j satisfy the radiation condition equation (13.9).

As a result, equation (14.80) becomes

$$0 = \int\limits_{S_b \cup S_s} \left(\phi_D \frac{\partial \phi_j}{\partial n} - \phi_j \frac{\partial \phi_D}{\partial n} \right) dS = \int\limits_{S_b} \left(\phi_D \frac{\partial \phi_j}{\partial n} - \phi_j \frac{\partial \phi_D}{\partial n} \right) dS + \int\limits_{S_s} \phi_j \frac{\partial \phi_I}{\partial n} dS,$$

(14.81)

since $\partial \phi_j / \partial n = 0$ if $x_j \in S_s$ and $\partial \phi_D / \partial n = -\partial \phi_I / \partial n$ on $x_j \in S_s$. The sea-floor condition is the result of $\partial \Phi_T / \partial n = 0$ on $x_j \in S_s$ for an arbitrary sea-floor geometry. If the sea floor is flat, obviously $\partial \phi_I / \partial n = \partial \phi_D / \partial n = 0$ will result. We can now rearrange equation (14.81) and write it as

$$\int\limits_{S_b} \phi_D \frac{\partial \phi_j}{\partial n} dS = - \int\limits_{S_b \cup S_s} \phi_j \frac{\partial \phi_I}{\partial n} dS,$$

(14.82)

because $\partial \phi_D / \partial n = -\partial \phi_I / \partial n$ on $x_j \in S_b$ also.

Finally then, equation (14.82) can be used in the force and moment integrals given by equation (14.78) to write

$$F_j = \Re \left\{ i \rho \omega e^{-i\omega t} \int\limits_{S_b} (\phi_I + \phi_D) n_j dS \right\}$$

$$= \Re \left\{ i \rho \omega e^{-i\omega t} \int\limits_{S_b \cup S_s} \left(\phi_I \frac{\partial \phi_j}{\partial n} - \phi_j \frac{\partial \phi_I}{\partial n} \right) dS \right\},$$

(14.83)

since $\partial \phi_j / \partial n = 0$ on $x_j \in S_s$. One can also apply Green's identity to the integral in equation (14.83) (by noting that the radiation condition is not applicable to the incoming potential) to obtain an equivalent expression for the diffraction forces and moments:

$$F_j = -\Re \left\{ i \rho \omega e^{-i\omega t} \int\limits_{S_\infty} \left(\phi_I \frac{\partial \phi_j}{\partial n} - \phi_j \frac{\partial \phi_I}{\partial n} \right) dS \right\}, \quad i = \sqrt{-1}, \quad j = 1,2,\ldots,6.$$

(14.84)

Equation (14.83) or equation (14.84) is known as the Haskind–Hanaoka formula or Haskind–Hanaoka relationship that gives the diffraction forces and moments by use of the radiation and incident potentials only. However, if one is interested in other physical quantities, e.g., the scattered surface elevation due to the diffraction potential or the pressure distribution on the body, then one must solve for ϕ_D explicitly. For elastic (or flexible) bodies, the Haskind–Hanaoka formula can also be proven (see Ertekin et al. (1995)).

Note that if the sea floor is flat or the water depth is infinite then equation (14.83) becomes

$$F_j = \Re \left\{ i \rho \omega e^{-i\omega t} \int\limits_{S_b} \left(\phi_I \frac{\partial \phi_j}{\partial n} - \phi_j \frac{\partial \phi_I}{\partial n} \right) dS \right\},$$

(14.85)

where the integral is now over the body surface only. Note that the accuracy of the diffraction forces calculated by determining ϕ_D and by use of equation (14.75) can be checked by use of equation (14.85) in a computer program if the radiation potentials, ϕ_j, have already been calculated.

14.5   Freely Floating Bodies

Let us first summarize the problem of a freely floating body in the *absence* of forward motion, but in the presence of incoming linear waves. We assume that the body makes "small" motions in six degrees of freedom, and that they can be written as

$$x_j = x_j^0 e^{-i\omega t}, \quad i = \sqrt{-1}, \quad j = 1, 2, \ldots, 6. \tag{14.86}$$

Each x_j, $j = 1,2,3$, refers to the translational displacements of surge, heave, and sway, respectively, and each x_j, $j = 4,5,6$, refers to the angular displacements (or rotations) of roll, yaw, and pitch, respectively, and x_j^0 denotes the complex amplitude of the motion (see also equation (14.67)).

The complex total potential due to the interaction of waves with the body can be written in a compact form as

$$\Phi_T = \sum_{j=0}^{7} \phi_j(x_1, x_2, x_3) e^{-i\omega t}, \tag{14.87}$$

where $\phi_0 \equiv \phi_I$ is the incoming potential, ϕ_j, $j = 1, 2, \ldots, 6$, are the radiation potentials, and $\phi_7 \equiv \phi_D$ is the diffraction potential, all being complex functions of the independent spatial variables.

Each potential in equation (14.87) must satisfy

$$\nabla^2 \phi_j(x_k) = 0, \quad x_k \in D,$$

$$\frac{\partial \phi_j(x_k)}{\partial x_2} - \frac{\omega^2}{g} \phi_j(x_k) = 0, \quad x_k \in S_f, \tag{14.88}$$

$$\frac{\partial \phi_j(x_k)}{\partial x_2} = 0, \quad x_k \in S_s,$$

for $j = 0, 1, 2, \ldots, 7$ and $x_k \equiv (x_1, x_2, x_3)$, see Figure 14.4. The sea-floor condition (as used here) implies that the water depth is constant (although this is not a necessary assumption in general).

In addition, we must have the following body-boundary conditions to be satisfied:

$$\frac{\partial \phi_j(x_k)}{\partial n} = n_j, \, j = 1, 2, \ldots, 6, \, x_k \in S_m; \qquad \frac{\partial \phi_7(x_k)}{\partial n} = -\frac{\partial \phi_0(x_k)}{\partial n}, \, x_k \in S_m. \tag{14.89}$$

And all ϕ_j, $j = 1, 2, 3, \ldots 7$, except the incident wave potential, must also satisfy the Sommerfeld radiation condition (see equation (13.9)).

The total pressure can be obtained from linearized Euler's integral:

$$p_T = -\rho g x_2 - \rho \frac{\partial \Phi_T}{\partial t}; \tag{14.90}$$

and the total force ($j = 1, 2, 3$) and moment ($j = 4, 5, 6$) are obtained from

$$F_{Tj} = - \int_{S_m} p_T \, n_j \, dS, \quad j = 1, 2, \ldots, 6. \tag{14.91}$$

The hydrostatic forces and moments due to the first term on the right-hand side of equation (14.90) can be written as

$$F_{Si} = -k_{ij}^{(S)} x_j, \quad i, j = 1, 2, \ldots, 6, \tag{14.92}$$

where $k_{ij}^{(S)}$ are the hydrostatic stiffness (or restoring) coefficients, e.g., $k_{22}^{(s)} = \rho g A_{WP}$ is the restoring coefficient in heave, where A_{WP} denotes the water-plane area of the body. Because there is no restoration in the horizontal plane, it is clear that $k_{ij}^{(s)} = 0$ if $i, j = 1, 3, 5$. Also, note that in a linear system and for a rigid body, $k_{ij}^{(S)} = k_{ji}^{(S)}$, i.e., the hydrostatic stiffness matrix (or tensor) is symmetric. For elastic (or deformable) bodies, this can also be shown (see Huang and Riggs (2000)).

The total forces and moments given by equation (14.91) include the hydrostatic, wave-exciting and radiation forces and moments. The wave-exciting forces can be written as

$$F_{Wj} = -\rho e^{-i\omega t} \int_{S_m} \left(\frac{\partial \phi_0}{\partial t} + \frac{\partial \phi_7}{\partial t} \right) n_j dS = A E_j e^{-i\omega t}, \quad j = 1, 2, \ldots, 6, \tag{14.93}$$

where E_j is the complex amplitude of the exciting force divided by the wave amplitude A. Note that equation (14.93) includes both the Froude–Krylov force ($\partial \phi_0 / \partial t$ term) and the scattering force ($\partial \phi_7 / \partial t$ term).

The hydrodynamic (or radiation) forces due to the motion of the body in each mode have been determined before by equation (14.69), i.e.,

$$F_{Ri} = -\mu_{ij} \ddot{x}_j - \lambda_{ij} \dot{x}_j, \quad i, j = 1, 2, \ldots, 6. \tag{14.94}$$

There may be mooring lines that are attached to the body to keep it in location. These mooring lines tend to restore the motion of the body, and thus, can be treated as restoring coefficients:

$$F_{Mi} = -k_{ij}^{(M)} x_j, \quad i, j = 1, 2, \ldots, 6. \tag{14.95}$$

And they can provide restoring (however small) in all modes of motion unlike the hydrostatic restoring.

We can now assemble the forces and moments to obtain

$$F_{Ti} = -(k_{ij}^{(S)} + k_{ij}^{(M)}) x_j - \mu_{ij} \ddot{x}_j - \lambda_{ij} \dot{x}_j + A E_i e^{-i\omega t}, \tag{14.96}$$

where we have also included the mooring line loads in equation (14.96) (they were not included in equation (14.91)).

We are now ready to consider Newton's equations which govern the motions of a body. These equations can be written as

$$F_{Ti} = m_{ij} \ddot{x}_j, \quad i, j = 1, 2, \ldots, 6, \tag{14.97}$$

where

$$m_{11} = m_{22} = m_{33} = m, \; m_{44} = I_1, \; m_{55} = I_2, \; m_{66} = I_3, \; m_{54} = m_{45} = -I_{12},$$

$$m_{46} = m_{64} = -I_{13}, \; m_{56} = m_{65} = -I_{23},$$

$$I_{jk} = \int_V \bar{x}_j \bar{x}_k \rho_B \, dV, \; I_1 = I_{22} + I_{33}, \; I_2 = I_{11} + I_{33}, \; I_3 = I_{11} + I_{22}, \quad (14.98)$$

where any unspecified mass or mass moment, m_{ij}, is zero provided that F_{Tj} refer to a coordinate system which is fixed in the mean position, S_m, of the body, and whose origin coincides with the center of gravity of the body. Here, ρ_B is the mass density of the body and I_{ij} are called the moment of inertia coefficients when $i = j$, and products of inertia when $i \neq j$ (see, e.g., Goldstein (1980)).

We can now set equation (14.96) equal to equation (14.97) and arrange the equation by moving some terms around to obtain

$$(m_{ij} + \mu_{ij})\ddot{x}_j + \lambda_{ij}\dot{x}_j + (k_{ij}^{(S)} + k_{ij}^{(M)})x_j = AE_i e^{-i\omega t}. \quad (14.99)$$

Considering equation (14.86), we can write the equations of motion given by equation (14.99) as

$$\frac{x_j^0}{A} = \{-\omega^2(m_{ij} + \mu_{ij}) - i\omega\lambda_{ij} + (k_{ij}^{(S)} + k_{ij}^{(M)})\}^{-1} E_i. \quad (14.100)$$

Recalling that x_j^0 is complex, we can write it as $x_j^0 = x_j^{0R} + ix_j^{0I} = |x_j| e^{-i\delta}$, where δ here is the phase angle of the motion relative to the incoming-wave crest. As a result, equation (14.100) becomes two sets of simultaneous 6×6 linear equations to be solved for the motion response, $|x_j^0|/A$. These motion responses are commonly called transfer functions. The square of a transfer function is sometimes known as the response amplitude operator (RAO).[1] RAOs are used in irregular-sea analysis to determine the random or stochastic response of a floating body in random waves and they will be discussed in Chapter 15. It is also noted that the term "transfer function" is also used for wave forces or moments per unit wave amplitude, A.

14.6   Concluding Remarks

It was mentioned in Chapter 13 that the problem of the interaction of a body with waves can be decomposed into a series of boundary-value problems once the assumption of linearity is made within the confines of the potential theory. In the case of a fixed obstruction, be it on the sea floor, submerged, or on the free surface, the total potential is decomposed into an incoming potential and a diffraction (or scattering) potential.

If, in addition, the body is oscillating harmonically, on or below the free surface, then an additional potential, called the radiation potential, must be determined. This

---

[1] But it is not uncommon that a transfer function, itself, is called an RAO, rather than its square.

potential basically represents the disturbance in the flow field due to the motion of the body. Since the diffraction problem already included the incoming and scattering potentials, the radiation potential is obtained by prescribing the motion of the body in each of the six degrees of freedom, one at a time, for the unit displacement or velocity of the body. In other words, if we can determine the potential (and therefore, the particle velocities and/or the pressure) for the unit displacement of the body, then the force that acts on the body due to its motion for arbitrary displacements can be written as the displacement times the force per unit displacement.

To determine the radiation potential, we need to review the kinematics of the rigid-body motion first. This is done in Section 7.1.2, where it is shown that the velocity of a body, equation (14.13), must be expressed as the sum of two vectors: (i) the rate of change of the position vector in time, because the body origin translates with respect to an earth-bound origin, and (ii) the rate of change of body unit base vectors, because the body-fixed coordinate system rotates with respect to the earth-bound coordinates (see equation (14.7)).

First, the unbounded-fluid case is considered to determine the forces and moments acting on a moving body. For convenience, the forces and moments are resolved in the body-fixed coordinate system. Expressing the absolute velocity in terms of the body coordinate system and by use of the body-boundary condition, it is shown that the velocity potential can be written as the product of two functions, one being dependent on time only and the other being dependent on the spatial coordinates only. The use of this decomposition, called the Kirchhoff decomposition, in the force and moment integrals, allowed us to define a second-order tensor called the added-mass tensor. It is shown that the added-mass tensor must be a symmetric tensor.

In the case of a body moving on or below the free surface, in addition to the body-boundary condition, the potential must also satisfy the free-surface and radiation conditions. The original Kirchhoff decomposition used in the unbounded-fluid case must now be modified to satisfy the radiation condition. When this is done, and the radiation potentials that are decomposed into their real and imaginary parts are substituted in the force and moment integrals, two distinct terms appear: one being proportional to the accelerations and the other being proportional to the velocities. The acceleration term is the added-mass term as before and the velocity term is called the (wave-making) damping term, given in terms of the second-order damping tensor. It can be shown that the damping coefficients are components of a symmetric tensor and that the diagonal components must be positive since (mechanical) energy input into the fluid is positive.

It is possible to obtain the scattering or diffraction force on a body (whether fixed or floating) without knowing the diffraction potential, by means of a remarkable formula called the Haskind–Hanaoka formula. This formula is discussed in Section 14.4. As a result of this formula, one can determine the diffraction forces and moments by use of the incoming and radiation potentials only. However, if one needs the diffraction pressure or particle velocities, then it is necessary that one fully solve the diffraction problem formulated in Chapter 13 to obtain the diffraction potential explicitly.

Finally, the equations of motion, for a rigid-body having six degrees of freedom, are formulated in Section 14.5. The linear set of equations, once solved, provides the three translational and three rotational displacements per unit wave amplitude. These responses are called the transfer functions or sometimes the response amplitude operators. The transfer functions can be used in an irregular-sea analysis (see Chapter 15) to determine the response of a body in a random-sea environment, provided that the linearity of the system (fluid and body, and the interaction between them) is preserved.

14.7   Self-Assessment

14.7.1

A three-dimensional body in an unbounded, ideal fluid has port-starboard symmetry ($\bar{x}_1 - \bar{x}_2$ plane is the symmetry plane). It is moving with a constant speed, $\bar{q}_1 = U$, along the $\bar{x}_1$ direction. At the same time, it rotates about the $\bar{x}_1$ axis with a velocity $\bar{q}_4 =$ constant. Determine the forces on the body acting in the $\bar{x}_1$, $\bar{x}_2$, and $\bar{x}_3$ directions. Comment on the results.

Notes:

$$F = -\frac{dB}{dt} - \Omega \times B, \quad B_i = \mu_{ij}\bar{q}_j, \ i = 1,2,3, \ j = 1,2,\ldots,6, \quad \Omega = \bar{q}_j \bar{e}_{j-3}, \ j = 4,5,6.$$

14.7.2

Consider a dipole and unsteady flow $U(t)$ which represents a flow around a stationary sphere in an unbounded fluid. The fluid is incompressible and inviscid, and the flow is irrotational. To obtain the case of a sphere moving in the negative x_1 direction in an otherwise calm fluid, add negative unsteady flow to the potential.

Calculate the added mass of the sphere whose radius is a in a fluid of density ρ, to show that μ_{11} is equal to one-half of the mass of the sphere, i.e., $\mu_{11} = (2\pi/3)\rho a^3$. Use the Kirchoff decomposition in obtaining ϕ_1,

Note:

$$\iint dS = \int_0^{2\pi} \int_0^{\pi} r^2 \sin\theta \, d\theta \, d\omega, \quad \int \cos^2\theta \sin\theta \, d\theta = -\frac{1}{3}\cos^3\theta.$$

14.7.3

Consider Figure 14.5 which shows a floating box, symmetric with respect to the x_2-x_3 plane. Show if the moment of added masses μ_{53} and μ_{51} should be zero or not. Note: $n_5 = x_3 n_1 - x_1 n_3$.

14.7.4

Consider a body in an unbounded, ideal fluid, and which is symmetric only with respect to the x_2-x_3 plane, i.e., there is fore-aft symmetry, but no other symmetry. By considering the components of the unit normal vector on the body,

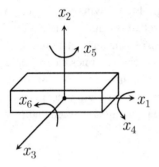

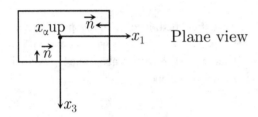

Figure 14.5 A floating box, symmetric with respect to the $x_2 - x_3$ plane.

e.g., $\bar{n}_1(\bar{x}_1, \bar{x}_2, \bar{x}_3) = -\bar{n}_1(-\bar{x}_1, \bar{x}_2, \bar{x}_3)$, etc., for all $\bar{n}_i$, $i = 1, 2, \ldots, 6$, examine the body-boundary condition

$$\frac{\partial \phi_i(\bar{x}_j)}{\partial n} = \bar{n}_j \frac{\partial \phi_i}{\partial \bar{x}_j} = \bar{n}_i, \quad \bar{x}_j \in S_b,$$

to determine which ϕ_i, $i = 1, \ldots, 6$, are symmetric (even function) and which ones are not (odd function). Then by considering the definition of the added-mass tensor, μ_{ij}, prove that $\mu_{12} = \mu_{13} = \mu_{14} = \mu_{52} = \mu_{53} = \mu_{54} = \mu_{62} = \mu_{63} = \mu_{64} = 0$.

14.7.5

Consider a two-dimensional, freely floating body, which is under the action of linear waves in an incompressible fluid. The flow is ideal.

Formulate the mathematical problem to be solved to determine the motions of the body, i.e., give the boundary conditions and the governing equation. Explain the decomposition of the total potential and the conditions each potential has to satisfy. What methods are available to solve for each potential? Also, state the equations of motion, and describe the Kirchoff decomposition.

14.7.6

A pontoon model (see Figure 14.6) is restricted to move along the x_2 coordinate only. Its heaving motion is due to monochromatic incoming waves (of frequency ω) generated by a wavemaker in the model tank. The pontoon model extends from one side wall

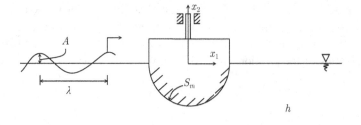

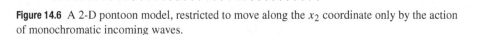

Figure 14.6 A 2-D pontoon model, restricted to move along the x_2 coordinate only by the action of monochromatic incoming waves.

of the tank to the other, so that the fluid flow is two-dimensional. The fluid is inviscid and incompressible, and the flow is irrotational. The wavelength over the water depth ratio is such that the water depth can be assumed infinite. Incoming wave slope is small, therefore the body motions are small so that linear theory is applicable.

For this problem, give the governing equation and the boundary conditions, and

1. Formulate the mathematical problem to be solved so that you can determine the heave displacement. Describe the decomposition of the potential and the resulting equations to be solved.
2. Give the equations of motion for this body and explain how you would calculate the forces acting on the body, i.e., explain the exciting forces, added-mass, damping, and stiffness coefficients. Also explain the Kirchoff decomposition of the potential.
3. Assuming that the pontoon has starboard-port symmetry, what can you say further about the added-mass and damping coefficients you discussed above?
4. Discuss the analytical and numerical methods available to obtain the velocity potential, and therefore the heave displacement.

14.7.7

Assume that we know the six-degree-of-freedom "small" motions of a rigid body given with respect to its center of gravity, G, located at x_{G1}, x_{G2}, x_{G3} (on or inside the body) measured with respect to a global (inertial) coordinate origin, O, fixed in space (say somewhere on the still-water plane). We also have a body-fixed coordinate system origin, on or inside the body, located at B. The coordinates of this point B measured in the inertial coordinate system are x_{B1}, x_{B2}, x_{B3}.

The motions (or 3 translational and 3 rotational displacements) of the body given with respect to G are x_{Gi}, $i = 1, 2, 3$, (surge, heave, and sway) and θ_{Gj}, $j = 1, 2, 3$ (roll, yaw, and pitch).

By keeping in mind that both points B and G are fixed on or inside a *rigid body*, and the motions are "small," show that the motions of the point B can be calculated by $x_{Bi} = x_{Gi} + \epsilon_{ijk}\theta_{Gj}r_k$, $i = 1,2,3$, and $\theta_{Bi} = \theta_{Gi}, i = 1,2,3$, where r_k are the components of the position vector that originates at G and ends at B, i.e., $r_k = x_{Bk} - x_{Gk}$, $k = 1,2,3$.

15 Irregular-Sea Analysis

15.1 Introduction

In Chapters 13 and 14, we have discussed the diffraction and radiation problems when water waves are regular and unidirectional, meaning they are monochromatic (long-crested, and with single frequency) waves. However, water waves are irregular most of the time, and they also are multidirectional in general. The irregularity of ocean waves is basically due to wave dispersion, i.e., longer waves move faster than shorter waves. The directionality of waves is mainly because of the variable direction of winds blowing over the ocean surface. Because of the randomness of wind magnitude and direction, ocean waves also are random, i.e., they are non-deterministic. This randomness of ocean waves prevents one from determining the future events, say at time t, in terms of some analytic expressions established deterministically.

Irregular-sea waves can be thought of as the sum of infinite number of sinusoidal waves (each of which has a different amplitude and frequency) whose phase angles are random. In general then, the amplitude of each component wave may be represented by $A(\omega, \gamma)$ which is a random variable itself. Here, ω is the angular wave frequency and γ is the heading angle of incoming waves. Because of this randomness of waves, a probabilistic approach is necessary to describe various parameters associated with a "confused sea." St. Denis and Pierson (1953) introduced first, the probabilistic description of confused seas in marine hydrodynamics involving ship motions. As an example of the superposition of regular waves of different (however, infinitesimal) heights and frequencies, consider Figure 15.1. Even this limited number of regular waves gives an irregular wave pattern when they are superposed. Furthermore, the resulting irregular shape is totally random, i.e., a slight change in wave amplitude, frequency, or phase of the waves will result in a different pattern for irregular waves (see Figure 15.2, from Faltinsen (1990), which also shows how the frequency-domain and time-domain representations of waves are related to each other in long-crested seas). Therefore, irregular waves cannot be identified by their shapes (surface elevation).

The surface elevation of a regular wave can be written as

$$\eta(x_1, x_3, t) = \mathcal{R}\{A e^{i(\mathbf{k} \cdot \mathbf{x} - \omega t)}\}, \quad i = \sqrt{-1}, \tag{15.1}$$

where $\mathbf{k}$ is the wave-number vector, A is the wave amplitude, and $\mathbf{k} \cdot \mathbf{x}$ contains the directionality of the wave. For a unidirectional wave, equation (15.1) is reduced to

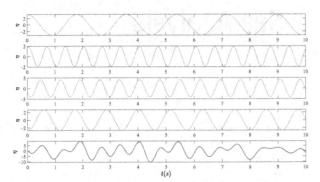

Figure 15.1 Superposition of four regular waves with different amplitudes and lengths, shifted randomly.

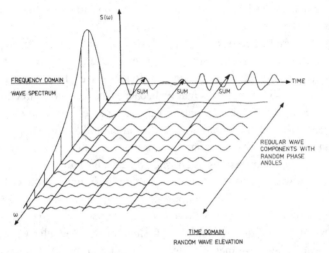

Figure 15.2 The relation between the frequency-domain and time-domain representations of waves in long-crested seas (from Faltinsen (1990)).

equation (12.39), i.e., $\eta(x_1,t) = A\cos(kx_1-\omega t)$. If we sum N of the waves in equation (15.1), we obtain

$$\eta(x_1,x_3,t) = \mathfrak{R}\left\{\sum_{n=1}^{N}\{A_n \exp(ik_n \cdot x - i\omega_n t)\}\right\}. \qquad (15.2)$$

Here A_n is a random function.

Let us consider the directionality of the waves next. Since $k = k_1 e_1 + k_3 e_3$, $x = x_1 e_1 + x_3 e_3$, $k_1 = k\cos\gamma$, $k_3 = k\sin\gamma$, $k = |k|$, where γ is the angle of wave incidence, as shown in Figure 15.3, we have

$$\eta(x_1,x_3,t) = \mathfrak{R}\left\{\sum_{n=1}^{N} A_n \exp\{i[k_n x_1 \cos(\gamma_n) + k_n x_3 \sin(\gamma_n)] - i\omega_n t\}\right\}. \qquad (15.3)$$

Because we cannot characterize an irregular sea by its shape, we need another criterion to base our approach on. This criterion is that the total (potential and kinetic) energy,

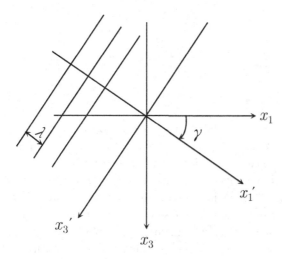

Figure 15.3 The definition of wave incidence angle (clockwise positive).

E, of an irregular wave train is the sum of the energies of all components of individual waves, i.e.,

$$E = \frac{\rho g}{2}\{A_1^2 + A_2^2 + \cdots\} = \frac{1}{2}\rho g \sum_{n=1}^{N} A_n^2, \tag{15.4}$$

where we have used equation (13.11). This concept will lead us to energy spectrum in which waves of many frequencies are present. In the limit, the number of individual wave components, N, in equation (15.3), tend to infinity, meaning the summation in equation (15.3) becomes an integral, i.e.,

$$\eta(x_1,x_3,t) = \Re\left\{\int_0^\infty \int_0^{2\pi} A(\omega,\gamma)\,\exp\{ik(\omega)(x_1\cos(\gamma) + x_3\sin(\gamma)) - i\omega t\}\,d\gamma\,d\omega\right\}. \tag{15.5}$$

Equation (15.5) admits all possible wave directions and frequencies.

15.2   Fourier Analysis

To see the relation between the wave elevation at time t and at a fixed location in space, (x_1, x_3), we need to discuss first the Fourier analysis of a random phenomenon.

If $f(t)$ is a periodic complex function, with a period T, then we may use Fourier series to represent $f(t)$, see also Section 4.3, i.e.,

$$f(t) = \frac{a_0}{2} + \sum_{n=1}^{\infty} a_n \cos\frac{2\pi nt}{T} + \sum_{n=1}^{\infty} b_n \sin\frac{2\pi nt}{T} = \sum_{n=-\infty}^{\infty} f_n e^{in\omega t}, \qquad \omega = \frac{2\pi}{T}, \tag{15.1}$$

where

$$f_0 = \frac{a_0}{2}, \quad f_n = \frac{a_n - ib_n}{2}, \quad f_{-n} = \frac{a_n + ib_n}{2} \quad \text{or} \quad f_n = \frac{1}{T} \int_{-\frac{T}{2}}^{\frac{T}{2}} f(t)e^{-in\omega t}\, dt.$$

(15.2)

The last equation in (15.2) can be obtained by multiplying equation (15.1) by $\cos(2\pi mt/T)$ and $\sin(2\pi mt/T)$ successively and integrating it with respect to t, making use of the orthogonality conditions, such as the one in equation (13.35).

Since $\eta(t)$ is not a periodic function, the above procedure may be applied to ocean waves if we assume that $T \to \infty$, i.e., periodicity is infinite. However, one needs to take a "sufficiently" long time interval for T so that Fourier representation is justified. We will, therefore, consider large time duration, $T = 2\pi/\Delta\omega$, where $\Delta\omega$ is "small." Then for $f(t)$, defined within $-T/2 < t < T/2$, and by use of equations (15.1) and (15.2), we have (see also Section 4.4)

$$f_T(t) = \sum_{n=-\infty}^{+\infty} e^{in\Delta\omega t} \left\{ \frac{\Delta\omega}{2\pi} \int_{-\frac{T}{2}}^{\frac{T}{2}} f_T(\tau)e^{-in\Delta\omega\tau}\, d\tau \right\}.$$

(15.3)

As $T \to \infty$ (or $\Delta\omega \to 0$), equation (15.3) becomes

$$f(t) = \frac{1}{2\pi} \int_{-\infty}^{\infty} e^{i\omega t} \left\{ \int_{-\infty}^{\infty} f(\tau)e^{-i\omega\tau}\, d\tau \right\} d\omega,$$

or

$$f(t) = \frac{1}{2\pi} \int_{-\infty}^{\infty} F(\omega)e^{i\omega t}\, d\omega,$$

(15.4)

where

$$F(\omega) = \int_{-\infty}^{\infty} f(\tau)e^{-i\omega\tau}\, d\tau.$$

(15.5)

The above two functions, $f(t)$ and $F(\omega)$, are the Fourier transform pairs, see Section 4.5. For the integral in equation (15.5) to exist, it is necessary that

$$\int_{-\infty}^{\infty} |f(t)|\, dt < \infty.$$

(15.6)

However, as $T \to \infty$, equation (15.6) is not finite when we replace $f(t)$ by $\eta(t)$. To avoid this difficulty, a box car function is introduced such that

$$C(t) = 1 \text{ if } -\frac{T}{2} < t < \frac{T}{2}, \quad C(t) = 0 \text{ if } t < -\frac{T}{2} \text{ or } t > \frac{T}{2}.$$

(15.7)

Now it is possible to define a truncated surface-elevation function by

$$\eta_T(t) = C(t)\eta(t),$$

(15.8)

and therefore, from equations (15.4) and (15.5), we have

$$\eta_T(t) = \frac{1}{2\pi} \int\limits_{-T/2}^{+T/2} F_T(\omega) e^{i\omega t} d\omega, \quad F_T(\omega) = \int\limits_{-T/2}^{+T/2} \eta_T(t) e^{-i\omega t} dt. \tag{15.9}$$

15.3   Spectral Density

One can now introduce the autocorrelation function defined by

$$R_T(\tau) = \frac{1}{T} \int\limits_{-\frac{T}{2}}^{\frac{T}{2}} \eta_T(t)\, \eta_T(t+\tau)\, dt. \tag{15.10}$$

If we substitute equation (15.9) in equation (15.10), we obtain

$$R_T(\tau) = \frac{2\pi}{T} \int\limits_{-\infty}^{\infty} e^{i\omega\tau} F_T(\omega)\, F_T^*(\omega) d\omega, \tag{15.11}$$

where $F_T^*(\omega)$ is the conjugate of the complex function $F_T(\omega)$. As $T \to \infty$, equation (15.11) can be written as

$$R_T(\tau) = \int\limits_{-\infty}^{\infty} \lim_{T\to\infty} \left\{ \frac{2\pi}{T} |F_T(\omega)|^2 \right\} e^{i\omega\tau} d\omega. \tag{15.12}$$

$R_T(0)$ is the mean square value of the function, $\eta(t)$, i.e., it is the time average of η^2:

$$R_T(0) = <\eta^2(t)> = \int\limits_{-\infty}^{\infty} \lim_{T\to\infty} \left\{ \frac{2\pi}{T} |F_T(\omega)|^2 \right\} d\omega = \int\limits_{-\infty}^{\infty} S_{\eta\eta}(\omega) d\omega. \tag{15.13}$$

$S_{\eta\eta}(\omega)$ is called the double-sided spectral density With this in mind, equation (15.12) becomes

$$R_T(\tau) = \int\limits_{-\infty}^{\infty} e^{i\omega\tau} S_{\eta\eta}(\omega) d\omega, \tag{15.14}$$

i.e., the autocorrelation function is the Fourier transform of the spectral density. Therefore, the inverse transform is

$$S_{\eta\eta}(\omega) = \frac{1}{2\pi} \int\limits_{-\infty}^{\infty} R_T(\tau) e^{-i\omega\tau} d\tau. \tag{15.15}$$

Several examples of spectral densities and autocorrelation functions are shown in Fig. 15.4.

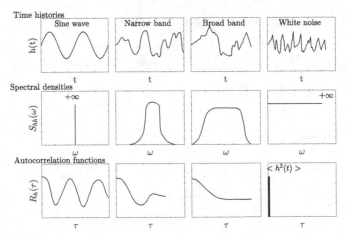

Figure 15.4 Some examples of time series, spectral densities, and autocorrelation functions.

Since we deal with $\omega \geq 0$, it is possible to define a one-sided spectrum by (note that $S_{\eta\eta}$ and R_T are real)

$$R_T(\tau) = \mathfrak{R}\left\{\int_{-\infty}^{+\infty} e^{i\omega\tau} S_{\eta\eta}(\omega)\,d\omega\right\} = \int_0^\infty 2S_{\eta\eta}(\omega)\cos(\omega\tau)d\omega = \int_0^\infty S(\omega)\cos(\omega\tau)d\omega,$$

$$(15.16)$$

where

$$S(\omega) = \frac{1}{2\pi}\int_0^\infty R_T(\tau)\cos(\omega\tau)d\tau. \tag{15.17}$$

Then, from equations (15.13) and (15.16),

$$<\eta^2(t)>=\int_0^\infty S(\omega)d\omega. \tag{15.18}$$

The average energy density for directional waves becomes

$$\bar{\eta}^2 =< \eta^2(t) >= \int_0^\infty \int_0^{2\pi} S(\omega,\gamma)\,d\gamma\,d\omega. \tag{15.19}$$

Since the energy is given by equation (15.4), the total mean energy of the wave system is

$$\bar{E} = \frac{1}{2}\rho g \int_0^\infty \int_0^{2\pi} S(\omega,\gamma)d\gamma d\omega. \tag{15.20}$$

Since the mean square value of a monochromatic wave is $A^2/2$, we can represent a random process by

$$\eta(t) = \lim_{\Delta\omega \to 0} \left\{ \sum_{n=1}^{N} \sqrt{2S(\omega_n)\Delta\omega} \cos(\omega_n t + \epsilon_n) \right\}, \quad N \to \infty, \qquad (15.21)$$

where ϵ_n is a random phase angle uniformly distributed in the interval 0 to 2π, and $\omega_n = n\Delta\omega$. Even though different choices of ϵ_n would produce different time histories, the resulting wave spectrum would be the same. Recalling equation (15.18), we see that the mean square value of $\eta(t)$ is the area under the spectral density versus frequency curve.

Let us consider the mean value of the surface elevation, $\eta(t)$:

$$< \eta(t) > = \frac{1}{T} \int_0^T \eta(t) dt, \qquad (15.22)$$

and the standard deviation, σ,

$$\sigma = \sqrt{\frac{1}{T} \int_0^T (\eta(t) - < \eta(t) >)^2 \, dt}. \qquad (15.23)$$

The standard deviation σ is a measure of how η deviates from the mean. Water surface elevation is assumed to have a Gaussian probability distribution with zero mean. Thus the variance, σ^2, becomes

$$\sigma^2 = < \eta^2(t) > = \frac{1}{T} \int_0^T \eta(t)^2 dt = \int_0^\infty S(\omega) d\omega = m_0, \qquad (15.24)$$

i.e., the area under the spectral density function is the variance of $\eta(t)$ and m_0 is known as the zeroth moment of the spectrum. In other words, the variance of $\eta(t)$ is equal to the standard deviation given by equation (15.23) if the probability distribution is Gaussian, i.e.,

$$\text{RMS} = \sqrt{< \eta^2(t) >} = \sqrt{m_0} = \sigma. \qquad (15.25)$$

15.4   Probability

Since we have already mentioned the Gaussian probability distribution, let us consider the concept of cumulative probability and probability density. Cumulative probability is the probability that a function, say $f(t)$, lies in the interval between f_1 and $f_1 + \Delta f = f_2$ (see Figure 15.5). The cumulative probability is then expressed by

$$P(f, f_1, f_2) = \text{Probability } \{f_1 \leq f(t) \leq f_1 + \Delta f\} = \frac{1}{T} \sum_{i=1}^{M} t_i, \qquad (15.26)$$

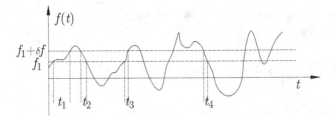

Figure 15.5 A time series used to define cumulative probability.

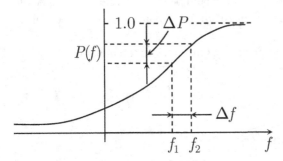

Figure 15.6 Cumulative probability.

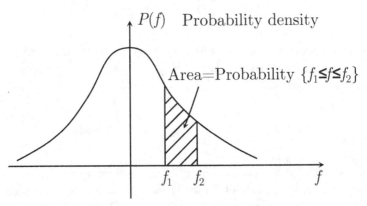

Figure 15.7 Probability density function.

where M is the number of occurrences. This function is shown in Figure 15.6. The slope of the function $P(f)$ is called the probability density function, $p(f)$, and is given by

$$p(f) = \frac{dP(f)}{df} = \lim_{\Delta f \to 0} \{\frac{P(f_1 + \Delta f) - P(f_1)}{\Delta f}\}. \tag{15.27}$$

Therefore, the area under the probability density curve, $p(f)$, gives the probability of $f(t)$ being between f_1 and $f_1 + \Delta f = f_2$, as shown in Figure 15.7.

The surface elevation is approximated very well by a Gaussian probability density function given by

$$p(\eta) = \frac{1}{\sigma \sqrt{2\pi}} \exp\left\{-\frac{\eta^2}{2\sigma^2}\right\}. \tag{15.28}$$

Because the mean value is assumed zero, equation (15.28) predicts that there are as many points above the SWL as there are below. However, if one examines the wave heights (that are always positive), one will see that wave heights follow, in general, the Rayleigh probability density function or distribution. This distribution, given by

$$p(H) = \frac{\pi}{2} \frac{H}{(\bar{H})^2} \exp\left\{-\frac{\pi}{4}\left(\frac{H}{\bar{H}}\right)^2\right\},$$

(15.29)

is for $H \geq 0$, and therefore, is not symmetric with respect to $H = 0$, unlike the Gaussian distribution of the surface elevation that may have a positive or negative value (see Figure 15.10). $\bar{H}$ in equation (15.29) is the mean wave height.

15.5   Transfer function and Spectral Parameters

The wave spectrum can directly be obtained from measured wave data. If, however, measured spectrum is not available, one would often use a formula for the spectrum. There are several well-known formulas that were obtained by fitting a formula to observational data. Among them are the Pierson–Moskowitz spectrum and Bretschneider spectrum (see, e.g., Chakrabarti (1987) and Faltinsen (1990) for more information). We discuss some of the spectral formulas in Section 15.9.

Next, let us discuss the relation between the wave spectrum and wave loads (or structure response or wave diffraction, etc.). Any physical system can be thought of a "black box" which converts an input to an output as shown in Figure 15.8. To show the relation between the output and the input, consider a function, $f = f(t)$, and multiply it by $\delta(t - \tau)$ at $t = \tau$, where $\delta(t - \tau)$ is the Dirac delta function, see also Section 4.6, that has the following properties:

$$\delta(t - \tau) = 0 \text{ if } t \neq \tau; \quad \delta(t - \tau) = \infty \text{ if } t = \tau; \quad \int_{-\infty}^{+\infty} \delta(t - \tau) = 1.$$

(15.30)

With these properties in mind, we then have

$$\int_{-\infty}^{+\infty} f(t)\delta(t - \tau)dt = f(\tau).$$

(15.31)

The Dirac delta function $\delta(t - \tau)$ represents a unit impulse. Since the relation between the input and output can be written as

$$y(t) = L[x(t)],$$

(15.32)

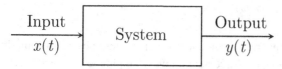

Input
$x(t)$

System

Output
$y(t)$

Figure 15.8 "Black box" representing a system that converts an "Input" to an "Output" when the system is deterministic.

where $L[\ ]$ is the linear operator of the system, we can denote the response (output) to a unit impulse (input) by

$$h(t,\tau) = L[\delta(t - \tau)]. \tag{15.33}$$

Suppose now that the input, $x(t)$, is arbitrary. By keeping equation (15.31) in mind, we can represent an arbitrary input as the sum of unit impulses modified by the value of the function at $t = \tau$, i.e.,

$$x(t) = \int_{-\infty}^{+\infty} \left\{ \int_{-\infty}^{+\infty} x(t)\delta(t - \tau)dt \right\} \delta(t - \tau)d\tau = \int_{-\infty}^{+\infty} x(\tau)\delta(t - \tau)d\tau. \tag{15.34}$$

By use of equations (15.32) through (15.34), we have

$$y(t) = \int_{-\infty}^{\infty} x(\tau)h(t,\tau)d\tau, \tag{15.35}$$

where, for a linear system, it can be shown that $h(t,\tau) = h(t - \tau) \equiv h(\xi)$. As a result, equation (15.35) becomes a convolution integral, i.e.,

$$y(t) = \int_{-\infty}^{+\infty} x(t - \xi)h(\xi)d\xi. \tag{15.36}$$

The Fourier transform pairs of the input (see equations (15.4) and (15.5)) can be written as

$$x(t) = \frac{1}{2\pi} \int_{-\infty}^{+\infty} X(\omega)e^{i\omega t}\,dt, \quad X(\omega) = \int_{-\infty}^{+\infty} x(t)e^{-i\omega t}\,dt. \tag{15.37}$$

Also,

$$Y(\omega) = \int_{-\infty}^{+\infty} y(t)e^{-i\omega t}\,dt = \int_{-\infty}^{+\infty} \left\{ \int_{-\infty}^{+\infty} x(t - \xi)h(\xi)d\xi \right\} e^{-i\omega t}\,dt$$

$$= \int_{-\infty}^{+\infty} e^{-i\omega\xi} \left\{ \int_{-\infty}^{+\infty} x(t-\xi)e^{-i\omega(t-\xi)}d(t-\xi) \right\} h(\xi)d\xi = \left\{ \int_{-\infty}^{+\infty} h(\xi)e^{-i\omega\xi}\,d\xi \right\} X(\omega). \tag{15.38}$$

Therefore, the output $Y(\omega)$ is related to the input $X(\omega)$ by the following relation for a linear system:

$$Y(\omega) = H(\omega)X(\omega), \quad H(\omega) = \int_{-\infty}^{+\infty} h(\xi)e^{-i\omega\xi}\,d\xi. \tag{15.39}$$

where $H(\omega)$ is called the transfer function or, sometimes, the response amplitude operator (RAO). For example, the force per unit wave amplitude on a pile can be written as $F(\omega)/A \equiv H(\omega) = Y(\omega)/X(\omega)$, where $X(\omega)$ represents the properties of the incoming waves and body geometry, both being supplied as input. Note that the square of the transfer function is sometimes called the RAO, defined by the square of the integral in equation (15.38). Obviously,

$$h(t) = \frac{1}{2\pi} \int_{-\infty}^{+\infty} H(\omega)e^{i\omega t}\, dt. \tag{15.40}$$

The above is the case for a regular deterministic system. What happens if the system is random? To see the answer to this question, consider the spectrum of the input

$$S_x(\omega) = \lim_{T \to \infty} \left\{ \frac{2\pi}{T} |X_T(\omega)|^2 \right\}, \tag{15.41}$$

where $X_T(\omega)$ is the Fourier transform of $X(t)$. Then the spectrum of the output will be (by use of equations (15.39) and (15.41))

$$S_y(\omega) = \lim_{T \to \infty} \left\{ \frac{2\pi}{T} |Y_T(\omega)|^2 \right\} = \lim_{T \to \infty} \left\{ \frac{2\pi}{T} |H(\omega)|^2 |X(\omega)|^2 \right\}$$

or

$$S_y(\omega) = |H(\omega)|^2 S_x(\omega), \tag{15.42}$$

where the modulus (or magnitude) of $H(\omega)$ is indicated since the transfer function, $H(\omega)$, can also be a complex function of the frequency. In other words, the output spectrum is linearly proportional to the square of the transfer function. Recall that equation (15.39) is true if the relation between the input and the output is linear. For example, if the force acting on a pile is 1,500 KN for a wave amplitude of 2 m, then the force will be 3,000 KN for a wave amplitude of 4 m. This is certainly true if we use the linear wave theory to calculate the wave force. A typical transfer function could be equal to force divided by wave amplitude. The force clearly varies with the angular frequency ω (rad/s) (the cyclic frequency, $f = \omega/2\pi$ is used sometimes). The "black box" shown earlier will then be as shown in Figure 15.9.

The transfer function, $H(\omega)$, can represent the wave force, wave run-up, surface-elevation amplitude, etc., as long as the system is linear. In other words, $L[\ ]$ is an operator such that for the input $x(t)$ and the output $y(t)$, and the relation $y(t) = L[x(t)]$ between them, we must have $L[x_1(t) + x_2(t)] = L[x_1(t)] + L[x_2(t)]$. Also, for any

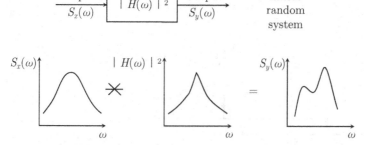

Figure 15.9 "Black box" representing a system that converts an "Input" to an "Output" when the system is linear and random, and the resulting spectral calculations.

constant α, $L[\alpha x(t)] = \alpha L[x(t)]$ needs to be satisfied. This is nothing but the definition of a linear operator. Here, $x_1(t)$ and $x_2(t)$ are the two inputs corresponding to the two outputs $y_1(t)$ and $y_2(t)$, respectively.

Let us go back to the output spectrum and relate it to various concepts such as the significant value of the output. If we assume that the wave height distribution follows the Rayleigh probability density function given by equation (15.29), one can show that the average of the $1/n$-th highest value of a function, $y(t)$, is equal to the abscissa of the centroid of the $1/n$-th part of the area under the Rayleigh probability density function, and is denoted by $\bar{y}^{1/n}$ as shown in Figure 15.10. In other words, the probability of y being between y^* and ∞ is $(Area)^{1/n}$. Therefore,

$$\bar{y}^{1/n} = \frac{\int_{y^*}^{\infty} y\, p(y)\, dy}{\int_{y^*}^{\infty} p(y)dy} = \frac{1}{n} \int_{y^*}^{\infty} y\, p(y)\, dy, \qquad (15.43)$$

where y^* is defined by

$$\frac{1}{n} = \int_0^{\infty} p(y)dy - \int_0^{y^*} p(y)dy = 1 - P(y^*), \qquad (15.44)$$

and $P(y^*)$ is the cumulative probability distribution function, see equation (15.26). Then for a narrow-banded spectrum (which means that most of the energy present in waves is concentrated around a rather small interval of wave frequencies) it can be shown, by use of the Rayleigh probability distribution, that for $n = 3$:

$$\bar{y}^{1/3} \cong 4.0\, \sqrt{m_0} = 4.0\, \sigma, \qquad (15.45)$$

where m_0 is the area under the spectrum curve (m_0 is also known as the zeroth moment of the wave spectrum), and σ is the root mean square (RMS) as given by equation (15.25). If, for instance, $y(t)$ is equal to $H(t)$, we have

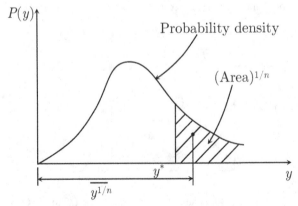

Figure 15.10 Raleigh probability density function and the definition of $\bar{y}^{1/n}$.

$$H_{1/3} = 4.0\sqrt{m_o} = 4.0\sqrt{\int_0^\infty S(\omega)d\omega},\qquad(15.46)$$

where $H_{1/3}$ is called the significant wave height and $S(\omega)$ is the wave spectrum. Note that if m_0 is taken as the area under the response spectrum, $y_{1/3}$ would give the significant height (double amplitude) of the force, moment, motion, etc., i.e., whatever the response (output) corresponds to.

If one is interested in the significant amplitude response, then, e.g.,

$$F_{1/3} = 2.0\sqrt{m_R} = 2.0\sqrt{\int_0^\infty S_R(\omega)\,d\omega} = 2.0\,\text{RMS}(\text{Force})\qquad(15.47)$$

will be the significant force amplitude, while $S_R(\omega)$ is the force amplitude response spectrum, i.e.,

$$S_R(\omega) = |H(\omega)|^2 S_W(\omega),\qquad(15.48)$$

where $H(\omega)$ is the force amplitude transfer function, $F(\omega)/A$, and $S_W(\omega)$ is the wave amplitude spectrum.

15.6   Short-Term Extreme Values

Once the significant response is known, it is possible to predict (if the Rayleigh distribution is valid) the short-term design extreme by the following formula[1]:

$$y_{\text{extreme}} = y_{1/3}\left\{\frac{1}{2}\log\frac{N}{0.01}\right\}^{1/2},\qquad(15.49)$$

where N is the number of waves expected to be encountered during a storm. For example, if the storm lasts for 3 hours and the average wave period is 15 seconds, then $N = 3 \times 60 \times 60/15 = 720$, and therefore, $y_{\text{extreme}} = 1.5584\,y_{1/3}$.

15.7   Encounter Frequency and Spectrum Conversion

The motion of waves can be measured with respect to either a fixed (or earth-bound) or a moving coordinate system. If a vessel has forward motion, the frequency (or period) measured in the moving system, attached to the vessel, is different from the absolute frequency of waves measured in a fixed coordinate system. Therefore, if the vessel motions are measured in a coordinate system steadily moving with the vessel, with a speed equal to vessel's forward speed, then one must use the encounter wave frequency measured in the moving coordinate system. The encounter frequency is similar in concept to the well-known one in physics. The same is also true when waves are riding on a current.

---

[1] For a derivation of this equation, see, e.g., Newman (1978, p. 319).

Consider a vessel moving with a steady forward speed U in the positive x_1 direction. The incoming waves make an angle of γ with the x_1 axis (see Figure 15.3). This wave-incidence angle is termed head waves if $\gamma = \pi$ and following waves if $\gamma = 0$. The relative frequency or the encounter frequency can be written as

$$w_e = \omega - kU\cos(\gamma), \quad \omega^2 = gk\tanh(kh). \tag{15.50}$$

In deep water and head seas, e.g., $k = \omega^2/g$, $w_e = \omega(1+\omega U/g)$. Equation (15.50) can be obtained by considering the phase speed of waves measured relative to an observer located on the vessel. Note that in a fixed coordinate system, the phase speed of waves is given by $c = \omega/k = \lambda/T$, where λ is the wavelength and T is the wave period. To an observer moving with the vessel, this phase speed must be modified by considering the component of the speed of the vessel (or the moving coordinate system) in the direction of the waves, i.e.,

$$c_e = c - U\cos(\gamma), \quad c_e = \frac{\lambda}{T_e}, \quad T_e = \frac{2\pi}{w_e}. \tag{15.51}$$

Equation (15.51) leads to equation (15.50) immediately. Note that it is implicit in equation (15.51) that the wavelength, which is an absolute physical quantity, remains the same in any coordinate system.

As a result of equation (15.50), the encounter wave spectrum must be calculated and used in irregular-sea analysis of vessel motions, bending moments, etc., or when waves and current are simultaneously present. To do this, one has to keep in mind that the energy content of the waves must remain the same no matter what coordinate system is used (this is called "Galilean Invariance"), including the steadily moving one. This obvious energy conservation principle leads, e.g., to the following relation:

$$S(w_e) \mid dw_e \mid = S(\omega) \mid d\omega \mid \Rightarrow S(w_e) = \frac{S(\omega)}{1 - \frac{2\omega U\cos(\gamma)}{g}}. \tag{15.52}$$

If one already has calculated, e.g., the motion transfer functions, as functions of the encounter frequency, the encounter wave spectrum can be used in equation (15.42) to calculate the encounter motion spectra, corresponding to each transfer function, in irregular waves.

The concept of relative frequency or encounter frequency may also be important in coastal engineering when one deals with wave–current interaction problems. For example, when one considers the refraction of waves by currents, the relative frequency must be used. This means that the frequency of waves is modified by the current magnitude (which is similar to vessel speed) and direction (which is similar to vessel heading). See, e.g., Ertekin, Liu, and Padmanabhan (1994).

Sometimes the period spectrum is preferred to frequency spectrum. The preference is a valid one since the wave period is a quantity that one can easily observe because it has the dimension of time, one of the fundamental dimensions. The use of a spectrum based on wave period is rather straightforward. Suppose a frequency spectrum is somehow available. We then have

$$S(T) \mid dT \mid = S(\omega) \mid d\omega \mid, \quad d\omega = -\frac{2\pi}{T^2}dT \Rightarrow S(T) = \frac{2\pi}{T^2}S(\omega). \tag{15.53}$$

Finally, it is common to see that cyclic wave frequency, $f = 1/T$ (Hz), is also used in calculations, especially in oceanography. In such cases, one can convert the angular-frequency spectrum into the cyclic-frequency spectrum by writing

$$S(f)df = S(\omega)d\omega, \quad \omega = 2\pi f \quad \Rightarrow \quad S(f) = 2\pi S(\omega). \tag{15.54}$$

15.8   Directional Wave Spectrum

So far, we have discussed the long-crested wave spectrum. The waves in the ocean are in fact short crested, meaning they move in different directions in general. Let us assume that this is the case, i.e., that the waves are multidirectional, and therefore, are short crested, with a dominant wave heading angle γ. The directional wave-energy spectral density can be written as

$$\bar{S}_W(\omega,\theta) = S_W(\omega)G(\theta), \quad G(\theta) = \frac{2}{\pi}\cos^2\theta, \quad -\frac{\pi}{2} \le \theta \le \frac{\pi}{2}, \tag{15.55}$$

where θ, the heading angle of each component wave, is measured from the axis of the dominant wave heading, and $G(\theta)$ is called the spreading function which, by assumption, is set to zero if $|\theta| > \pi/2$. Note that there exist spreading functions that are different from equation (15.55). This is because of a better fit of a particular spreading function to observational data at a specific ocean site.

Indicating the variance of the response in short-crested waves by $\bar{\sigma}^2$, we have

$$\bar{\sigma}^2 = \int_{-\pi}^{\pi} \int_0^{\infty} H^2(\omega,\theta)\bar{S}_W(\omega,\theta)d\omega d\theta. \tag{15.56}$$

By use of equation (15.55), the variance for the dominant wave-heading angle, γ, can be written as

$$\begin{aligned}
\sigma_\gamma^2 &= \int_{-\pi/2}^{\pi/2} \int_0^{\infty} H^2(\omega,\gamma+\theta)S_W(\omega)G(\theta)d\omega d\theta \\
&= \int_{-\pi/2}^{\pi/2} \sigma^2(\gamma+\theta)G(\theta)d\theta = \frac{1}{4}\int_{-\pi/2}^{\pi/2} R_{1/3}^2(\gamma+\theta)G(\theta)d\theta.
\end{aligned} \tag{15.57}$$

Therefore, the *amplitude* of the significant response becomes

$$R_{1/3}(\gamma) = 2.0\sigma_\gamma = \left\{ \int_{-\pi/2}^{\pi/2} R_{1/3}^2(\gamma+\theta)G(\theta)d\theta \right\}^{1/2}. \tag{15.58}$$

Equation (15.58) requires that the significant response, $R_{1/3}(\gamma+\theta)$, in long-crested waves is obtained for a range of wave-heading angles, θ, to determine the significant response, $R_{1/3}(\gamma)$, in multidirectional waves. Note that we use here the significant amplitude of the response, rather than the significant height (or double amplitude) of the response, and therefore, in equation (15.58) the RMS of the response is multiplied by 2.0, not by 4.0 as in equation (15.46).

$R_{1/3}$ can represent the significant value of the motion, force, moment, etc., depending on what the transfer function $H(\omega,\theta)$ represents. The long-crested wave spectrum, $S_W(\omega)$, can be given by one of the spectrum equations, such as the Bretschneider spectrum (see Section 15.9), or it can be a measured spectrum. See, e.g., Du and Ertekin (1991) for some example calculations of directional wave response.

15.9   Some Wave-Spectra Formulas

Wave spectrum for a given location is, in general, not available from observational data. As a result, we must use one or more of a number of formulas developed for estimating wave spectrum. Here we summarize some of them, and the reader is referred to other works for more detailed analysis of and references on the subject, e.g., Sarpkaya and Isaacson (1981) or Chakrabarti (1987).

15.9.1

The Bretschneider spectrum is based on the significant wave height, H_s, and peak wave (angular) frequency, $\omega_p = 2\pi/T_p$, where T_p is the peak period of the wave spectrum, i.e., it is a two-parameter spectrum. The wave spectrum is for fully developed seas:

$$S(\omega) = \frac{5H_s^2}{16\omega_p} \frac{1}{(\omega/\omega_p)^5} exp\left\{ -\frac{5}{4}\left(\frac{\omega}{\omega_p}\right)^{-4}\right\}. \tag{15.59}$$

The dimension of the wave spectrum is L^2T.

An example of a Bretschneider spectrum is shown in Figure 15.16.

15.9.2

The Pierson–Moskowitz (P-M) spectrum is based on the wind speed U_w (m/s) alone, i.e., it is a one-parameter spectrum, and is given by

$$S(\omega) = \frac{\alpha g^2}{\omega^5} \exp\left\{ -\frac{B}{\omega^4}\right\}, \tag{15.60}$$

where $\alpha = 8.1 \times 10^{-3}$ is the Phillips' constant, $B = 0.74(g/U_w)^4$, and g is the gravitational acceleration.

It is possible to express the P-M spectrum in terms of the significant wave height, H_s, that is the average height of the highest one-third of the waves. Recall that H_s is given by equation (15.46):

$$H_s = 4.0\sqrt{m_o} = 4.0\sqrt{\int_0^\infty S(\omega)d\omega} = \frac{2U_w^2}{g}\sqrt{\frac{\alpha}{0.74}} \quad \text{or} \quad U_w^{-4} = \frac{0.044}{g^2(H_s)^2},$$

$$\tag{15.61}$$

where we used equation (15.60). Therefore, we can now write the one-parameter P-M spectrum in terms of H_s :

$$S(\omega) = \frac{8.1 \times 10^{-3} g^2}{\omega^5} \exp\left\{-0.032 \frac{g^2}{\omega^4 (H_s)^2}\right\}. \qquad (15.62)$$

Figure 15.19 shows a comparison of the Bretschneider and the one-parameter P-M spectrum.

15.10 Concluding Remarks

Ocean waves are in general random and directional in nature. The study of such surface waves would require us to do calculations by decomposing a random wave into its sinusoidal components, or by constructing a random wave by summing up sinusoidal waves of random amplitudes and directions and phase angles.

In this chapter, this approach is discussed in terms of the energy content of the waves. This leads to the concept of wave spectrum and autocorrelation function. Various statistical and probabilistic quantities are introduced in the study of wave spectrum. We also introduced the encounter frequency spectrum as the bodies may be in motion or there may be a current present in addition to the waves that the body encounters.

Both the long-crested wave spectrum and directional wave spectrum are introduced. These are followed by listing of some spectrum formulas used in industry. These formulas are partially based on observations.

15.11 Self-Assessment

15.11.1

The horizontal force on a pile is given by the transfer function shown in Figure 15.11. The wave spectrum is shown in Figure 15.12.

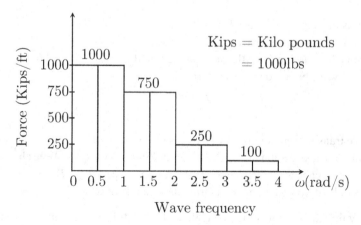

Figure 15.11 Horizontal-force transfer function.

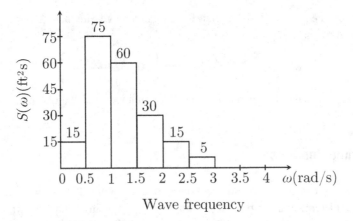

Figure 15.12 Incoming-wave spectrum.

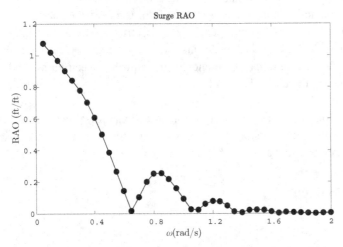

Figure 15.13 Surge response amplitude operator (or transfer function) as a function of angular wave frequency.

1. What is the significant wave height?
2. What is the significant wave-force amplitude?
3. What is the extreme force that acts on the pile if the storm lasts 3 hours and the average wave period during the storm is 9 s.

15.11.2

The surge transfer function (or RAO), that is the surge amplitude of motion over wave amplitude, for a semisubmersible is shown in Figure 15.13. The Bretschneider spectrum of significant wave height H_s = 9 ft and peak period of T_p = 8 s (see Figure 15.14) will be used to determine the irregular-sea response.

1. Verify that the significant wave height of H_s = 9 ft is correct for the given spectrum in Figure 15.12.

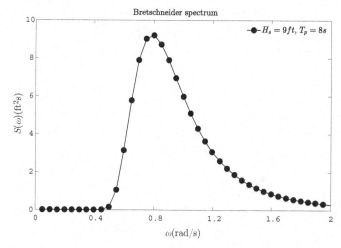

Figure 15.14 Bretschneider spectrum for significant wave height of 9 ft and peak period of 8 s.

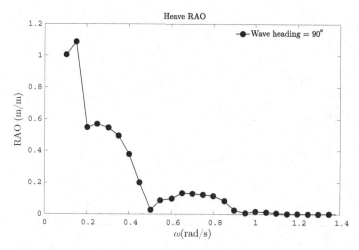

Figure 15.15 Heave transfer function (or RAO) for a semisubmersible.

2. Determine the extreme surge response amplitude during a storm that lasts one week with an average wave period of 8 s.

15.11.3

The heave transfer function (or RAO), that is the heave amplitude of motion over wave amplitude, for a semisubmersible is shown in Figure 15.15 for wave heading angle of 90° (beam seas). The Bretschneider spectrum of significant wave height $H_s = 15.25$ m and peak period of $T_p = 20$ s (see Figure 15.16) will be used to determine the irregular-sea response.

Determine the extreme heave response amplitude during a storm that lasts one week with an average wave period of 12 s.

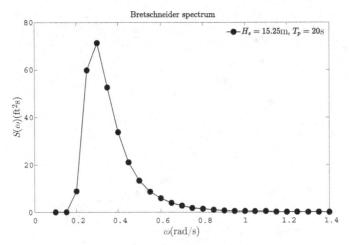

Figure 15.16 Bretschneider spectrum for a significant wave height of 15.25 m and a peak period of 20 s.

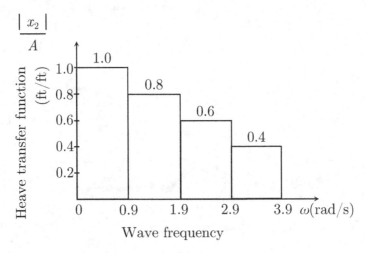

Figure 15.17 Heave motion transfer function.

15.11.4

After solving the potential problem for a series of wave frequencies, ω, by one of the methods you described in answering Prob. 14.7.6, you would like to obtain the heave response in irregular waves. Because the waves are random, the resulting heave response is also random. Let us suppose that you obtained the heave response transfer function as shown in Figure 15.17, where $|x_2|$ is the heave motion amplitude and A is the incoming-wave amplitude, and that you analyzed the time series of the random surface elevation of the incoming waves and calculated the incoming-wave spectrum shown in Figure 15.18.

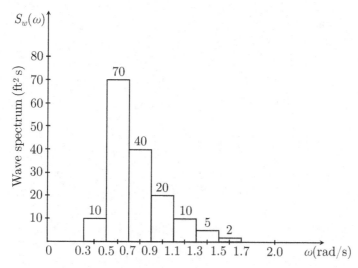

Figure 15.18 Incoming-wave spectrum.

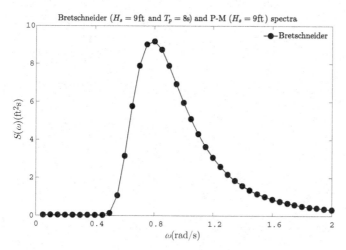

Figure 15.19 Bretschneider and P-M spectra.

1. What is the significant wave height?
2. What is the significant heave amplitude?
3. What is the extreme heave amplitude if the storm lasts 2 hours and the average wave period is 10 s.

15.11.5

Starting with equation (15.60), derive equation (15.62). Similar to the comparison of the two spectra results shown in Figure 15.19, calculate and plot the two spectra for $H_s = 9$ ft and $T_p = 14$ s. Prove that your calculations are correct by finding the area under the spectra and checking if $H_s = 4.0 \sqrt{m_0}$ or not. Use any numerical integration algorithm that you prefer.

Part V

Variational Methods

Here we provide a brief introduction to the interesting field of variational calculus. For moving bodies and flows, the modeling process begins with descriptions of energy rather than forces. Often, complicated dynamics among interacting bodies can be modeled more easily with variational techniques than if we were to apply Newton's laws to determine the overall motion.

16   Introduction to Analytical Dynamics

In this chapter we discuss how some methods from Analytical Dynamics can be applied to advantage in marine applications. These techniques are particularly valuable in problems involving multiple connected bodies, flexible structures, and rigid bodies connected with flexible structures. Such situations arise frequently in the marine world, particularly in operation of shipboard machinery and cargo handling equipment, ship-to-ship material transfer, mooring lines, ships and offshore structures equipped with a drill pipe or marine riser, wave energy converters, oceanographic moorings, etc.

Analytical Dynamics represents an alternative to the Vector Dynamics approach based on Newton's laws. Applications of Newton's laws begin with free-body and kinetic diagrams and a recognition of all of the forces acting on bodies (including any constraint forces, connection forces, etc.). Force and moment equilibrium equations require the use of three-dimensional vectors, and motion is determined when we determine displacements of chosen points as functions of time in response to any forces and moments acting on the bodies. Forces are the fundamental entity in Newton's mechanics. Analytical Mechanics on the other hand rests on the use of kinetic and potential energies as fundamental. Considerable benefit results from the use of scalars as opposed to vectors, for interacting multiple bodies can be included in a single formulation simply by adding their energies together. Additionally, rather than being confined to using Cartesian rectangular or cylindrical or spherical coordinates to define forces and motions, we are able to work with "generalized coordinates" in Analytical Mechanics where the coordinates can be chosen to be inherently compatible with any constraints or connections within the system (Lanczos 1970).

The basic principle of Analytical Dynamics is that the equations of motion for a system simply describe that space–time trajectory which minimizes a particular attribute (e.g., time of travel, distance to travel, etc.) for a system, though more generally, it simply asks that the total "action" (which is typically a function of the kinetic and potential energies of a system) be minimized for any realizable motion. That minimization is of a functional rather than a function. A functional is a function of functions over a trajectory whose value only depends on the end points of the trajectory. These ideas fall in the realm of Calculus of Variations or Variational Calculus. Analytical Dynamics relies on the Calculus of Variations or Variational Calculus for arriving at the equations of motion in any given situation. Here we review the mathematical foundations of Analytical Dynamics and consider its application in marine situations.

The foundations of Analytical Dynamics were developed by Euler and Lagrange, and subsequently extended by Hamilton. Though the early formulations of Euler and Lagrange limited application to conservative systems (i.e., with the sum of kinetic and potential energies conserved), Hamilton's work extended applications to non-conservative systems. Use of the extended Hamilton's principle enables application to systems with most types of external forces not expressible using a potential function. We proceed in a gradual manner below, initially considering just static systems,

16.1   Shape of a Hanging Chain or a Cable

Here we consider the shape taken by a chain or a cable supported at its two ends. The only force acting over the chain is its own weight. This problem thus defines the foundations of more complex problems involving mooring lines in the deep ocean. We will find that the shape taken by a hanging cable is that of a catenary (Meirovitch 2010). Consider a cable suspended between two points A and B with coordinates (x_1, y_1), and $x_2, y_2)$, respectively. Figure 16.1 shows a schematic.

The length of the cable is prescribed as L. Suppose that the cable has a constant cross-sectional area a and is homogeneous with a mass density ρ. Using the path variable s to define the length coordinate along the cable, let us consider a small elemental length ds whose mass is $\rho a ds$. The potential energy dV of the element ds relative to the datum $y = 0$ can be written as

$$dV = \rho a g y ds, \tag{16.1}$$

where y defines the vertical coordinate of the element ds. The potential energy for the entire cable can be found by integrating dV over the length of the cable between the two points A and B:

$$V = \rho g a \int_A^B y ds. \tag{16.2}$$

An arbitrary shape of the cable can be described by specifying y as a function of x, i.e., using $y(x)$. The variational principle of Analytical Mechanics states that the shape of the cable as described by $y(x)$ will be one that minimizes the potential energy V. Since the points y_1 and y_2 are defined and fixed, V is thus a functional, and our aim

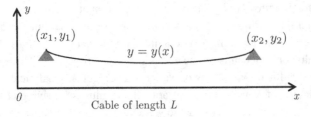

Figure 16.1 A cable of prescribed length L suspended between two supports.

is to determine a function $y(x)$ that minimizes the functional V essentially by looking at its value as evaluated between the two end points A and B. However, the cable also has a fixed length L, and therefore the function $y(x)$ to be found must be constrained by the prescribed length. Letting L be

$$L = \int_A^B ds. \tag{16.3}$$

To incorporate the constraint within our minimization procedure, we use a Lagrange's constant λ as

$$V^* = V - \int_A^B \lambda ds. \tag{16.4}$$

In order to find $y(x)$, we must express the integral such that x rather than s is the integration variable. Recognizing that

$$ds = \sqrt{dx^2 + dy^2}, \tag{16.5}$$

where dx and dy represent small quantities, in the limit, we can write

$$ds = \sqrt{1 + \left(\frac{dy}{dx}\right)^2} \, dx \equiv= \sqrt{1 + y'^2} dx. \tag{16.6}$$

Substitution into equation (16.4) leads to

$$V^{ast} = \rho g a \int_{x_1}^{x_2} y \sqrt{1 + y'^2} dx - \int_{x_1}^{x_2} \lambda \sqrt{1 + y'^2} dx. \tag{16.7}$$

This expression can be simplified to the form

$$V^* = \int_{x_1}^{x_2} (y - \lambda) \sqrt{1 + y'^2}, \tag{16.8}$$

where we have absorbed the multiplier $\rho g a$ into λ. We are led to the statement that the cable will take the shape $y(x)$ for which

$$\delta V^* = \delta \int_{x_1}^{x_2} F^*(y, y'; \lambda) dx = 0, \tag{16.9}$$

where F^* represents the integrand of equation (16.7). Further, δ denotes the first variation of V^*. Since x_1 and x_2 are prescribed points,

$$\delta V^* = \int_{x_1}^{x_2} \delta F^*(y, y^*; \lambda) dx$$

$$= \int_{x_1}^{x_2} \left(\frac{\partial F^*}{\partial y} \delta y + \frac{\partial F^*}{\partial y'} \delta y' + \frac{\partial F^*}{\partial \lambda} \delta \lambda \right) dx. \tag{16.10}$$

Using integration by parts on the term $\delta y'$ in equation (16.10), we find

$$\delta V^* = \int_{x_1}^{x_2} \frac{\partial F^*}{\partial y}\delta y + \frac{\partial F^*}{\partial y'}\delta y\Big|_{x_1}^{x_2} - \int_{x_1}^{x_2} \frac{d}{dx}\frac{\partial F^{*1}}{\partial y'}\delta y dx, + \int_{x_1}^{x_2} \frac{\partial F^*}{\partial \lambda}\delta\lambda dx.$$

$$= \int_{x_1}^{x_2} \left(\frac{\partial F^*}{\partial y} - \frac{d}{dx}\frac{\partial F^*}{\partial y'}\right)\delta y dx + \int_{x_1}^{x_2} \frac{\partial F^*}{\partial \lambda}\delta\lambda dx. \tag{16.11}$$

Since x_1 and x_2 are prescribed, $\delta y = 0$ at both x_1 and x_2. This implies that for an arbitrary choice of x, the functional V^* is minimized when (i) the term within the parentheses in the integrand vanishes, and when (ii) for an arbitrary choice of λ, the term attached to $\delta\lambda$ equals zero. In other words, when

$$\left(\frac{\partial F^*}{\partial y} - \frac{d}{dx}\frac{\partial F^{*1}}{\partial y'}\right) = 0,$$

$$\frac{\partial F^*}{\partial \lambda} = 0. \tag{16.12}$$

The first of equations (16.12) is the Euler–Lagrange equation for this problem, and its solution subject to the second of equations (16.12) gives the shape $y(x)$ of the cable. With variational extremization, the path to the solution is often made smoother by algebraic manipulations. We see here that the function F^* has an explicit dependence on the shape y and y'. It has no explicit dependence on the independent variable x. For such functions, Weinstock (1952) has shown that the condition

$$\frac{d}{dx}\left(y'\frac{\partial f}{\partial y'} - f\right) = 0. \tag{16.13}$$

In equation (16.13), f represents a function such as F^* in equation (16.12). One integration of equation (16.13) in our case leads to the result

$$y'\frac{\partial F^*}{\partial y'} - F^* = C. \tag{16.14}$$

The end result is that

$$(y - \lambda)y'' - y'^2 - 1 = 0. \tag{16.15}$$

The solution for y leads to the catenary form through (Meirovitch 2010)

$$y(x) = \lambda + A\cosh\frac{x + B}{A}. \tag{16.16}$$

The constants can be determined using the boundary conditions. However, it is suffi-cient here to understand that the shape a chain or cable that is freely hanging between two points would take is a catenary. We see from the literature on mooring cables that they too take the form of a catenary when supporting a buoy or another offshore platform. In such cases, one end of the cable is connected to a point on the floating buoy or platform structure, and the other end to an anchor on the seafloor. This basic understanding makes it easier to see how other forces, such as current and wave loads, would affect the dynamics of the cable and the platform it supports.

16.2   Dynamics of a Beam

The variational treatment is well suited for analyzing dynamic situations. As in most dynamics analysis situations, it should first be verified that the system is in static equilibrium. Here we focus on just the dynamics. Whereas potential energy considerations alone were sufficient for understanding the shape of a hanging chain or cable, when there is interest in quantifying possible oscillatory behavior of a flexible structure such as a cable or chain, or other structural members, it is necessary to consider both potential and kinetic energy, as well as any time-dependent external forces involved in the dynamics.

Here we consider a beam undergoing transverse flexural oscillations. This treatment applies to finite beams under different support conditions (i.e., boundary conditions) such as pinned at each end (pinned–pinned), cantilevered or fixed at one end (fixed–free), fixed at both ends (fixed–fixed), free at both ends (free–free), etc. Ships can sometimes be treated as free–free beams with variable cross-sectional dimensions. Hydrodynamic forces (e.g., due to waves) can be added in as external forces by invoking the extended Hamilton principle. We consider a beam whose length dimension is much greater than width and depth, so that bending oscillations occur in a single plane, i.e., in a Cartesian coordinate system xyz, the xz plane, with x being the only independent variable needed, and $w(x,t)$ denoting the deflection in the z direction. The beam is assumed to be homogeneous [i.e., has the same density ρ (mass per unit volume)] and Young's modulus E throughout. Further, we let $A(x)$ and $I(x)$ denote the beam sectional area and sectional second moment of area, respectively. The origin of the coordinate system is at the left-hand end of the beam, and the beam length is L.

To begin, we assume that the beam deflection w is small, so that its curvature can be approximated as the second derivative with respect to x, namely, w_{xx}. We consider a small element of length dx along the beam length. The potential energy for the small element can be expressed as

$$dV = \frac{1}{2}EI(x)w_{xx}^2 dx. \tag{16.17}$$

Integrating over the length of the beam L, we have

$$V = \frac{1}{2}\int_0^L EI(x)w_{xx}^2 dx. \tag{16.18}$$

The kinetic energy of the small element is

$$dT = \frac{1}{2}\rho A(x)\dot{w}^2 dx, \tag{16.19}$$

which leads to

$$T = \frac{1}{2}\int_0^L \rho A(x)\dot{w}^2 dx. \tag{16.20}$$

Based on Hamilton's principle, we define a Lagrangian $\mathcal{L}$ for the beam thus modeled as,

$$\mathcal{L} = T - V. \tag{16.21}$$

The action variable that describes the dynamic transition of the beam from time $t = 0$ to $t = T$ equals the integral of the Lagrangian $\mathcal{L}$ from 0 to T; or,

$$J = \int_0^T \mathcal{L} dt = \int_0^T \int_0^L \frac{1}{2} \left(\rho A(x) \dot{w}^2 - EI(x) w_{xx}^2 \right) dx dt. \tag{16.22}$$

The motion $w(x,t)$ resulting from the kinetic and potential energy functions is such as to minimize the action functional J. The necessary condition for minimum J is

$$\delta J = \delta \int_0^T \mathcal{L} dt. \tag{16.23}$$

Next, for convenience, we introduce a function $F = F(w_{xx}, \dot{w})$ as

$$F(w_{xx}, \dot{w}) = \frac{1}{2} \left(\rho A(x) \dot{w}^2 - EI(x) w_{xx}^2 \right). \tag{16.24}$$

Since the integration limits 0 and T and 0 and L are prescribed and known,

$$\delta J = \int_0^T \int_0^L \delta \left(F(w_{xx}, \dot{w}) \right) dt dx. \tag{16.25}$$

Equation (16.25) implies that

$$\delta J = \int_0^T \int_0^L \left(\frac{\partial F}{\partial \dot{w}} \delta \dot{w} + \frac{\partial F}{\partial w_{xx}} \delta w_{xx} \right) dx dt. \tag{16.26}$$

Since it is the motion $w(x,t)$ we need to solve for, we must express the variation δJ in terms of the variation δw. Thus, we need to use one integration by parts on the first term in equation (16.26), and two successive integrations by parts on the second. The net result is

$$\delta J = -\int_0^T \frac{\partial F}{\partial w_{xx}} \delta w_x \Big|_0^L + \int_0^T \frac{\partial}{\partial x} \left(\frac{\partial F}{\partial w_{xx}} \right) \delta w \Big|_0^L$$

$$+ \int_0^L \frac{\partial F}{\partial \dot{w}} \delta w \Big|_0^T - \int_0^T \int_0^L \left[\frac{\partial F}{\partial t} \left(\frac{\partial F}{\partial \dot{w}} \right) + \frac{\partial^2}{\partial x^2} \left(\frac{\partial F}{\partial w_{xx}} \right) \right]. \tag{16.27}$$

Considering the last term first, we argue that the double integral is zero if the integrand is identically zero over $(0,T)$ and $(0,L)$. Thus,

$$\frac{\partial F}{\partial t} \left(\frac{\partial F}{\partial \dot{w}} \right) + \frac{\partial^2}{\partial x^2} \left(\frac{\partial F}{\partial w_{xx}} \right) = 0. \tag{16.28}$$

Equation (16.28) is the Euler–Lagrange equation for the motion of the beam. The rest of the terms are used to define the boundary conditions that the motion must satisfy. We consider an example that would enable the most straightforward illustration. Specifically, let our beam be simply supported, i.e., be pinned at both ends. The pin joints with their support bases do not allow the two ends to move. They do allow rotation, however, and hence, they do not support moments. Thus, any bending moments the beam might experience as a result of its loads must vanish at the two ends. Substituting the expression for F in equation (16.24) into equation (16.28), we can see that the Euler–Lagrange equation leads to

$$\rho A(x)\frac{\partial^2 w}{\partial x^2} + E\frac{\partial^2}{\partial x^2}\left(I(x)\frac{\partial w^2}{\partial x^2}\right) = 0. \tag{16.29}$$

Next let us consider the boundary terms in equation (16.27). The second boundary term in δw is zero because $\delta w = 0$ at both ends of the beam, $w(x,t)$ being known to be zero at $x = 0$ and at $x = L$. These boundary conditions are prescribed or known by virtue of the kinematic constraints at the two pinned joints. On the other hand, the first boundary term is with respect to δw_x. As pointed out above, the beam is free to rotate at its two ends, so its slope w_x is not known *a priori*, but rather, is determined by the flexural dynamics we are in the process of analyzing. For the first boundary term to be zero at both ends,

$$\frac{\partial F}{\partial w_{xx}} = EI(x)\frac{\partial^2 w}{\partial x^2} = 0, \text{ at } x = 0, \, x = L. \tag{16.30}$$

Note that equation (16.30) simply expresses the condition that the bending moments at the two ends of the pinned–pinned beam are zero, as indicated in the discussion in the paragraph above equation (16.29). The third term in equation (16.27) represents the two temporal boundaries, i.e., initial and terminal conditions. If the initial condition on $w(x,t)$ is specified, then $\delta w = 0$ at $t = 0$. If the terminal condition is also prescribed, as in control of flexible robot arms, then $\delta w = 0$ at $t = T$ as well. If $x(x,t)$ is not known *a priori*, however, then

$$\frac{\partial F}{\partial \dot{w}} = \rho A(x)\dot{w} = 0, \, t = T. \tag{16.31}$$

In other words, the velocity of the beam must be zero at $t = T$. The two end conditions in the time domain are, therefore,

$$w(x,0) = 0, \quad \dot{w}(x,T) = 0. \tag{16.32}$$

If we now consider a beam with a constant cross section, then both A and I are independent of x. Equation (16.29) then becomes

$$\rho A\frac{\partial^2 w}{\partial t^2} + EI\frac{\partial^4 w}{\partial x^4} = 0. \tag{16.33}$$

Using the methods we reviewed in Chapter 7 (starting with a variable separation $w(x,t) = W(x)T(t)$), we find the functions $W(x)$ and $T(t)$ (16.29) to be

$$W(x) = C_1 \cos \beta x + C_2 \sin \beta x + C_3 \cosh \beta x + C_4 \sinh \beta x,$$
$$T(t) = A \cos \omega t + B \sin \omega t. \tag{16.34}$$

Along the way, we find that

$$\omega = \beta^2 \sqrt{\frac{EI}{\rho A}}. \tag{16.35}$$

Applying 3 of the 4 the boundary conditions at $x = 0$ and $x = L$ first, we find that all but the constants C_2 are in fact zero. The boundary condition at $x = L$ then leads to the conclusion that there is a countable infinity of parameters β_n and ω_n, where

$$\beta_n = \frac{n\pi}{L}, \quad n = 1, 2, \dots. \tag{16.36}$$

This leads to

$$\omega_n = \beta_n^2 \sqrt{\frac{EI}{\rho A}} = (\beta_n L)^2 \sqrt{\frac{EI}{\rho A L^4}}. \tag{16.37}$$

The second equality in equation (16.37) helps in relating the more easily determined $\beta_n L$ quantity directly with the beam natural frequency ω_n. β_n are the eigenvalues for the pinned–pinned beam, with $\sin(\beta_n x)$ being the corresponding eigenfunctions. Different boundary conditions will lead to different eigenvalues and eigenfunctions, and by extension, different natural frequencies. It is interesting to see how closely the spatial functions in x are related to the temporal functions in t. One could use this fact to alter the natural frequencies of beam vibrations by introducing small spatial changes (e.g., addition of a point mass, etc.). The complete natural response can be written as

$$w(x,t) = \sum_{n=1}^{\infty} (A_n \cos \omega_n t + B_n \sin \omega_n t) \sin \beta_n x, \tag{16.38}$$

where the constants attached to the spatial eigenfunctions are absorbed into A_n and B_n (which can be found by projecting the initial conditions along each eigenfunction $\sin \beta_n x$. It should be added that having the eigenfunctions in hand for specific boundary conditions, one can find the forced response of beams under external forcing, as long as the assumptions of linearity still remain valid. Indeed, by the extended Hamilton principle, the response of the beam in equation (16.33) to a distributed load $q(x,t)$ can be written as

$$\rho A \frac{\partial^2 w}{\partial t^2} + EI \frac{\partial^4 w}{\partial x^4} = q(x,t) \tag{16.39}$$

To find the forced response, we would express the forced solution as

$$w(x,t) = \sum_{n=1}^{\infty} \zeta_n(t) \phi_n(x). \tag{16.40}$$

ϕ_n denotes the nth eigenfunction. The forcing function could similarly be expanded as

$$q(x,t) = \sum_{n=1}^{\infty} q_n(t)\phi_n(x). \qquad (16.41)$$

Using orthogonality of eigenfunctions, individual terms ζ_n can be evaluated using the temporal differential equation that results for each n. Often, an expansion up to a finite N is sufficient.

16.3 Lagrange's Equations

In this section, we will review the very general technique for writing down the equations of motion for a system that may involve multiple connected bodies. The "generalized" coordinates to describe the motion are chosen to be compatible with any constraints among the variables that are introduced by the kinematic relations among the way different connected bodies move. Doing so enables us to derive and solve the equations of motion using physically meaningful coordinates while avoiding solving for the reaction forces between connected bodies. As an example, consider the motion of a marble on a hoop that rotates about a vertical line and is supported by a pivot on a table. We can use an angular position coordinate θ to denote the position of the marble along the hoop, knowing that it always maintains a constant distance (i.e., the radius of the hoop) from the hoop center (and the table top). We can further use an angle ϕ to represent the rotation of the hoop about the vertical. The kinematic constraint that the marble always remains on the hoop is automatically satisfied by the chosen generalized coordinate θ for the marble position. The constraint that the hoop rotates about the vertical is automatically satisfied by the generalized coordinate ϕ used to denote the hoop's angular rotation.

16.3.1 Pendulum on an Oscillating Support

Here we consider a pendulum whose support is free to oscillate vertically. In a somewhat simplistic sense, this problem models the dynamics of a load suspended from a ship-board crane while the ship executes heave oscillations. We use θ and s as our generalized coordinates to describe pendulum oscillation and support oscillation, respectively, and require that our initial conditions be $\theta(0) = 0$ and $s(0) = 0$ (see Figure 16.2). The pendulum length is L with a mass m, and the origin is set below the lowest point of the pendulum, so that the vertical coordinate y for the pendulum at any time is $y(t) = s(t) + L(1 - \cos\theta)$. The horizontal coordinate for the pendulum is $x(t) = L\sin\theta$ for any $\theta(t)$.

The potential energy of the pendulum can be expressed as

$$V = mgy = mg[s + L(1 - \cos\theta)], \qquad (16.42)$$

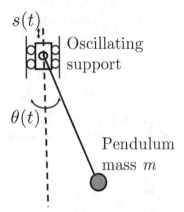

Figure 16.2 Pendulum oscillations in the presence of vertical oscillations of its support.

while its kinetic energy is given by

$$T = \frac{1}{2}m(\dot{x}^2 + \dot{y}^2).$$ (16.43)

We have

$$\dot{x} = (L\cos\theta)\dot{\theta}, \quad \dot{y} = (-L\sin\theta)\dot{\theta} + \dot{s}.$$ (16.44)

Substitution of the quantities in equation (16.44) into equation (16.43) leads to

$$T = \frac{1}{2}m\left[L^2\dot{\theta}^2 - 2L\dot{s}\dot{\theta}\sin\theta + \dot{s}^2\right].$$ (16.45)

Now forming the Lagrangian $\mathcal{L}$ for the system,

$$\mathcal{L} = T - V = \frac{1}{2}m\left[L^2\dot{\theta}^2 - 2L\dot{s}\dot{\theta}\sin\theta + \dot{s}^2\right] - mg[s + L(1 - \cos\theta)],$$ (16.46)

we can represent the total action required for taking the system from time $t = 0$ to the present, say $t = T$ as

$$J = \int_0^T \mathcal{L}(\theta, \dot{\theta}, s, \dot{s})dt.$$ (16.47)

The action functional J for the known integration limits is minimized when

$$\delta J = \int_0^T \delta\mathcal{L}dt = 0.$$ (16.48)

Realizing that $L = L(\theta, \dot{\theta}, s, \dot{s})$, we can verify that the necessary condition $\delta J = 0$ leads to

$$\frac{\partial L}{\partial \dot{\theta}} \delta\theta + \int_0^T \left[\frac{\partial L}{\partial \theta} - \frac{d}{dt}\left(\frac{\partial L}{\partial \dot{\theta}} \right) \right] \delta\theta \, dt$$

$$+ \frac{\partial L}{\partial \dot{s}} \delta s + \int_0^T \left[\frac{\partial L}{\partial s} - \frac{d}{dt}\left(\frac{\partial L}{\partial \dot{s}} \right) \right] \delta s \, dt = 0. \tag{16.49}$$

For arbitrary functions θ, s, the two integrals are only zero when the two integrands in the two integrals above are identically zero. The boundary terms must also be zero, and since we are working with prescribed initial conditions,

$$\delta\theta(t = 0) = 0, \text{ and } \delta s(t = 0) = 0, \tag{16.50}$$

and

$$\frac{\partial L}{\partial \dot{s}}(t = T) = 0 \Rightarrow \dot{s}(T) = 0, \text{ and } \frac{\partial L}{\partial \dot{\theta}}(t = T) = 0, \Rightarrow \dot{\theta}(T) = 0. \tag{16.51}$$

As argued above, for the two integrals to be zero,

$$\frac{d}{dt}\left(\frac{\partial L}{\partial \dot{s}} \right) - \frac{\partial L}{\partial s} = 0,$$

$$\frac{d}{dt}\left(\frac{\partial L}{\partial \dot{\theta}} \right) - \frac{\partial L}{\partial \theta} = 0. \tag{16.52}$$

If we add a rotary motor at the pendulum pivot and a linear motor to drive the linear oscillation, with the rotary motor applying a torque M_θ, and the linear motor applying a force f_s, the extended Hamilton principle would allow us to write

$$\frac{d}{dt}\left(\frac{\partial L}{\partial \dot{s}} \right) - \frac{\partial L}{\partial s} = M_\theta,$$

$$\frac{d}{dt}\left(\frac{\partial L}{\partial \dot{\theta}} \right) - \frac{\partial L}{\partial \theta} = f_s. \tag{16.53}$$

Equations (16.52) and (16.53) are referred to as Lagrange's equations of motion, and they essentially lead to the dynamic equations of motion in the generalized coordinates (θ, s). Using equation (16.46) to substitute for $\mathcal{L}$ in equations (16.53), we arrive at

$$mL^2\ddot{\theta} - mL\sin\theta \ddot{s} - mL\dot{s}\dot{\theta}\cos\theta + mgL\sin\theta = M_\theta,$$

$$m\ddot{s} - mL\sin\theta \ddot{\theta} - mL\cos\theta \dot{\theta}^2 + mg = f_s. \tag{16.54}$$

Equations (16.54) are nonlinear ordinary differential equations. They can be expressed in state-space form and their equilibrium points can be evaluated following the

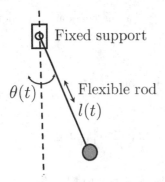

Figure 16.3 Pendulum oscillations in the presence of flexibility of the pendulum rod.

techniques described in Greenberg (1978), Braun (1991), and other texts that discuss qualitative analysis of nonlinear differential equations. In addition, they can be solved numerically using a software package such as MATLAB. An evaluation of the numerical solutions together with findings from the qualitative analysis can lead to meaningful insights into the behavior of the system. Alternatively, for prescribed support oscillations s, sufficiently general conclusions that apply to "parametrically excited oscillations" can be made, as discussed in Nayfeh (1973), where the pendulum oscillations can be described using the Mathieu equation.

16.3.2   Pendulum on a Flexible Rod

Next let us consider the problem of a pendulum swinging about a fixed support, while suspended by a rod that is flexible, i.e., extensible. Figure 16.3 depicts this situation. Effectively, this implies that while the pendulum is oscillating about the pivot, the rod that supports it is free to expand and contract within its elastic limits.

We may now again use two generalized coordinates: (i) the angular position of the pendulum θ and (ii) the axial deformation of the support rod l. The undeformed initial length of the support rod is now denoted as L. The mass of the pendulum is m, and the rod itself is assumed to be massless. However, its axial flexibility is captured using an equivalent stiffness with a spring constant k. We assume the rod oscillations to be small enough to be within the linear region of the elastic curve for the rod material. We take a counterclockwise θ as positive, while the rod deflection l is taken to be positive when compressive.

Relative to a datum that coincides with the lowest point for an undeflected pendulum and undeflected rod, the position vector $\mathbf{r}$ for the pendulum position at any point in its trajectory can be written as

$$\mathbf{r} = (L - l)\sin\theta\,\mathbf{i} + (L - l)(1 - \cos\theta)\,\mathbf{j}. \tag{16.55}$$

The pendulum velocity $\dot{\mathbf{r}}$ can be found as

$$\dot{\mathbf{r}} = -\left[\dot{l}\sin\theta + (L - l)\cos\theta\dot{\theta}\right]\mathbf{i} + \left[-\dot{l}(1 - \cos\theta) + (L - l)\sin\theta\dot{\theta}\right]\mathbf{j}. \tag{16.56}$$

The kinetic energy of the pendulum mass m can now be expressed using

$$T = \frac{1}{2}m\dot{\mathbf{r}} \cdot \dot{\mathbf{r}}. \tag{16.57}$$

There are now two components to the pendulum potential energy: (i) that due to gravity and (ii) that due to spring compression. The potential energy due to gravity is given by

$$V_1 = mg(L - l)(1 - \cos\theta). \tag{16.58}$$

The potential energy component due to spring stiffness is

$$V_2 = \frac{1}{2}kl^2. \tag{16.59}$$

The Lagrangian for the motion can be written as

$$\mathcal{L} = \frac{1}{2}m\left[\left(\dot{l}\sin\theta + (L - l)\cos\theta\dot{\theta}\right)^2 + \left(-\dot{l}(1 - \cos\theta) + (L - l)\sin\theta\dot{\theta}\right)^2\right]$$
$$- mg(L - l)(1 - \cos\theta) - \frac{1}{2}kl^2. \tag{16.60}$$

The generalized coordinates here being (θ, l), with no external forcing, the Lagrange's equations for this problem can be written as

$$\frac{d}{dt}\left(\frac{\partial\mathcal{L}}{\partial\dot{\theta}}\right) - \frac{\partial\mathcal{L}}{\partial\theta} = 0,$$

$$\frac{d}{dt}\left(\frac{\partial\mathcal{L}}{\partial\dot{l}}\right) - \frac{\partial\mathcal{L}}{\partial l} = 0. \tag{16.61}$$

We can see the algebra getting rather tedious at this point, even though we just have two generalized coordinates. Working through the first differentiations, we find

$$\frac{\partial\mathcal{L}}{\partial\theta} = m\left[\left(\dot{l}\sin\theta + (L - l)\cos\theta\dot{\theta}\right)\left(\dot{l}\cos\theta - (L - l)\sin\theta\dot{\theta}\right)\right]$$
$$+ m\left[\left(-\dot{l}\sin\theta + (L - l)\cos\theta\dot{\theta}\right)\right]$$
$$- mg(L - l)\sin\theta. \tag{16.62}$$

Further,

$$\frac{\partial\mathcal{L}}{\partial\dot{\theta}} = m\left(\dot{l}\sin\theta + (L - l)\cos\theta\dot{\theta}\right)\left((L - l)\cos\theta\right)$$
$$+ m\left(-\dot{l}(1 - \cos\theta) + (L - l)\sin\theta\dot{\theta}\right)(L - l)\sin\theta. \tag{16.63}$$

In a similar manner, we arrive at

$$\frac{\partial\mathcal{L}}{\partial l} = m\left[\left(\dot{l}\sin\theta + (L - l)\cos\theta\dot{\theta}\right) + \left(-\dot{l}(1 - \cos\theta) + (L - l)\sin\theta\dot{\theta}(-\sin\theta\dot{\theta})\right)\right]$$
$$+ mg(1 - \cos\theta) + kl, \tag{16.64}$$

and

$$\frac{\partial \mathcal{L}}{\partial \dot{l}} = m \left[(\dot{l} \sin\theta + (L - l) \cos\theta \dot{\theta}) \sin\theta + (-\dot{l}(1 - \cos\theta) + (L - l)\sin\theta\dot{\theta})(1 - \cos\theta) \right].$$

(16.65)

We leave the second differentiation as an exercise for the reader, if only to show that the treatment can get rather tedious as the number of generalized coordinates grows, particularly when we have linear and rotational motions occurring together. We also should point out that the equations of motion that result from this treatment are best derived without any simplifying assumptions such as linearity in the earlier stages. Once the final equations are available, one is able to linearize consistently across all of the terms using Taylor expansions for instance.

As we saw in the previous example, the Lagrange's equations may be written directly in applications, and are particularly useful when different bodies perform coupled motions, as demonstrated in Korde (1990), where Lagrange's equations are used to derive the equations of motion for a rolling barge with an internal cylindrical mass that in turns rolls over a circular track within the barge.

16.4   Concluding Remarks

We discussed an important and useful mathematical technique for solving problems that involve forces and motions. Indeed, the calculus of variations encompass a wide variety of problems wherein the objective is to determine a function that "extremizes" a functional, which is a function of a function that takes on a minimum value, for instance, when it represents a minimum-potential energy shape of a hanging chain, or when it represents the "action" corresponding to the dynamics of moving body or bodies. This approach allows us to choose motion variables that are compatible with any kinematic constraints that the motion of the moving bodies must satisfy. We used the action minimization approach to derive the equations of motion for a vibrating beam, and then also used it to describe the dynamics of a pendulum whose support oscillates in the vertical direction. As a further example of the power of the variational approach, we arrived at the equations of motion of a pendulum swinging about a fixed pivot, but on a flexible support rod. We used the calculus of variations approach to derive the Lagrange's equations for the pendulum problems. Oftentimes, it is convenient to write the Lagrange's equations directly and use the resulting equations of motion to evaluate the motions of a system.

16.5   Self-Assessment

16.5.1

Derive the equation of motion for a pendulum with a swiveling support, so that the pendulum oscillation can be described by two generalized coordinates, θ and ϕ. Assume that the pendulum rod is inextensible.

16.5.2

Using as many degrees as necessary, derive the equations of motion for a rectangular crane barge in beam seas, with the crane carrying a large load that is swinging in response to the wave excitation.

16.5.3

Assume that a ship is moored at its bow to a floating dock by means of a thick composite cable with non-negligible mass. Derive the equation of motion for the cable, assuming that the floating dock oscillations are negligible.

References

Abramowitz, M. and I. Stegun (1972). *Handbook of Mathematical Functions*. New York: Dover Publications. Reissue of 1964 version by National Bureau of Standards.

Adiutori, E. (2005). FOURIER: Is this French mathematician the true father of modern engineering. *Mechanical Engineering Vol. 127*(8), 30–31.

Batchelor, G. K. (1967). *An Introduction to Fluid Dynamics*. Cambridge University Press. Call No. (QA 911.B33), xviii+615pp.

Blevins, R. D. (1977). *Flow-Induced Vibration*. New York: Van Nostrand Reinhold Co. Call No. (TA 355.B52), xiii+363pp.

Braun, M. (1991). *Differential Equations and Their Applications* (4th ed.). New York: Springer-Verlag.

Brebbia, C. A. (1978). *The Boundary Element Method for Engineers*. New York: John Wiley & Sons. Call No. (TA 335.B73).

Brown, J. and R. Churchill (2008). *Fourier Series and Boundary Value Problems* (7th ed.). New York: McGraw-Hill.

Buckingham, E. (1914). On physically similar systems: Illustrations of the use of dimensional equations. *Physical Review Vol. 4*(4), 345–376.

Caro, C. G., T. J. Pedley, R. C. Schroter, and W. A. Seed (2012). *The Mechanics of the Circulation*. Cambridge University Press.

Chakrabarti, S. K. (1987). *Hydrodynamics of Offshore Structures*. Computational Mechanics Publications. New York: Springer-Verlag. Call No. (TC 1665.C43), xvii+440pp.

Chakrabarti, S. K. (1994). *Offshore Structure Modeling*. New Jersey: World Scientific, xx+470pp.

Chakrabarti, S. K. (2002). *The Theory and Practise of Hydrodynamics and Vibration*. New Jersey: World Scientific.

Chakrabarti, S. K. and W. A. Tam (1975). Interaction of waves with a large vertical cylinder. *Journal of Ship Research Vol. 19*, 23–33.

Comstock, J. P. E. (1967). *Principles of Naval Architecture*. New York: The Society of Naval Architects and Marine Engineers (SNAME).

Currie, I. G. (1974). *Fundamental Mechanics of Fluids*. New York: McGraw-Hill. Call No. (QA 901.C8), xiv+441pp.

Dean, R. G. and R. A. Dalrymple (1991). *Water Wave Mechanics for Engineers and Scientists*. Singapore: World Scientific.

Delaney, N. and N. Soreasen (1953, November). *Low-Speed Drag of Cylinders of Various Shapes*. Washington, DC: National Advisory Committee for Aeronautics, NACA, Technical Note 3038.

Dillingham, J. (1984). Recent experience in model-scale simulation of tension leg platforms. *Marine Technology Vol. 21*(2), 186–200.

Du, S. X. and R. C. Ertekin (1991, October). Dynamic response analysis of a flexibly joined, multi-module very large floating structure. In *Proceedings of Oceans '91 Conference*, Volume 3, Honolulu, pp. 1286–1293. IEEE.

Ertekin, R. C. (1984, May). *Soliton Generation by Moving Disturbances in Shallow Water: Theory, Computation and Experiment.* Ph.D. thesis, University of California, Berkeley, 352pp.

Ertekin, R. C. and J. W. Kim (1999, September 22–24). *Third International Workshop on Very Large Floating Structures, VLFS '99*, Volumes I and II. University of Hawaii, 916pp.

Ertekin, R. C., Y. Z. Liu, and B. Padmanabhan (1994). Interaction of incoming waves with a steady intake-pipe flow. *Journal of Offshore Mechanics and Arctic Engineering Vol. 116*(4), 214–220.

Ertekin, R. C. and B. Padmanabhan (1994). Graphical aid in potential flow problems. *Computers in Education Journal Vol. IV*(4), 24–31.

Ertekin, R. C., H. R. Riggs, X. L. Che, and S. X. Du (1993). Efficient methods for hydroelastic analysis of very large floating structures. *Journal of Ship Research Vol. 37*(1), 58–76.

Ertekin, R. C. and G. Rodenbusch (2016). *Wave, Current and Wind Loads.* Handbook of Ocean Engineering, Part D, Chapter 35. Heidelberg: Springer.

Ertekin, R. C. and H. Sundararaghavan (1995). The calculation of the instability criterion for a uniform viscous flow past an oil boom. *Journal of Offshore Mechanics and Arctic Engineering Vol. 117*, 24–29.

Ertekin, R. C., S. Q. Wang, X. L. Che, and H. R. Riggs (1995). On the application of the Haskind-Hanaoka relations to hydroelasticity problems. *Marine Structures Vol. 8*(6), 617–629.

Ertekin, R. C. and Y. Xu (1994). Preliminary assessment of the wave-energy resource using observed wave and wind data. *Energy Vol. 19*(7), 729–738.

Falnes, J. (1995). On non-causal impulse response functions related to propagating water waves. *Applied Ocean Research Vol. 17*(6), 379–389.

Faltinsen, O. M. (1990). *Sea Loads on Ships and Offshore Structures.* Ocean Technology Series. Cambridge, UK: Cambridge University Press. Call No. (TC 1665.F35), viii+328pp.

Friedland, B. (1986). *Control System Design: An Introduction to State-Space Methods.* New York: McGraw-Hill. Reissued by Dover, 2005.

Fung, Y. C. (1977). *A First Course in Continuum Mechanics.* Englewood Cliffs: Prentice Hall. Call No. (QA808.2.F85), xii+340pp.

Goldstein, H. (1980). *Classical Mechanics* (2nd ed.). London: Addison-Wesley. Call No. (QA 805.G6 1980), xiv+672pp.

Gradshteyn, I. S. and I. M. Ryzhik (1980). *Table of Integrals, Series, and Products* (4th ed.). New York: Academic Press. Call No. (QA 55.G6613 1979), xlv+1160pp.

Gradshteyn, I. S. and I. M. Ryzhik (1994). *Table of Integrals, Series, and Products, Ed. Alan Jeffrey* (5th ed.). San Diego, CA: Academic Press.

Graff, K. (1991). *Wave Motion in Elastic Solids.* New York: Dover Publications. Reissue of 1975 version by Oxford University Press.

Greenberg, M. D. (1978). *Foundations of Applied Mathematics.* Englewood Cliffs: Prentice Hall.

Greenberg, M. D. (1998). *Advanced Engineering Mathematics* (2nd ed.). Upper Saddle River: Prentice Hall/Pearson.

Happel, J. and H. Brenner (1965). *Low Reynolds Number Hydrodynamics.* Englewood Cliffs: Prentice Hall. Call No. (QA 929. H35), xiii+553pp.

Harms, V. W. (1979, May). Diffraction of water waves by isolated structures. *Journal of the Waterway Port Coastal and Ocean Division Vol. 105*(WW2), 131–147.

Havelock, T. H. (1940). The pressure of water waves upon a fixed obstacle. *Proceedings of the Royal Society of London. Series A, Mathematical and Physical Sciences Vol. 175*(Ser. A), 409–421.

Hogben, N. and R. G. Standing (1974). Wave loads on large bodies. *Proceedings of the International Symposium on Dynamics of Marine Vehicles and Structures in Waves*, University College London.

Howarth, L. (1938). On the solution of the laminar boundary layer equations. *Proceedings of the Royal Society A, London Vol. 164*(A), 547–579.

Huang, L. L. and H. R. Riggs (2000). The hydrostatic stiffness of flexible floating structures for linear hydroelasticity. *Marine Structures Vol. 13*, 91–106.

Hughes, S. A. (1993). *Physical Models and Laboratory Techniques in Coastal Engineering*. New Jersey: World Scientific, xv+568pp.

Kennard, E. H. (1967, February). *Irrotational Flow of Frictionless Fluids, Mostly Invariable Density*. Washington, DC, AD 653463: David Taylor Model Basin Report No. 2299, xxiv+410pp.

Keulegan, G. H. and L. H. Carpenter (1958). Forces on cylinders and plates in an oscillating fluid. *Journal of Research of the National Bureau of Standards Vol. 60*(5), 423–440.

Korde, U. A. (1990). Study of a wave energy device for possible application in communication and spacecraft propulsion. *Ocean Engineering Vol. 17*(6), 587–599.

Korde, U. A. and J. V. Ringwood (2016). *Hydrodynamic Control of Wave Energy Devices*. Cambridge, UK: Cambridge University Press.

Koterayama, W. (1984). Wave forces acting on a vertical circular cylinder with a constant forward velocity. *Ocean Engineering Vol. 11*(4), 363–379.

Kreyszig, E., H. Kreyszig, and E. Norminton (2011). *Advanced Engineering Mathematics* (10th ed.). New York: John Wiley & Sons.

Krolikowski, L. and T. Gay (1980). An improved linearization technique of frequency domain riser analysis. In *Proceedings of Offshore Technology Conference*, Houston, TX, pp. 341–353. OTC.

Lamb, H. (1932). *Hydrodynamics*. (6th ed.). Cambridge, UK: Cambridge University Press. Reissued by Dover, NY, 1945.

Lanczos, C. (1970). *The Variational Principles of Mechanics*. Toronto, Canada: University of Toronto Press. Reissued by Dover, NY, 1986.

Landweber, L. (1957). Generalization of the logarithmic law of the boundary layer on a flat plate. *Schiffstechnik Vol. 4*(21), 110–113.

Lee, J. J. (1971). Wave induced oscillations in harbors of arbitrary geometry. *Journal of Fluid Mechanics Vol. 45*, 375–394.

Levi-Civita, T. (1925). Determination rigoureuse des ondes permanentes d'ampleur finie, *Mathematische Annalen Vol. 93*, 264–314.

MacCamy, R. C. and R. A. Fuchs (1954, December). *Wave Forces on Piles: A Diffraction Theory*. Tech. Memo. No. 69. Beach Erosion Board. Army Corps of Engineers, 17pp.

Mei, C. C. (1989). *The Applied Dynamics of Ocean Surface Waves*. New Jersey: World Scientific, xx+740pp.

Mei, C. C. (1992). *The Applied Dynamics of Ocean Surface Waves*. Singapore: World Scientific. Chapter 7.

Meirovitch, L. (2010). *Methods of Analytical Dynamics*. Reissue by Dover, McGraw-Hill, 1970.

Meriam, J. and L. Kraige (1997). *Dynamics: Engineering Mechanics* (4th ed.). New York: John Wiley & Sons.

Milne-Thompson, L. M. (1968). *Theoretical Hydrodynamics*. London: MacMillan Press. Call No. (QA 911.M5), xxii+4 plates+743pp.

Mogridge, G. R. and W. W. Jamieson (1975). Wave forces on a circular caisson: Theory and experiment. *Canadian Journal of Civil Engineering Vol. 2*, 540–548.

Morison, J. R., M. P. O'Brien, J. W. Johnson, and S. A. Schaaf (1950). The force exerted by surface piles. *Petroleum Transactions Vol. 189*, 149–154.

Naito, S. and S. Nakamura (1985). Wave energy absorption in irregular waves by feedforward control system. In D. Evans and A. de O. Falcão (Eds.), *Proceedings of IUTAM Symposium on Hydrodynamics of Wave Energy Utilization*, pp. 269–280. Berlin: Springer-Verlag.

Nayfeh, A. (1973). *Perturbation Methods*. New York: John Wiley & Sons.

Newman, J. N. (1976). The interaction of stationary vessels with regular waves. In *Proceedings of 11th Symposium on Naval Hydrodynamics*, London, pp. 491–501.

Newman, J. N. (1978). *Marine Hydrodynamics*. Cambridge: MIT Press. Second Printing, Call No. (VM 156.N48), ix+402pp.

Nikuradse, J. (1942). Laminare reibungsschichten an der längsangetrömten platte. *Zentrale f. wiss. Berichtswesen Monograph*.

Omer, G. and H. Hall (1949). The scattering of a tsunami by a circular island. *Journal of Seismological Society of America Vol. 39*(4), 257–260.

Paulling, J. R. (1979a). An equivalent linear representation of the forces exerted on the OTEC CW pipe by combined effects of waves and current. In *Ocean Engineering for OTEC; 1980 Energy-Sources Technology Conference and Exhibition*, New York: American Society of Mechanical Engineers, pp. 21–28.

Paulling, J. R. (1979b). Frequency-domain analysis of OTEC CW pipe and platform dynamics. In *Offshore Technology Conference*, pp. 1641–1651.

Potter, M. C. and J. F. Foss (1982). *Fluid Mechanics*. Okemos, MI: Great Lakes Press. Call No. (TA 357.P72), x+588pp.

Prandtl, L. and E. W. F. Durand (1935). *The Mechanics of Viscous Fluids in Aerodynamic Theory*, Vol. III (Dover Publications, Inc., 1963). Call No. (TL 570.D931a), VI Plates + 34–208pp.

Press, W., S. Teukolsky, W. Vetterling, and B. Flannery (1992). *Numerical Recipes in FOR-TRAN: The Art of Scientific Computing*, Volume Second Edition. Cambridge University Press. Call No. (QA297.N866).

Rahman, M. (1988). *The Hydrodynamics of Waves and Tides, with Applications*. Boston: Computational Mechanics, 322pp.

Rawson, K. J. and E. C. Tupper (1984). *Basic Ship Theory* (3rd ed.), Vols. 1 and 2. London: Longman.

Richardson, S. M. (1989). *Fluid Mechanics*. New York: Hemisphere Pub. Co. Call No. (TA 357.R53), x+314pp.

Roach, G. N. (1982). *Green's Functions* (2nd ed.). New York: Cambridge University Press. Call No. (QA 379.R6), xiv+325pp.

Sarpkaya, T. (1977, December). In line and transverse forces on cylinders in oscillatory flow at high reynolds numbers. *Journal of Ship Research Vol. 21*(4), 200–216.

Sarpkaya, T. and M. Isaacson (1981). *Mechanics of Wave Forces on Offshore Structures*. New York: Van Nostrand Reinhold Co. Call No. (TC 1650.S26), xiv+651pp.

Saunders, H. E. (1957). *Hydrodynamics in Ship Design: Vol. 1*. New York: Society of Naval Architects and Marine Engineers.

Schlichting, H. (1968). *Boundary Layer Theory* (6th ed.). New York: McGraw-Hill. Call No. (TL 574.B6 S283), xix+748pp.

Schmidt, R. and K. Housen (1995). Problem solving with dimensional analysis. *The Industrial Physicist Vol. 1*(1), 21–24.

Schwartz, L. W. (1974). Computer extension and analytic continuation of Stokes' expansion for gravity waves. *Journal of Fluid Mechanics Vol. 62*, 553–578.

Sedov, L. I. (1959). *Similarity and Dimensional Methods in Mechanics*. New York: Academic Press. Call No. (QC131.Se29s), xvi+363pp.

Sokolnikoff, I. S. and Redheffer, R. M. (1966). *Mathematics of Physics and Modern Engineering* (2nd ed.). New York: McGraw-Hill. Call No. (QA 401.S64), xi+752pp.

Sommerfeld, A. (1949). *Partial Differential Equations in Physics*, Vol. 1 of *Pure and Applied Mathematics*. New York: Academic Press. Call No. (QA 3.P8), xi+335pp.

St. Denis, M. and W. J. Pierson (1953). On the motions of ships in confused seas. *Transactions: The Society of Naval Architects and Marine Engineers Vol. 61*, 280–357.

Stakgold, I. (1979). *Green's Functions and Boundary Value Problems*. New York: John Wiley & Sons. Call No. (QA 379.S72), xv+638pp.

Steele, C. (1994). Beams with and without elastic foundation. Mechanical Engineering, Stanford University.

Stokes, G. G. (1883). *Mathematical and Physical Papers*, Vol. 2. Cambridge University Press.

Stokes, G. G. (1849). *On the Theory of Oscillatory Waves*, Vol. 8, Also Mathematical and Physical Papers, Vol. 1. Cambridge (1880): Transactions of the Cambridge Philosophical Society. Call No. (GC 211.2 .M44).

Sundararaghavan, H. and R. C. Ertekin (1997). Near-boom oil-slick instability criterion in viscous flows and the influence of free-surface boundary conditions. *Journal of Offshore Mechanics and Arctic Engineering Vol. 119*, 26–33.

Thomson, W. (1887). On stationary waves in flowing water. *Philosophical Magazine 22* (Lord Kelvin).

Townsend, A. A. (1954, March). Turbulent friction on a flat plate. In *Proceedings of the 7th International Congress on Ship Hydrodynamics*. Oslo, Norway.

Tulin, M. P. (1999, September). *Hydroelastic Scaling*, Vol. II. Proceedings of 3rd International Workshop on Very Large Floating Structures, VLFS 1999.

Van Dyke, M. (1975). *Perturbation Methods in Fluid Mechanics*. Stanford, CA: The Parabolic Press. Call No. (QA 911.V3), ix+271pp.

Wehausen, J. V. (1971). The motion of floating bodies. *Annual Review of Fluid Mechanics Vol. 3*, 237–268.

Wehausen, J. V. and E. V. Laitone (1960). *Surface Waves*, Vol. 9 of *in Handbuch der Physik, Ed. S. Flugge*. Berlin: Springer-Verlag. Call No. (QC 21.H191), 446–778pp.

Wehausen, J. V., W. C. Webster, and R. W. Yeung (2016). *Hydrodynamics of Ships and Offshore Systems*. Lecture Notes for Course ME241. Berkeley, CA: University of California. Revised Edition, July.

Weinstock, R. (1952). *Calculus of Variations*. New York: McGraw-Hill. Reissued by Dover, NY, 1974.

White, F. M. (1974). *Viscous Fluid Flow*. New York: McGraw-Hill. Call No. (QA929.W48), xix+725pp.

Wiegel, R. L. (1964). *Oceanographical Engineering*. Englewood Cliffs: Prentice Hall. Call No. (GC 201.W5), xi+532pp.

Wilcox, D. C. (1993). *Turbulence Modeling for CFD*. La Canada, California: DCW Industries, Inc.

Wylie, C. R. (1975). *Advanced Engineering Mathematics*. New York: McGraw-Hill. Call No. (QA 401.W9), xii+937pp.

Yih, C. S. (1977). *Fluid Mechanics*. Michigan: West River Press. Call No. (QA 901.Y54), xviii+622pp.

Author Index

Index

Printed in the United States
by Baker & Taylor Publisher Services